Mechanical Engineering Reference Manual

Mechanical Engineering Reference Manual

Edited by **Bilroy Lockhart**

NY RESEARCH PRESS
New York

Published by NY Research Press,
23 West, 55th Street, Suite 816,
New York, NY 10019, USA
www.nyresearchpress.com

Mechanical Engineering Reference Manual
Edited by Bilroy Lockhart

International Standard Book Number: 978-1-63238-318-1 (Hardback)

Printed in China.

Contents

Preface

Mechanical engineering is a branch of engineering that applies the principles and tools of engineering and its related fields to analyze, design, manufacture and maintain mechanical systems. It also involves the production and usage of mechanical and heat power for the design, manufacturing and operation of machines and tools. Mechanical engineering is one of the oldest and expansive of the engineering disciplines. This engineering field requires knowledge of core scientific and engineering concepts like mechanics, thermodynamics, kinematics, structural analysis, materials science and electricity. Mechanical engineers use these core principles along with other tools and methodologies to design and examine manufacturing plants, industrial equipment and machinery, medical devices, transport systems, heating and cooling systems, aircraft, robotics and weapons. There are various technologies being developed as a result of advancements in the field of mechanical engineering. This discipline is constantly and rapidly pushing itself towards technological developments and advancements in order to produce safer, cheaper, and more efficient machines and mechanical systems. Mechanical engineers can be found in almost all aspects of industry and thus enjoy innumerable opportunities. A major asset of an education in mechanical engineering is the flexibility it gives for future employment opportunities for skilled graduates.

This book is an attempt to compile and collate all available research on mechanical engineering under one project. I am grateful to those who put their hard work, effort and expertise into these research projects as well as those who were supportive in this endeavor. I wish to acknowledge the never-ending efforts of the contributors of this book to present the most up-to-date data on all the topics presented in the book. I also wish to thank my publisher for giving me this unparalleled opportunity and supporting me throughout the editing process.

Editor

1

Numerical Investigation of Entropy Generation in Unsteady MHD Generalized Couette Flow with Variable Electrical Conductivity

T. Chinyoka[1] and O. D. Makinde[2]

[1] *Center for Research in Computational and Applied Mechanics, University of Cape Town, Rondebosch 7701, South Africa*
[2] *Institute for Advanced Research in Mathematical Modelling and Computations, Cape Peninsula University of Technology, P.O. Box 1906, Bellville 7535, South Africa*

Correspondence should be addressed to T. Chinyoka; tchinyok@vt.edu

Academic Editors: S. Park and A. Szekrenyes

The thermodynamic second law analysis is utilized to investigate the inherent irreversibility in an unsteady hydromagnetic generalized Couette flow with variable electrical conductivity in the presence of induced electric field. Based on some simplified assumption, the model nonlinear governing equations are obtained and solved numerically using semidiscretization finite difference techniques. Effects of various thermophysical parameters on the fluid velocity, temperature, current density, skin friction, the Nusselt number, entropy generation number, and the Bejan number are presented graphically and discussed quantitatively.

1. Introduction

Investigation of the flow of electrically conducting fluids in porous geometries is of particular importance due to the widespread prevalence in a number of industrial applications [1–4]. These include applications to geothermal reservoirs, nuclear reactor cooling, Magnetohydrodynamic (MHD) marine propulsion, electronic packaging, microelectronic device operations, thermal insulation, and petroleum reservoirs. Preliminary experimental evidence suggests promising future applications in the field of metallurgy, in particular in the MHD stirring of molten metals and in magnetic-levitation casting. Experimental and theoretical investigations of MHD flow in various geometries and under various conditions remain a topic of ongoing interest [5]. Similar investigations not including the effects of electrical conductivity are summarized in [6–12].

The cornerstone in the field of heat transfer and thermal design is the second law analysis and its design-related concept of entropy generation minimization. The foundation of knowledge of entropy production goes back to Clausius and Kelvin's studies on the irreversible aspects of the second law of thermodynamics. Since then, the theories based on these foundations have rapidly developed. However, the entropy production resulting from combined effects of velocity and temperature gradients has remained untreated by classical thermodynamics, which has motivated many researchers to conduct analyses of fundamental and applied engineering problems based on the second law analyses. Entropy generation is associated with thermodynamic irreversibility, which is common in all types of heat transfer processes, namely, conduction, convection, and radiation. In thermodynamical analysis, the fundamental principle remains the improvement of the relevant thermal systems to mitigate against energy losses and hence fully optimize the energy resources.

The purpose of this paper is to analyze the second law of thermodynamics with respect to inherent irreversibility in an unsteady flow of a viscous, incompressible, and electrically conducting fluid. The fluid flows through a channel with isothermal walls and a transversely imposed magnetic field. Expressions for the dimensionless velocity, temperature, current density, wall shear stress, all heat transfer, Bejan number, and entropy generation number are presented. These flow quantities are discussed qualitatively with respect to the

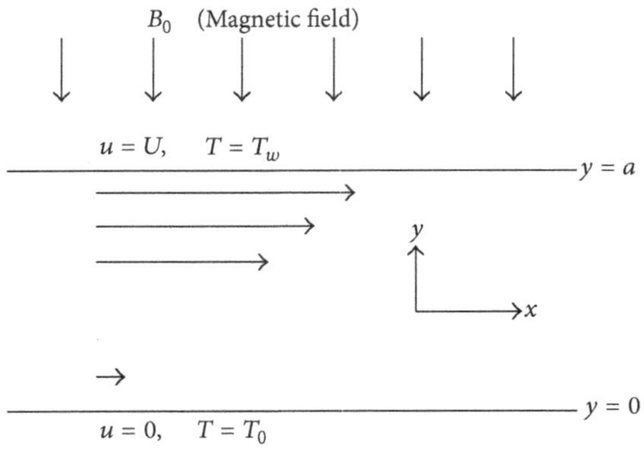

FIGURE 1: Geometry of the problem.

embedded parameters. The mathematical formulation of the problem is established in Section 2. In Section 3, the semi-implicit finite difference technique is implemented for the solution process of the coupled nonlinear problem. Graphical results are presented and discussed qualitatively and quantitatively with respect to various parameters embedded in the system in Section 4.

2. Mathematical Model

The unsteady hydromagnetic generalized Couette flow of a viscous conducting incompressible fluid is considered in the presence of an imposed transverse magnetic field of strength B_0 taking into account the induced electric field E_z. The induced magnetic field is assumed to be small compared with the applied magnetic field and is neglected. Initially, at time $\bar{t} \leq 0$, it is assumed that the fluid is stationary with temperature T_i. At $\bar{t} > 0$, the fluid is subjected to a constant axial pressure gradient while the upper plate moves with a uniform velocity U. The plates' surface temperatures are non-uniform with temperature T_0 at the lower plate and T_w at the upper moving plate such that $T_w > T_0$ as shown in Figure 1.

The temperature-dependent variable electrical conductivity is given as [5]

$$\bar{\sigma} = \sigma_w \left(\frac{T - T_0}{T_w - T_0} \right)^{\lambda}, \tag{1}$$

where σ_w is the fluid electrical conductivity at the upper plate and λ is the electrical conductivity variation index. Under the above assumptions, the dimensionless governing equations for the momentum and energy balance can be expressed as

$$\begin{aligned} \mathrm{Re}\frac{\partial w}{\partial t} &= G - \mathrm{Ha}\left(\mathrm{Le} + w\right)\theta^{\lambda} + \frac{\partial^2 w}{\partial \eta^2}, \\ \mathrm{Re}\,\mathrm{Pr}\frac{\partial \theta}{\partial t} &= \frac{\partial^2 \theta}{\partial \eta^2} + \mathrm{Ec}\,\mathrm{Pr}\left(\frac{\partial w}{\partial \eta}\right)^2 + \mathrm{Ec}\,\mathrm{Pr}\,\mathrm{Ha}(\mathrm{Le} + w)^2\theta^{\lambda}, \end{aligned} \tag{2}$$

where θ is the dimensionless temperature, w is the dimensionless velocity, t is the dimensionless time, Re, Le, Ha, Ec, Pr, and G are the moving upper plate Reynolds number, the electric field loading parameter, magnetic field parameter, the Eckert number, the Prandtl number, and the pressure-gradient parameter. The appropriate initial and boundary conditions in dimensionless form are given as follows:

$$\begin{aligned} w(\eta, 0) &= 0, \qquad \theta(\eta, 0) = \theta_i, \\ w(0, t) &= 0, \qquad \theta(0, t) = 0, \end{aligned} \tag{3}$$

$$w(1, t) = 0, \qquad \theta(1, t) = 1. \tag{4}$$

The following quantities have been utilized in order to obtain the dimensionless governing equations ((2) and (3)):

$$\begin{aligned} &\eta = \frac{y}{a}, \qquad t = \frac{U\bar{t}}{a}, \qquad X = \frac{x}{a}, \qquad w = \frac{u}{U}, \\ &\theta = \frac{T - T_0}{T_w - T_0}, \qquad P = \frac{\bar{P}a}{\mu U}, \qquad \theta_i = \frac{T_i - T_0}{T_w - T_0}, \\ &\mathrm{Re} = \frac{\rho h U}{\mu}, \qquad \mathrm{Pr} = \frac{\mu c_p}{k}, \qquad G = -\frac{dP}{dX}, \\ &\mathrm{Ec} = \frac{U^2}{c_p\left(T_w - T_0\right)}, \qquad \mathrm{Le} = \frac{E_z}{UB_0}, \qquad \mathrm{Ha} = \frac{\sigma_w B_0^2 a^2}{\mu}, \end{aligned} \tag{5}$$

where T is the temperature, U is the uniform velocity of the upper plate, a is the channel width, k is the thermal conductivity, $\bar{t}$ is the time, μ is the dynamic viscosity coefficient, c_p is the specific heat at constant pressure, and P is the fluid pressure. Other quantities of interest are the skin-friction coefficient (C_f) and the Nusselt number (Nu) which are given as

$$C_f = \left.\frac{\partial w}{\partial \eta}\right|_{\eta=0,1}, \qquad \mathrm{Nu} = -\left.\frac{\partial \theta}{\partial \eta}\right|_{\eta=0,1}. \tag{6}$$

The current density and the total current generated within the MHD flow system are given as

$$\begin{aligned} J_z &= \theta^{\lambda}\left(\mathrm{Le} + w\right) \quad \text{(current density)}, \\ I_T &= \int_0^1 \theta^{\lambda}\left(\mathrm{Le} + w\right) d\eta \quad \text{(total current)}. \end{aligned} \tag{7}$$

It is important to note that, for a short circuit configuration where there is no electric field loading, Le = 0.

3. Entropy Analysis

Hydromagnetic generalized Couette flow is inherently irreversible. This may be due to the exchange of energy between the conducting fluid and the moving plate surface. According to [13], the local volumetric rate of entropy generation for a viscous incompressible conducting fluid in the presence of magnetic field and induced electric field is given as

$$E_G = \frac{k}{T_0^2}\left(\frac{\partial T}{\partial \bar{y}}\right) + \frac{\bar{\mu}}{T_0}\left(\frac{\partial \bar{u}}{\partial \bar{y}}\right)^2 + \frac{\bar{\sigma}\left(E_z + uB_0\right)}{T_0}. \tag{8}$$

The first term in (8) describes the heat transfer irreversibility while the second and third terms represent irreversibility due

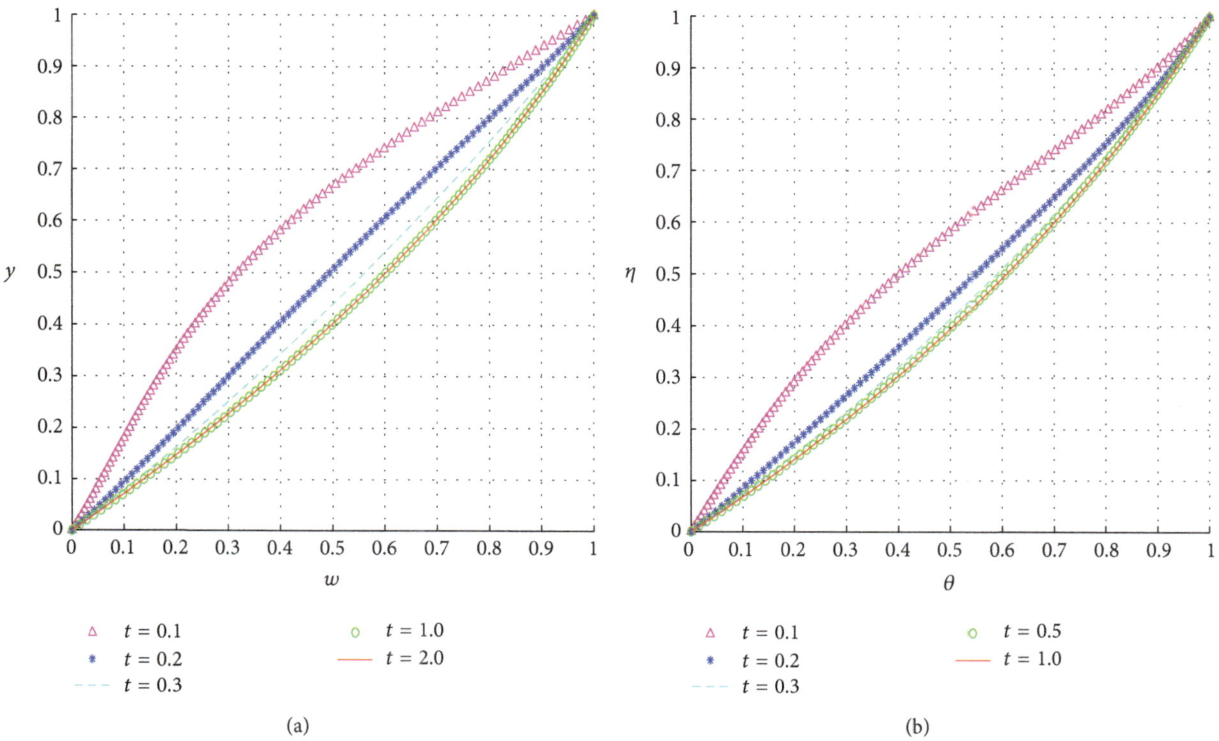

FIGURE 2: Transient and steady-state profiles.

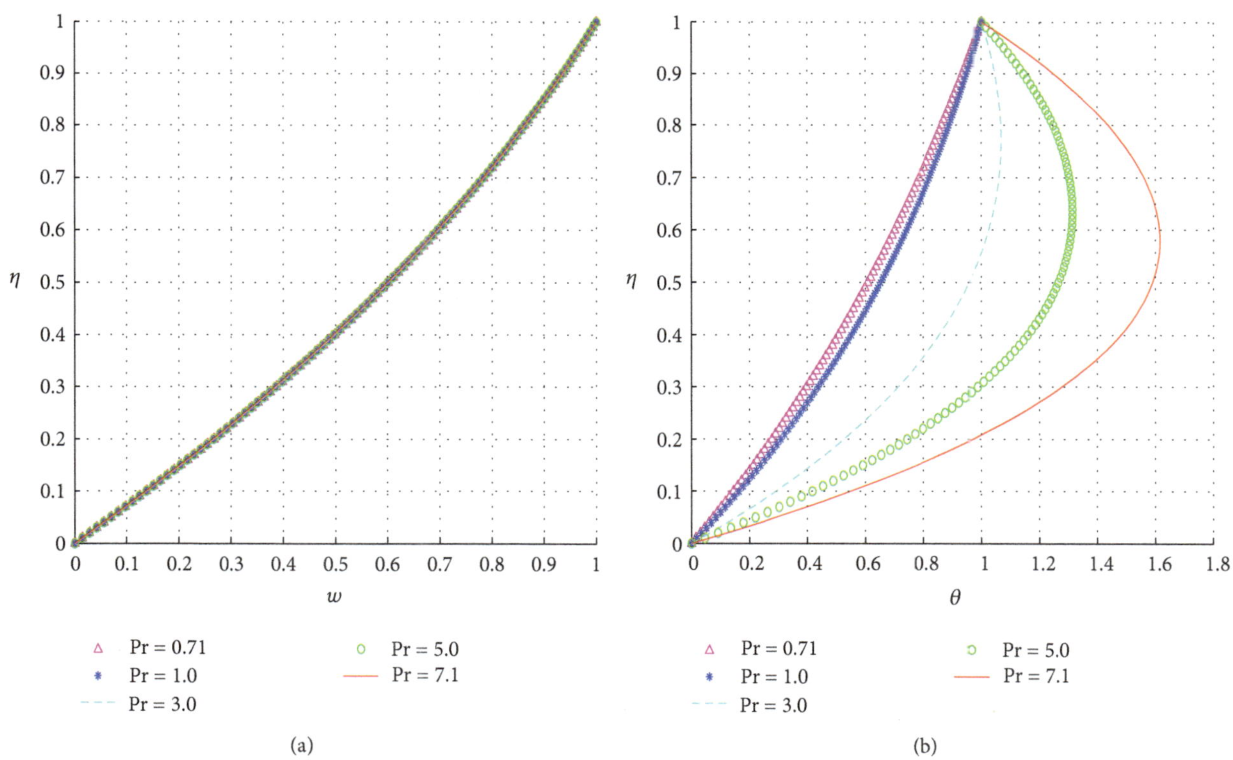

FIGURE 3: Effects of the Prandtl number, Pr.

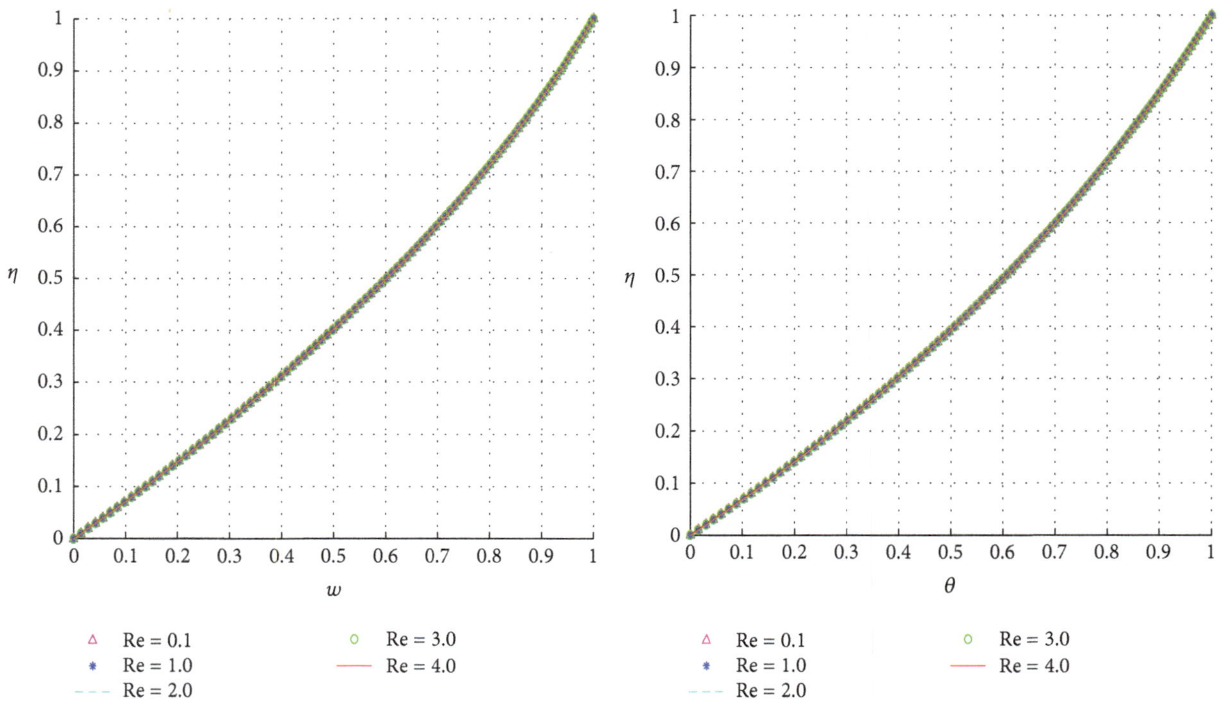

(a) (b)

FIGURE 4: Effects of the Reynolds number, Re.

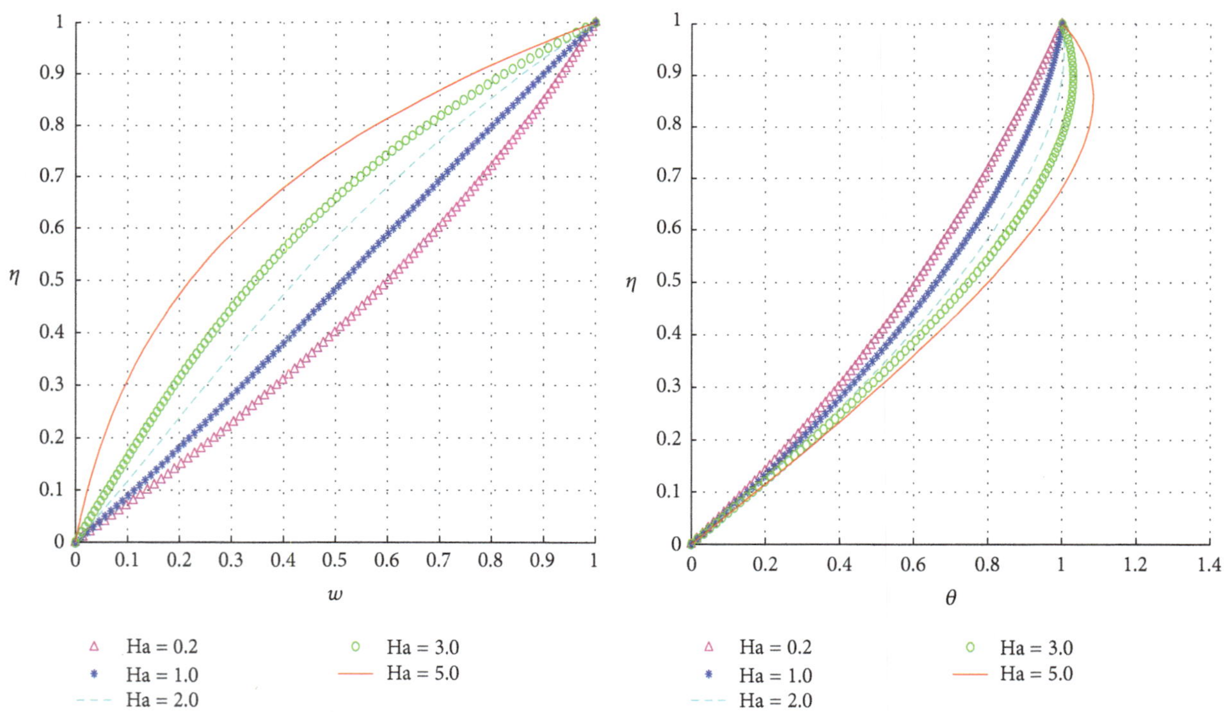

(a) (b)

FIGURE 5: Effects of Ha.

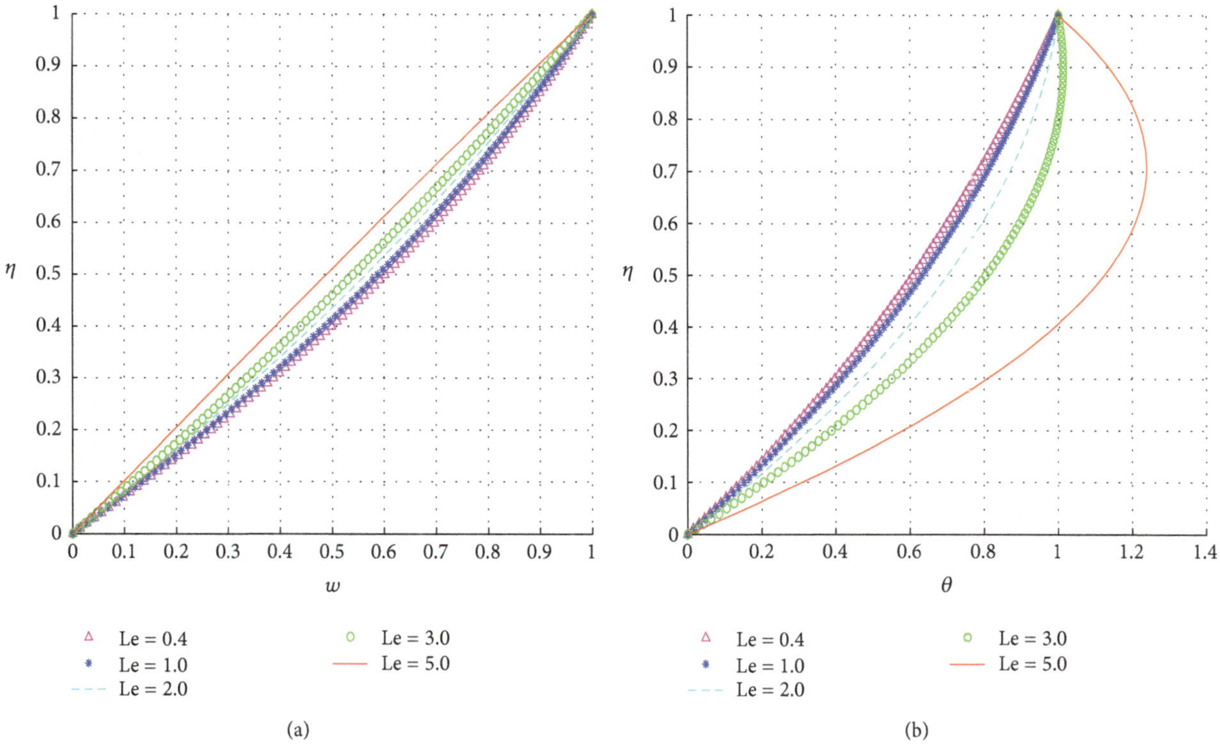

FIGURE 6: Effects of Le.

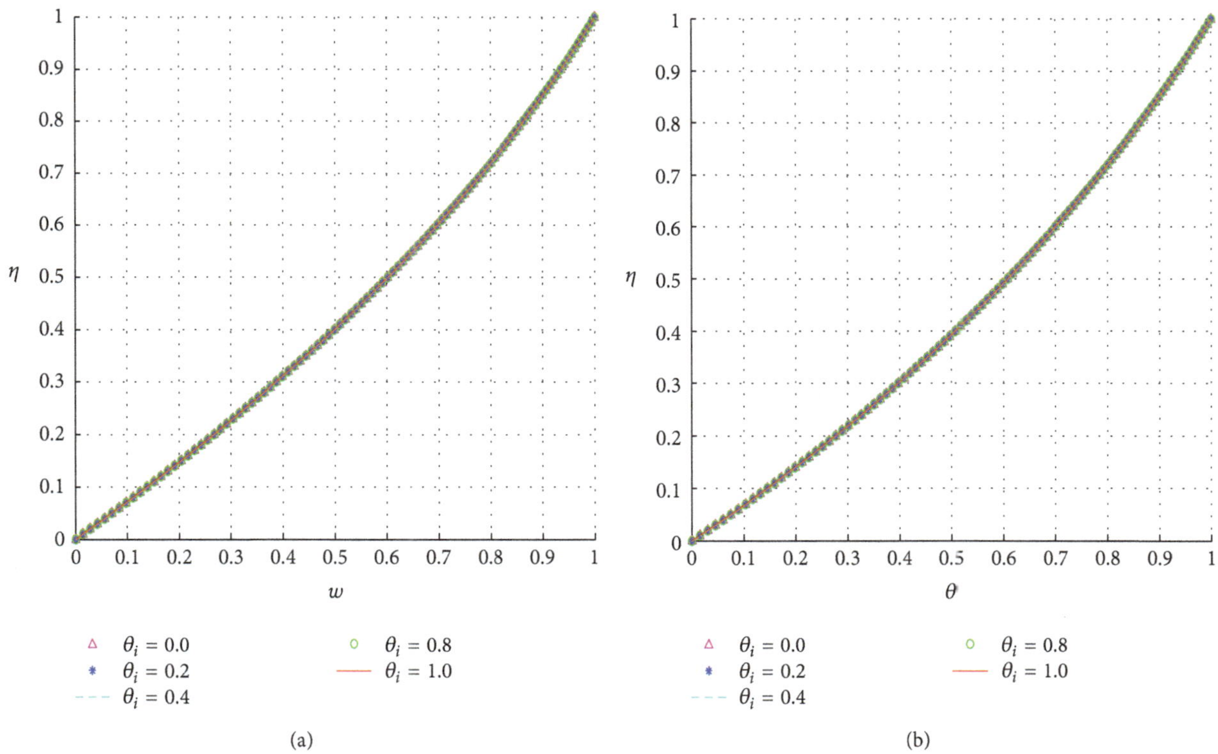

FIGURE 7: Effects of the initial fluid temperature, θ_i.

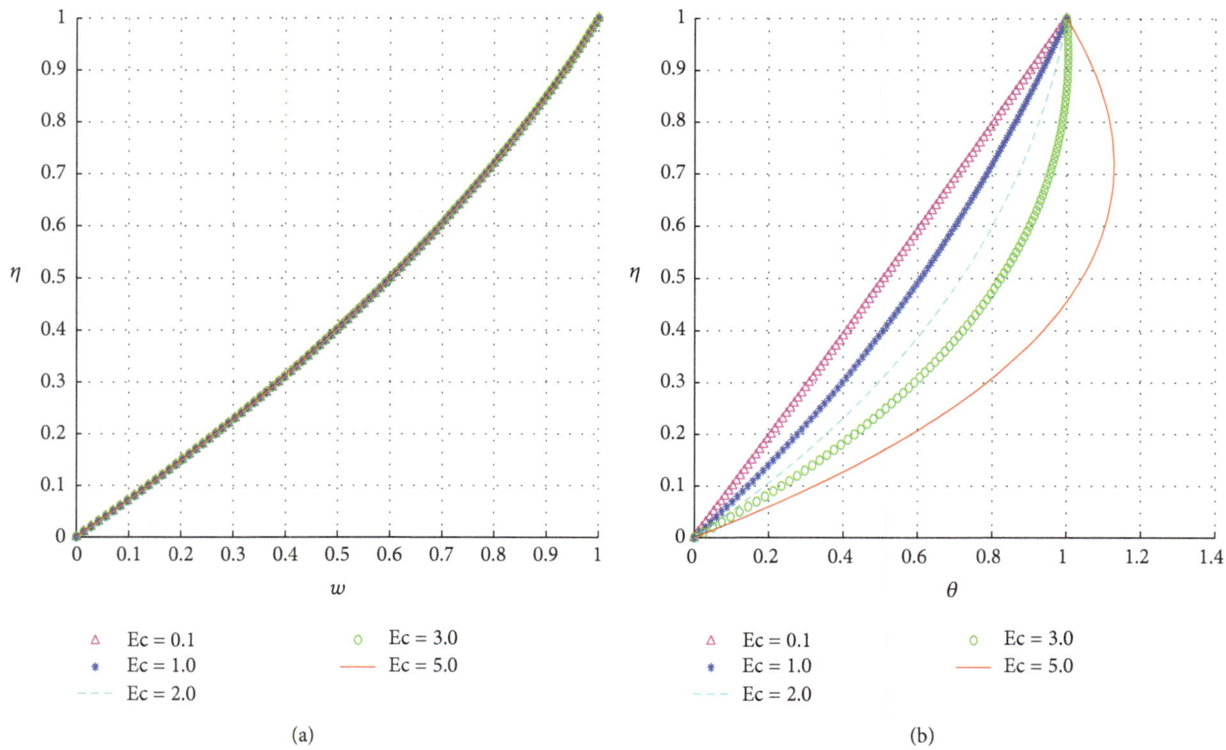

FIGURE 8: Effects of the Eckert number, Ec.

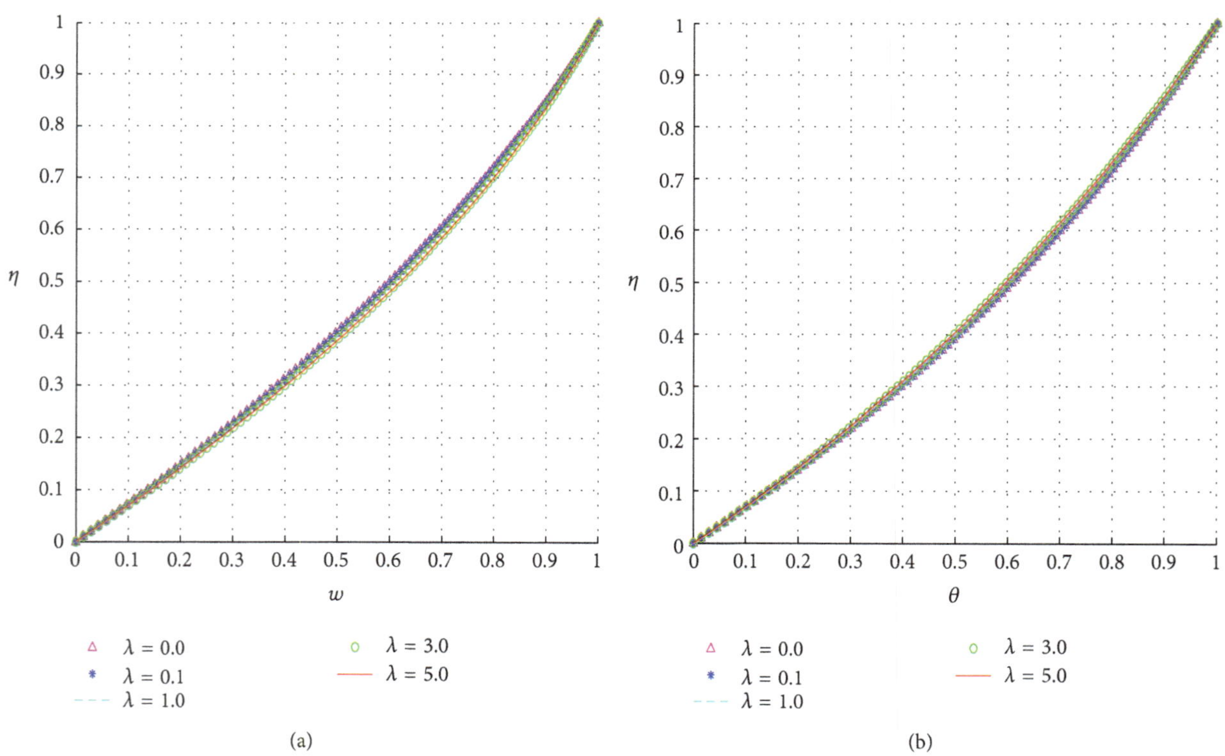

FIGURE 9: Effects of the electrical conductivity variation index, λ.

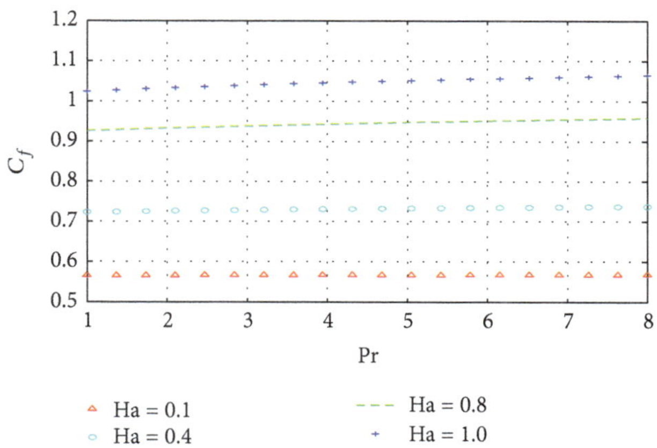

FIGURE 10: Variation of wall shear stress with Pr and Ha.

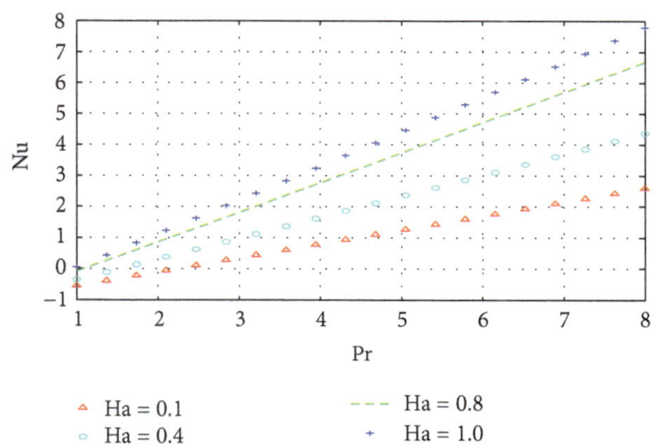

FIGURE 11: Variation of wall shear stress with Pr and Ha.

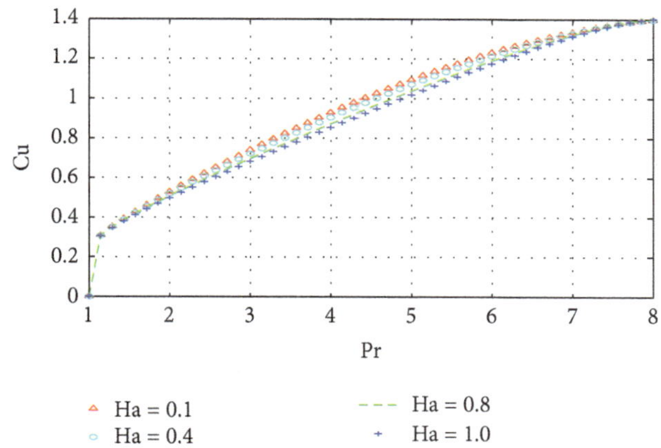

FIGURE 12: Variation of current density with Pr and Ha.

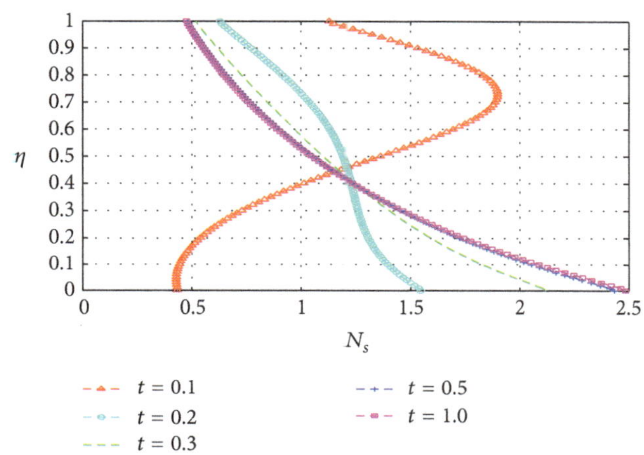

FIGURE 13: Variation of entropy generation rate with η and t.

to fluid friction and magnetic field, respectively. Using (6), the dimensionless form of local entropy generation rate in (8) is given as

$$N_s = \frac{a^2 T_0^2 E_G}{k(T_w - T_0)^2} = \left(\frac{\partial\theta}{\partial\eta}\right)^2 + \frac{\text{Br}}{\Omega}\left[\left(\frac{\partial w}{\partial\eta}\right)^2 + \text{Ha}(\text{Le} + w)^2\theta^\lambda\right], \quad (9)$$

where $\Omega = (T_w - T_0)/T_0$ is the temperature difference parameter and $\text{Br} = \text{Ec}\,\text{Pr}$ is the Brinkman number. The Bejan number, Be, is defined as

$$\text{Be} = \frac{N_1}{N_s} = \frac{1}{1+\Phi}, \quad (10)$$

where $N_s = N_1 + N_2$,

$$N_1 = \left(\frac{\partial\theta}{\partial r}\right)^2,$$

$$N_2 = \frac{\text{Br}}{\Omega}\left[\left(\frac{\partial w}{\partial\eta}\right)^2 + \text{Ha}(\text{Le} + w)^2\theta^\lambda\right], \quad (11)$$

$$\Phi = \frac{N_2}{N_1}.$$

N_1 represents the irreversibility due to heat transfer, N_2 represents fluid friction and magnetic field irreversibility, and Φ is the irreversibility ratio.

The Bejan number (Be) as shown in (10) has the range $0 \leq \text{Be} \leq 1$. If Be = 0, then the irreversibility is dominated by the combined effects of fluid friction and magnetic fields, but if Be = 1, then the irreversibility due to heat transfer dominates the flow system by the virtue of finite temperature differences.

4. Numerical Solution

Our numerical algorithm is based on semi-implicit finite difference schemes [14–19]. Implicit terms are taken at the intermediate time level $(N + \xi)$, where $0 \leq \xi \leq 1$. The discretization of the governing equations is based on a linear Cartesian mesh and uniform grid on which finite differences are taken. We approximate both the second and first spatial derivatives with second-order central differences. The equations corresponding to the first and last grid points

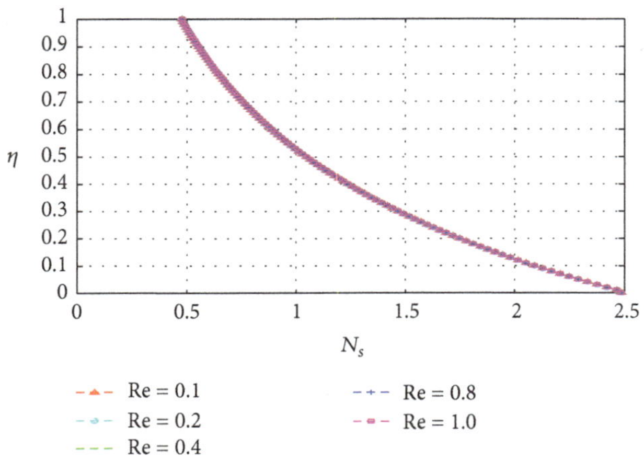

FIGURE 14: Variation of entropy generation rate with η and Re.

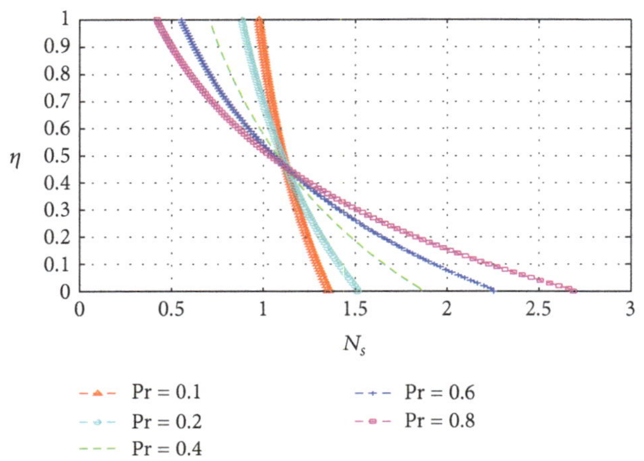

FIGURE 15: Variation of entropy generation rate with η and Pr.

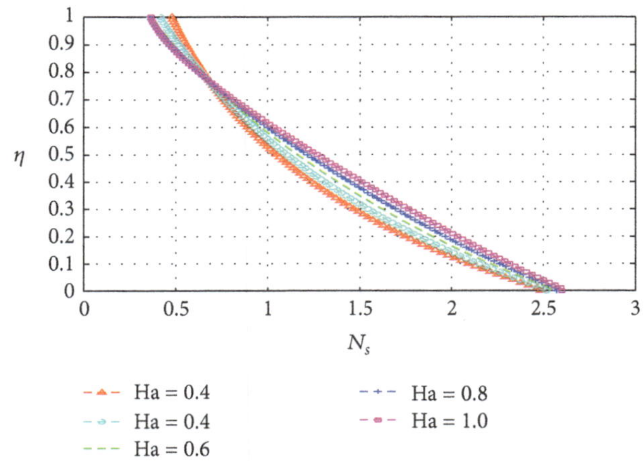

FIGURE 16: Variation of entropy generation rate with η and Ha.

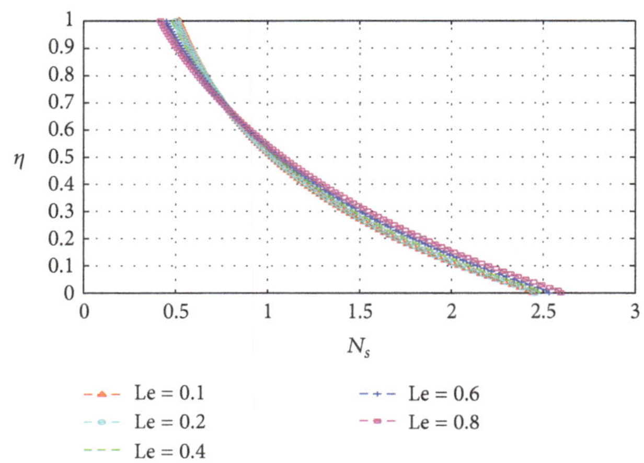

FIGURE 17: Variation of entropy generation rate with η and Le.

are modified to incorporate the boundary conditions. The semi-implicit scheme for the velocity component reads:

$$\operatorname{Re}\frac{\partial w}{\partial t} = G - \operatorname{Ha}\left(\operatorname{Le} + w^{(N+\xi)}\right)\left(\theta^{\lambda}\right)^{(N)} + \frac{\partial^2}{\partial \eta^2} w^{(N+\xi)}. \tag{12}$$

In (12), it is understood that $\partial \# / \partial t := (\#^{(N+1)} - \#^{(N)})/\Delta t$. The equation for $w^{(N+1)}$ then becomes

$$-r_1 w_{j-1}^{(N+1)} + \left(\operatorname{Re} + 2r_1 + \operatorname{Ha}\Delta t\left[\theta^{\lambda}\right]^{(N)}\right) \times w_j^{(N+1)} - r_1 w_{j+1}^{(N+1)} = \text{explicit terms}, \tag{13}$$

where

$$r_1 = \xi \frac{\Delta t}{\Delta \eta^2}. \tag{14}$$

The solution procedure for $w^{(N+1)}$ thus reduces to inversion of tridiagonal matrices, which is an advantage over a fully implicit scheme. The semi-implicit integration scheme for the temperature equation is similar to that for the velocity component. Unmixed second partial derivatives of the temperature are treated implicitly as follows:

$$\operatorname{Re}\operatorname{Pr}\frac{\partial \theta}{\partial t} = \frac{\partial^2}{\partial \eta^2}\theta^{(N+\xi)} + \operatorname{Ec}\operatorname{Pr} \times \left[\left(\frac{\partial w}{\partial \eta}\right)^2 + \operatorname{Ha}(\operatorname{Le} + w)^2\theta^{\lambda}\right]^{(N)}. \tag{15}$$

The equation for $\theta^{(N+1)}$ thus becomes

$$-r_1 \theta_{j-1}^{(N+1)} + (\operatorname{Re}\operatorname{Pr} + 2r_1)\,\theta_j^{(N+1)} - r_1 \theta_{j+1}^{(N+1)} = \text{explicit terms}. \tag{16}$$

The solution procedure again reduces to inversion of tridiagonal matrices. The schemes ((13) and (16)) were checked for consistency. For $\xi = 1$, these are first order accurate in time but second order in space. The schemes in [14] have $\xi = 1/2$, which improves the accuracy in time to second order. As in [14–18] we, however, use $\xi = 1$ here so that we are free to choose larger time steps and still converge to the steady solutions.

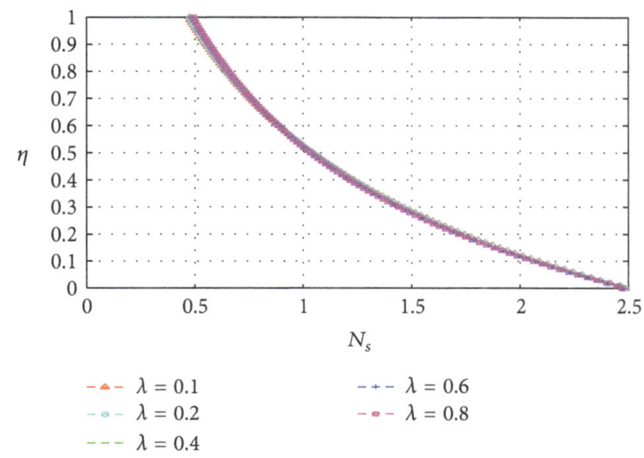

FIGURE 18: Variation of entropy generation rate with η and λ.

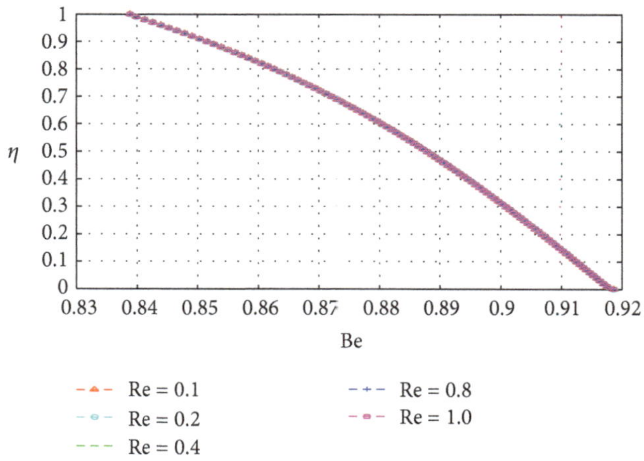

FIGURE 20: Variation of the Bejan number with η and Re.

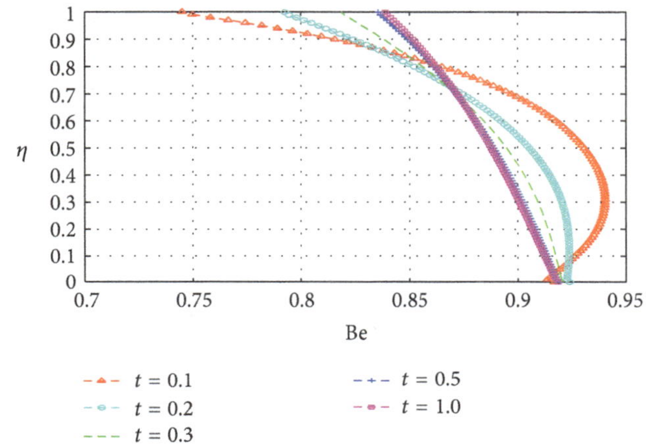

FIGURE 19: Variation of the Bejan number with η and t.

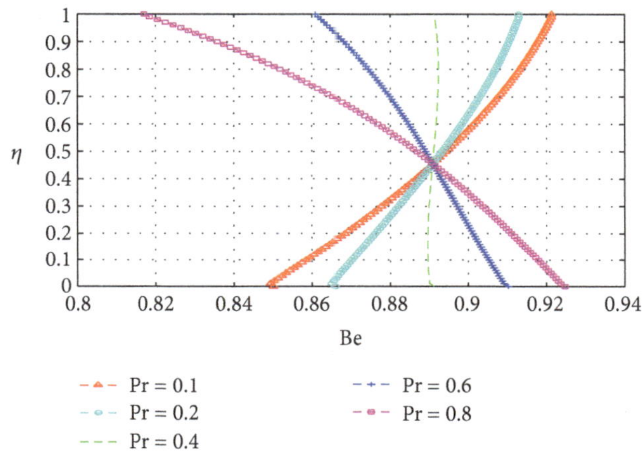

FIGURE 21: Variation of the Bejan number with η and Pr.

5. Results and Discussion

Unless otherwise stated, we employ the following parameter values:

$\text{Re} = 1$, $\text{Pr} = 0.71$, $\text{Ec} = 1$, $\text{Ha} = 0.2$, $\text{Le} = 0.4$, $\lambda = 0.1$, $\theta_i = 0$, $\Delta\eta = 0.01$, $\Delta t = 0.01$, $G = 1$, and $t = 10$.

These will be the default values in this work. In the succeeding graphical, if any of these parameter values is not explicitly mentioned, it will be understood that such parameters take on the default values.

5.1. Transient and Steady-State Flow Profiles. We display the transient solutions in Figure 2. The figures show a transient increase in both fluid velocity, Figure 2(a), and temperature, Figure 2(b), until a steady state is reached.

5.1.1. Parameter Dependence of Solutions. The response of the velocity and temperature to varying values of the Prandtl number (Pr) is illustrated in Figure 3.

Larger values of the Prandtl number correspondingly increase the strength of the heat sources in the temperature equation, and hence this strength in turn increases the overall fluid temperature as clearly illustrated in Figure 3(b). Due to the implicit relationship between the Prandtl number and the velocity field via the magnetic field terms, we notice as expected that increases in the Prandtl number have limited effects on the fluid velocities as illustrated in Figure 3(a).

The response of the velocity and temperature to varying values of the Reynolds number (Re) is illustrated in Figure 4.

The Reynolds number being coupled to the time-dependent terms shows limited effect on both the steady fluid velocity and temperature as shown in Figures 4(a) and 4(b).

The response of the velocity and temperature to varying values of the magnetic field parameter (Ha) is illustrated in Figure 5.

Larger values of Ha correspondingly increase the strength of the heat sources in the temperature equation, and hence, this strength in turn increases the overall fluid temperature as clearly illustrated in Figure 5(b). On the other hand, higher values of Ha decrease the strength of the source terms in the velocity equation, and hence this decreased strength in turn increases the overall fluid temperature as clearly illustrated in Figure 5(a).

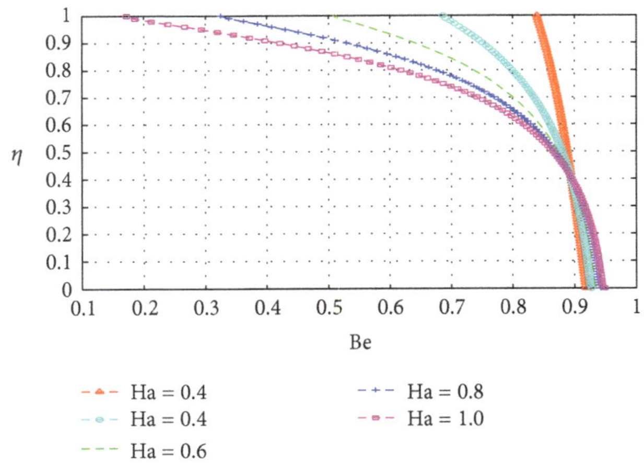

FIGURE 22: Variation of the Bejan number with η and Ha.

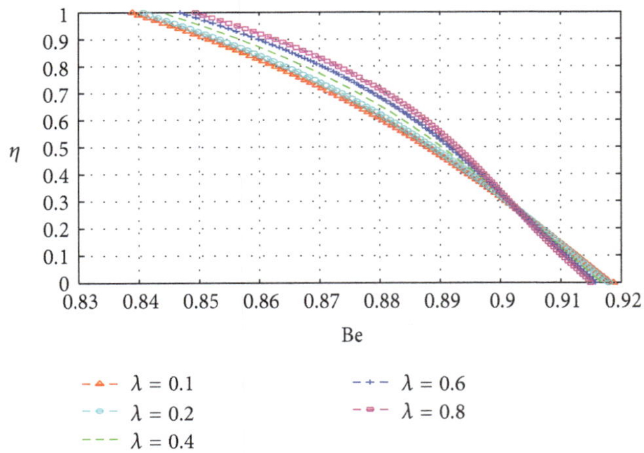

FIGURE 24: Variation of the Bejan number with η and λ.

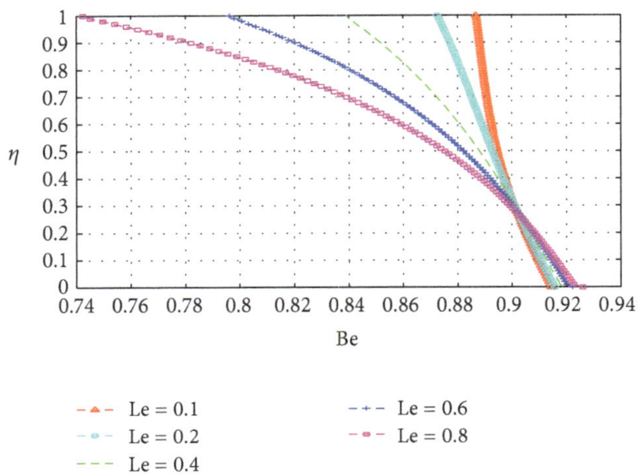

FIGURE 23: Variation of the Bejan number with η and Le.

The effects of Le on both the velocity and temperature profiles are similar to those of Ha as illustrated in Figures 6(a) and 6(b).

Figures 7(a) and 7(b) show that the values of the initial temperature have limited effect on the steady-state flow profiles.

The response of the velocity and temperature to varying values of the Eckert number (Ec) are illustrated in Figure 8.

The effects of the Eckert number are similar to those for the Prandtl number. The effects of the electrical conductivity variation index on the velocity and temperature profiles are illustrated in Figure 9.

As expected, since $0 \leq \theta \leq 1$, an increase in the electrical conductivity variation index (λ) correspondingly decreases the values of θ^{λ}. The source terms in the temperature equation will thus decrease in magnitude, whereas those in the velocity equation become less negative. This explains the slight decreases noticed in the fluid temperature, Figure 9(b), as well as the slight increases in the fluid velocity, Figure 9(a).

5.2. Wall Shear Stress and Wall Heat Transfer Rate. The wall shear stress and wall heat transfer rates are illustrated in Figures 10 and 11, respectively, with varying Pr and Ha.

The results in Figures 10 and 11 simply summarize those in the previous section on parameter dependence of solutions. In particular, the behavior of the velocity and temperature with respect to certain parameters can be summarized in terms of the behavior of their respective gradients. The results of the current density similarly reflect the corresponding results on parameter dependence as shown in Figure 12.

5.3. Entropy Generation. In this section, we plot the entropy generation rate (N_s) across the channel under varying parameter conditions. Except for Figure 13, all graphs are otherwise drawn at the time $t = 10$.

Figures 13, 14, 15, 16, 17, and 18 show the expected results for N_s. In particular, parameters that increase the velocity and temperature gradients also increase the entropy generation rate and vice versa. In Figures 13–18, the values of the entropy generation rate expectedly vary across the channel including the walls in response to the evolution of the velocity and temperature gradients with increasing parameter values.

5.4. The Bejan Number. In this section, we plot the Bejan number (Be) across the channel under varying parameter conditions. The analysis in this section is similar to that for the previous section with N_s now replaced by Be.

Figures 19, 20, 21, 22, 23, and 24 show as expected that parameters which increase the entropy generation rate will correspondingly decrease the Bejan number and vice versa. In the vicinity of the walls, the strength of the fluid parameters will determine which mode of irreversibility dominates over the other.

6. Conclusion

We computationally investigate the inherent irreversibility in an unsteady hydromagnetic generalized Couette flow with variable electrical conductivity in the presence of induced electric field. We also notice that, due to the nature of the source terms, the fluid velocity and temperature will each decrease (resp., increase) with a corresponding increase in the parameters that decrease/increase the magnitudes of the source terms. We have also demonstrated computationally that parameters which increase the entropy generation rate will correspondingly decrease the Bejan number and vice versa. In particular, as the flow profiles vary in shape from linear to "parabolic" in response to varying parameter values, the velocity/temperature gradients correspondingly change in magnitude leading to noticeable effects in the entropy generation rates and Bejan numbers.

References

[1] R. Moreau, *Magnetohydrodynamics*, Kluwer Academic Publishers, Dordrecht, The Netherlands, 1990.

[2] J. Hartmann and F. Lazarus, "Hg-dynamics I. Theory of the laminar flow of an electrically conductive liquid in a homogeneous magnetic field," *Kongelige Danske Videnskabernes Selskab. Matematisk-Fysiske Meddelelser*, vol. 15, no. 6, pp. 1–28, 1937.

[3] A. Bejan, "Second-law analysis in heat transfer and thermal design," *Advances in Heat Transfer*, vol. 15, pp. 1–58, 1982.

[4] A. Bejan, *Entropy Genaration Minimization*, CRC, Boca Raton, Fla, USA, 1996.

[5] O. D. Makinde and O. O. Onyejekwe, "A numerical study of MHD generalized Couette flow and heat transfer with variable viscosity and electrical conductivity," *Journal of Magnetism and Magnetic Materials*, vol. 323, no. 22, pp. 2757–2763, 2011.

[6] O. Haddad, M. Abuzaid, and M. Al-Nimr, "Entropy generation due to laminar incompressible forced convection flow through parallel-plates microchannel," *Entropy*, vol. 6, no. 5, pp. 413–426, 2004.

[7] O. D. Makinde and A. Aziz, "Second law analysis for a variable viscosity plane Poiseuille flow with asymmetric convective cooling," *Computers and Mathematics with Applications*, vol. 60, no. 11, pp. 3012–3019, 2010.

[8] O. D. Makinde, "Thermodynamic second law analysis for a gravity-driven variable viscosity liquid film along an inclined heated plate with convective cooling," *Journal of Mechanical Science and Technology*, vol. 24, no. 4, pp. 899–908, 2010.

[9] A. Z. Sahin and R. Ben-Mansour, "Entropy generation in laminar fluid flow through a circular pipe," *Entropy*, vol. 5, no. 5, pp. 404–416, 2003.

[10] A. Z. Şahin, "Entropy generation in turbulent liquid flow through a smooth duct subjected to constant wall temperature," *International Journal of Heat and Mass Transfer*, vol. 43, no. 8, pp. 1469–1478, 2000.

[11] O. D. Makinde and A. S. Eegunjobi, "Effects of convective heating on entropy generation rate in a channel with permeable walls," *Entropy*, vol. 15, pp. 220–233, 2013.

[12] A. S. Eegunjobi and O. D. Makinde, "Effects of Navier slip on entropy generation in a porous channel with suction/injection," *Journal of Thermal Science and Technology*, vol. 7, no. 4, pp. 522–535, 2012.

[13] L. C. Wood, *Thermodynamics of Fluid Systems*, Oxford University Press, Oxford, UK, 1975.

[14] T. Chinyoka, "Computational dynamics of a thermally decomposable viscoelastic lubricant under shear," *Journal of Fluids Engineering*, vol. 130, no. 12, Article ID 121201, 7 pages, 2008.

[15] T. Chinyoka and O. D. Makinde, "Viscoelastic modeling of the diffusion of polymeric pollutants injected into a pipe flow," *Acta Mechanica Sinica*, vol. 29, no. 2, pp. 166–178, 2013.

[16] T. Chinyoka and O. D. Makinde, "Unsteady hydromagnetic flow of a reactive variable viscosity third-grade fluid in a channel with convective cooling," *International Journal for Numerical Methods in Fluids*, vol. 69, no. 2, pp. 353–365, 2012.

[17] T. Chinyoka, "Poiseuille flow of reactive Phan-Thien-Tanner liquids in 1D channel flow," *Journal of Heat Transfer*, vol. 132, no. 11, Article ID 111701, 7 pages, 2010.

[18] T. Chinyoka, "Suction-injection control of shear banding in non-isothermal and exothermic channel flow of Johnson-Segalman liquids," *Journal of Fluids Engineering*, vol. 133, no. 7, Article ID 071205, 12 pages, 2011.

[19] I. E. Ireka and T. Chinyoka, "Non-isothermal flow of a Johnson-Segalman liquid in a lubricated pipe with wall slip," *Journal of Non-Newtonian Fluid Mechanics*, vol. 192, pp. 20–28, 2013.

2

Fuzzy Performance between Surface Fitting and Energy Distribution in Turbulence Runner

Zhongwei Liang,[1] Xiaochu Liu,[1] Bangyan Ye,[2] and Richard Kars Brauwer[3]

[1] *School of Mechanical and Electrical Engineering, Guangzhou University, Guangzhou 510006, China*
[2] *School of Mechanical and Automotive Engineering, South China University of Technology, Guangzhou 510640, China*
[3] *Department of Information Engineering, Indian Institute of Technology (IIT), Kanpur 208023, India*

Correspondence should be addressed to Zhongwei Liang, lzwstalin@126.com

Academic Editors: A. Szekrenyes and H. Zhang

Because the application of surface fitting algorithms exerts a considerable fuzzy influence on the mathematical features of kinetic energy distribution, their relation mechanism in different external conditional parameters must be quantitatively analyzed. Through determining the kinetic energy value of each selected representative position coordinate point by calculating kinetic energy parameters, several typical algorithms of complicated surface fitting are applied for constructing microkinetic energy distribution surface models in the objective turbulence runner with those obtained kinetic energy values. On the base of calculating the newly proposed mathematical features, we construct fuzzy evaluation data sequence and present a new three-dimensional fuzzy quantitative evaluation method; then the value change tendencies of kinetic energy distribution surface features can be clearly quantified, and the fuzzy performance mechanism discipline between the performance results of surface fitting algorithms, the spatial features of turbulence kinetic energy distribution surface, and their respective environmental parameter conditions can be quantitatively analyzed in detail, which results in the acquirement of final conclusions concerning the inherent turbulence kinetic energy distribution performance mechanism and its mathematical relation. A further turbulence energy quantitative study can be ensured.

1. Introduction

Through searching the academic literature published in recent years we learn that resulting from the rapid progress of turbulence science, the complexity and spatial meaning of objective fluid characteristic have already broken the traditional sense, and the precise turbulence monitoring has touched upon energy distribution domain in its flow runner; furthermore, we found that the optimization design and arrange deployment on reversing valve runner wall play an important influence effect on the following model construction of kinetic energy distribution in practice. Although surface fitting provides crucial theoretical foundations for precision machining, the performance assessment between surface fitting algorithms and turbulence kinetic energy distribution in different parameter conditions still remains unstudied and needs further detailed investigations.

In the research area of microturbulence energy distribution, some original papers have been published in the topic of energy spectra measurement on a given complicated turbulence fluid section during the past several years, which provide a new investigation idea for turbulence fluid structure modeling [1–3]. For example, there are some researchers who focus on the comparison of the liquid energy spectra and velocity probability density functions with experimental data obtained by phase-sensitive constant-temperature anemometry [4]. Simultaneously, Liu et al. [5] studied the statistical properties of complex fluid field networks which were constructed from energy distribution in three-dimensional fully developed turbulence runner by using the visibility algorithm. All these works provide original ideas and science references for our further researches. In the subject of microfluid section feature analysis and kinetic energy distribution modeling, Panidis [6] has investigated the topic of turbulent flow field kinetic energy generated due to the interaction of grid turbulence pressure in a vertical channel of rectangular cross-section. More relevant theoretical progress can also be found in [7–9]. It can be seen that these traditional research results still keep a considerable distance from microfluid section kinetic energy characteristics and their respective fitting algorithms that being paid attention to, which become our research interest in this paper. With the help of three-dimensional fluid modeling, Ahmed [10] employed laser Doppler velocimeter to measure and model the three-dimensional flow properties of a confined, isothermal, swirling flow field in an axisymmetric sudden expansion research combustor. Fujiwara et al. [11] investigated the statistical characteristics of spatial distribution fluctuations of kinetic energies of each component wave and its time derivative in wave turbulence for a Hamiltonian system with a nondecay type dispersion relation. We can also learn other similar investigations and representative results from [12–14]. Most of them focused on the fluid mechanics analysis in a narrow theoretical sense which describes the studied turbulence energy distribution properties without any detailed consideration on its three-dimensional structure features, the mutual fuzzy performance mechanisms between fluid energy distribution models and their respective surface fitting algorithms, leaving these difficult problems unsolved which should be further investigated.

This paper is structured as follows. Section 1 outlines the importance and necessity of three-dimensional fuzzy evaluation of the influence mechanism between surface fitting and turbulence kinetic energy distribution in a reversing valve runner; Section 2 explains the detailed computation process and theoretical basis of turbulence kinetic energy; Section 3 describes some newly proposed mathematical features for accurately describing energy distribution surface models. Section 4 presents an experimental process and illustrates the computed results; then a detailed fuzzy mutual-influence analysis and quantitative data evaluation can be made in Section 5 with the newly proposed three-dimensional fuzzy evaluation method after performance comparison and suggestion presentation in the specific experimental process Section 6 concludes this paper as required.

2. Turbulence Kinetic Energy

Consider the fact that the inner structure of reversing valve runner is characterized by complicated topography, ultra-miniature size, and high-pressure closed condition, which explains the difficulties of describing the dynamic fluid properties. A microcomputerized runner model simulating a high-pressure reversing valve was obtained by computer-aided design, and then its solid model in 1 : 1 size scale can easily be produced by rapid forming/fast molding when using transparent high-polymer PVC material for clearly demonstrating the inner microtopography of target runner space and its subsequent turbulence moving process promptly.

In this paper, we use fluid function as the most frequently-used model to compute turbulence kinetic energy according to its inherent mechanism. In order to clearly describe the detailed computation condition and dynamic process of kinetic energy, we place microfluid sensors on the inner wall of valve runner, which contributes to the acquirement of turbulence parameter (time duration, fluid velocity, viscosity, pressure, temperature, concentration, density, and things like that vary with time and space) from different measurement positions; afterwards, we define turbulence kinetic energy K in the ith sampling time interval, by recording turbulence flow signal such as velocities, motion vectors, and flow directions in each instantaneous monitor position; it is supposed that the statistic time moments should be denoted as $T = \{t_1, t_2, t_3, \ldots, t_n\}$. With the amount of sampling levels being denoted as n, then turbulence average velocity in one presumptive time interval $\overline{\mu_i}$ can be computed as [15–17]

$$\overline{\mu_i} = \frac{1}{n}\sum_{i=1}^{n}\left[\frac{1}{T}\int_0^T \mu_i(t)dt\right]. \tag{1}$$

Here $\mu_i(t)$ denotes the turbulence instantaneous velocity in the time moment of t which locates in the ith sampling time interval. Thus turbulence intensity μ_i' in one presumptive time interval can be computed as

$$\mu_i' = \frac{1}{n}\sum_{i=1}^{n}\left[\frac{1}{T}\int_0^T \left[\mu_i(t) - \overline{\mu_i}\right]^2 dt\right]^{1/2}. \tag{2}$$

Here the dissipation rating ε is defined as

$$\varepsilon = \mu_t\left[\frac{\partial \mu_i}{\partial x}\right]^2 = C_D\frac{K^{3/2}}{\lambda}. \tag{3}$$

Turbulence kinetic energy K of the ith sampling time interval can be defined by following parametric equations:

$$\begin{aligned}
\rho\frac{\partial K}{\partial t} + \rho\mu_i\frac{\partial K}{\partial x_i} &= \frac{\partial}{\partial x_i}\left[\left(\mu + \frac{\mu_t}{\sigma_\varepsilon}\right)\frac{\partial K}{\partial x_i}\right] \\
&\quad + \mu_t\frac{\partial \mu_t}{\partial x_i}\left(\frac{\partial \mu_t}{\partial x_i} + \frac{\partial \mu_t}{\partial y_i} + \frac{\partial \mu_t}{\partial z_i}\right) - \rho\varepsilon, \\
\rho\frac{\partial K}{\partial t} + \rho\mu_i\frac{\partial K}{\partial y_i} &= \frac{\partial}{\partial y_i}\left[\left(\mu + \frac{\mu_t}{\sigma_\varepsilon}\right)\frac{\partial K}{\partial y_i}\right] \\
&\quad + \mu_t\frac{\partial \mu_t}{\partial y_i}\left(\frac{\partial \mu_t}{\partial x_i} + \frac{\partial \mu_t}{\partial y_i} + \frac{\partial \mu_t}{\partial z_i}\right) - \rho\varepsilon, \\
\rho\frac{\partial K}{\partial t} + \rho\mu_k\frac{\partial K}{\partial z_i} &= \frac{\partial}{\partial z_i}\left[\left(\mu + \frac{\mu_t}{\sigma_\varepsilon}\right)\frac{\partial K}{\partial z_i}\right] \\
&\quad + \mu_t\frac{\partial \mu_t}{\partial z_i}\left(\frac{\partial \mu_t}{\partial x_i} + \frac{\partial \mu_t}{\partial y_i} + \frac{\partial \mu_t}{\partial z_i}\right) - \rho\varepsilon.
\end{aligned} \tag{4}$$

Coefficient of the turbulence's viscosity μ_t is given as

$$\mu_t = C_\mu'\rho K^{1/2}L = \left(C_\mu' + C_D\right)\rho K^2\frac{L}{C_D K^{3/2}} = \frac{C_\mu'\rho K^2}{\varepsilon}. \tag{5}$$

Here C_μ' is an empirical factor, K is the kinetic energy value to be computed, and L is the length scale of turbulence movement. When integrating both sides of the differential equations, turbulence kinetic energy K can be finally obtained as

$$\begin{aligned} K &= \frac{1}{2}(\overline{\mu_i \mu_t}) \\ &= \left(\frac{1}{2n}\sum_{i=1}^{n}\left[\frac{1}{T}\int_0^T [\mu_i(t) - \overline{\mu_i}]^2 dt\right]^{1/2}\right) \times \frac{C_\mu{}' \rho}{\varepsilon} \\ &\quad \times \frac{1}{n\mu'^2}\sum_{i=1}^{n}\left[\frac{1}{T-\tau}\int_0^{T-\tau} [\mu_i(t)\mu_i(t+\tau)] dt\right] \\ &= \frac{C_\mu{}' \rho}{2\varepsilon n^2 \mu'^2}\left(\sum_{i=1}^{n}\left\{\left[\frac{1}{T}\int_0^T [\mu_i(t) - \overline{\mu_i}]^2 dt\right]^{1/2}\right.\right. \\ &\quad \left.\left. \times\left[\frac{1}{T-\tau}\int_0^{T-\tau} [\mu_i(t)\mu_i(t+\tau)] dt\right]\right\}\right) \\ &= \frac{C_\mu{}' \rho}{2\varepsilon n^2 \mu'^2 T(T-\tau)} \\ &\quad \times \sum_{i=1}^{n}\left\{\left[\int_0^T [\mu_i(t) - \overline{\mu_i}]^2\right]^{1/2} [\mu_i(t)\mu_i(t+\tau)]\right\} dt. \end{aligned} \tag{6}$$

This result can be regarded as the computed turbulence kinetic energy in reversing valve runner as the objective target [18, 19].

3. Mathematical Features of Energy Distribution Surface

As we use several typical surface fitting algorithms in this experiment, such as surface of NURBS, energy optimization modeling, B-spline of quasiuniform bicubic, trigonometry Bernstein-Bezier, and scattered data interpolation, the following mathematical features are newly proposed for describing our fitting results.

Feature 1. Consider

$$\begin{aligned} \varphi &= \alpha_1 \sum_{i=1}^{m} W_{ui}{}^2 + \beta_1 \sum_{i=1}^{m} W_{uui}{}^2 \\ &\quad + \alpha_2 \sum_{j=1}^{n} W_{vj}{}^2 + \beta_2 \sum_{j=1}^{n} W_{vvj}{}^2 \\ &\quad + \alpha_1\alpha_2\beta_1\beta_2 \sum_{i=1}^{m}\sum_{j=1}^{n} W_{uivj}{}^2 \\ &\quad - 2f(u,v)W. \end{aligned} \tag{7}$$

Here W is one constructed surface in the form of B-spline primary function; $W_u, W_v, W_{uu}, W_{vv}, W_{uv}$ are the partial derivatives of the objective fitted surface W in the first order, second order, and hybrid state of u, v axes, respectively; $\alpha_1, \alpha_2, \beta_1, \beta_2$ are given parameters, and $f(u,v)$ is a given function of surface vector, m, n are the order numbers of surface vector of u, v axes [20].

Feature 2. Consider

$$\begin{aligned} \zeta &= \sum_{i=1}^{m}\int_\Omega S_u(u_i)^2 du + \sum_{i=1}^{m}\sum_{i=1}^{m}\iint_{i\in\Omega} S_{uu}(u_{ii})^2 du\, du \\ &\quad + \sum_{j=1}^{n}\int_\Omega S_v(v_j)^2 dv + \sum_{j=1}^{n}\sum_{j=1}^{n}\iint_{j\in\Omega} S_{vv}(v_{jj})^2 dv\, dv \\ &\quad + \sum_{i=1}^{m}\sum_{j=1}^{n}\iint_\Omega S_{uv}(u_i v_j)^2 du\, dv. \end{aligned} \tag{8}$$

Here $S_u(u_i)$, $S_{uu}(u_{ii})$, $S_v(v_j)$, $S_{vv}(v_{jj})$, $S_{uv}(u_i v_j)$ are the first order, second order, and hybrid derivatives of surface $f(u,v)$ in u, v axes.

Feature 3. Consider

$$\begin{aligned} \rho &= -2 \sum_{i=0,m_u}^{u}\sum_{j=0,m_v}^{v} V_{i,j} \oiint_{i,j\in\Omega} N_{i,s_u}(u) \\ &\quad * N_{j,s_v}(v) N_{i,j}(uv) f(u,v) du\, dv. \end{aligned} \tag{9}$$

Here $N_{i,s_u}(u), N_{j,s_v}(v), N_{i,j}(uv)$ are the boundary control B-spline surface in u, v, uv axes, respectively; $f(u,v)$ is a given vector function, with $V_{i,j}$ denoting the transitional vector obtained from the surface external load.

Feature 4. Consider

$$Z_{nm} = \frac{n+1}{\pi}\int_0^1\int_0^{2\pi} R_{nm}(\gamma) e^{jm\theta} f(\gamma,\theta)\gamma\, d\gamma\, d\theta. \tag{10}$$

Here n is a positive integer or zero, m is an integer and $n - |m|$ = even number, with $|m| \le n$; r is vector length from an origin point to a given target control point of surface $(x,y,z) : \gamma = \sqrt{x^2 + y^2 + z^2}$, $-1 < x, y, z < 1$.

Feature 5. Consider

$$\begin{aligned} \varepsilon_{n_1,n_2} &= \frac{\max_{(u,v)V} |P(u,v) - P_{n_1,n_2}(u,v)|}{|P(u,v)|} \times 100\%; \\ \overline{S}_{n_1,n_2} &= 1 - \frac{\varepsilon - \varepsilon_{n_1,n_2}}{\varepsilon_{n_1-1,n_2-1} - \varepsilon_{n_1,n_2}}. \end{aligned} \tag{11}$$

Here $\beta_{n-1}(t)$ is the high-frequency surface obtained from the 1st order wavelet decomposition of $f_n(t)$.

4. Experiment and Computation

Figure 1 illustrates the constructed three-dimensional model of one given high-pressure reversing valve runner (Type No.

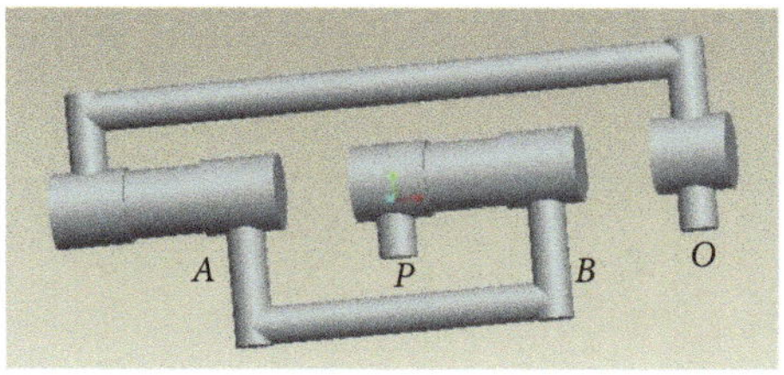

FIGURE 1: The constructed three dimensional model of target runner.

D5-02-2B-AC-A01) by using PRO-E software, with its spatial structure gridded in fluent system. Figure 2 denotes the distribution characteristics and change processes of turbulence kinetic energy in it, the values of kinetic energy illustrated by different color sections in the left column. First the required turbulence field is simulated with SNQ-1TX-140 microturbulence generator, and a produced PVC transparent valve runner is applied for clearly observing the detailed flow process. The specific experimental condition can be defined as follows: flow quantity is 10–20 Min/L, working pressure is higher than 20–30 MPa, flow velocity of flow field exit is faster than 10–30 cm/s, the spatial arrangement of valve runner is 150 mm × 150 mm × 30 mm, together with the experimental time duration being kept as long as 2–4 hours; all these condition parameters require precision adjustment in the interest of energy distribution modeling.

As Reynolds number $\mathrm{Re} = uh/v$ is defined as 4700~4900, Figure 3 denotes the gridded fluid runner, and Figure 4 shows the turbulence imaging result. Through adopting finite volume method (FVM) in a staggered grid we implement a discretized data process on turbulence equation set. By positioning those monitor points that show key fluid parameters such as pressure P, dissipation rating ε at the center of grid boundary, and the monitor points of flow velocity μ on the grid boundary, we use a power function to parameterize the whole duration of data processing.

The exit boundary pressures of turbulence field are supposed as identical to those of external environment, whose normal gradient value is normally determined as zero. For the purpose of describing the boundary influences emerging from turbulence field wall, we assume they are from a nonslip condition. Namely, the three-dimensional motion velocities at the objective positions of turbulence monitoring points $\mu_i(U, V, R)$ are defined as $\Delta S_{pU} = -A_{\text{cell}}\Gamma_{\text{wall}}/\delta_p$.

Here ΔS_{pU} denotes the corrected value of an original item, A_{cell} denotes the area of a boundary grid which parallels a flow field section, and Γ_{wall} denotes an effective exchanging coefficient of velocity components that normal to the runner wall [21].

Turbulence motion parameters such as flow velocities, pressures, and pressure intensities are calculated or measured at each grid monitory point in different boundary conditions; we compute turbulence kinetic energy K with one detailed representative calculating process in k-ε model, as described by (1)–(6). Table 1 shows the computation process and result of comparison of turbulence parameter in different experimental condition by using k-ε model.

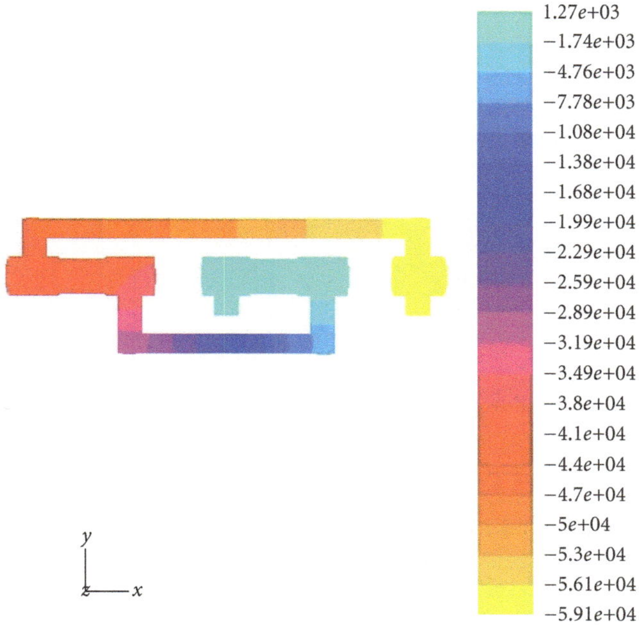

FIGURE 2: The distribution characteristics and change processes of turbulence kinetic energy in a high-pressure valve runner.

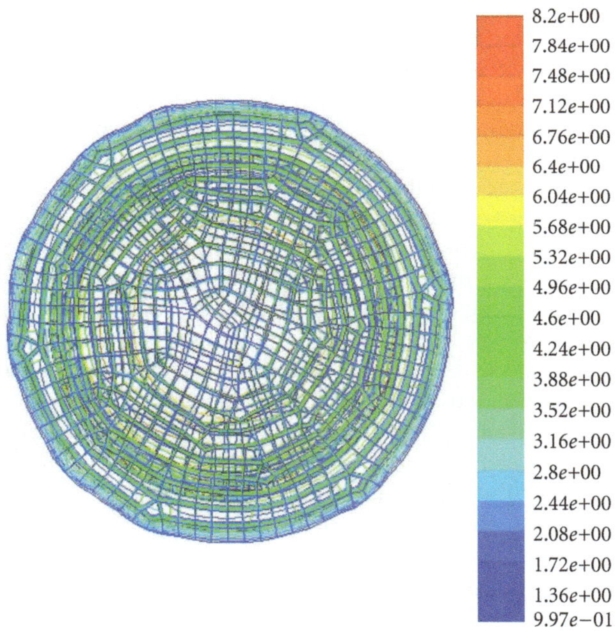

FIGURE 3: The gridded fluid section in the high-pressure reversing valve runner.

Simultaneously an instantaneous flow tester from the state key laboratory for hydraulic control technology hosted at Guangzhou University is used for directly measuring turbulence velocities, motion directions, and determining kinetic energy on the above-mentioned section point, which facilitates the comparison between those measured results and computed ones in this table; through inspection of the deviation value between K and K_{measured} this newly proposed

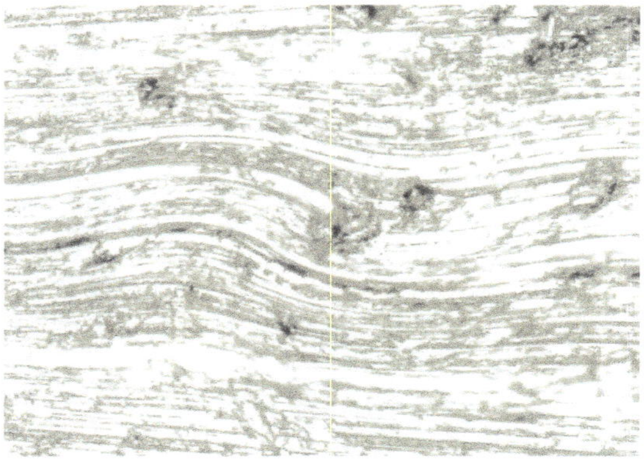

Figure 4: Turbulence in the runner of a high-pressure reversing valve.

calculation method of turbulence kinetic energy can be verified [22].

Afterwards, in the desire of calibrating the distribution of those obtained kinetic energy values on the objective section, we mesh the whole section plane into 100×80 points in x-axis and$y$-axis, respectively, with consideration of practical conditions and precision requirements, as shown in Figure 3.

For describing the respective microfluid kinetic energy distribution on the objective runner, those inflection points representing their particular energy value with a symbolized significance are chosen as the surface control points, such as the coordinate points with boundary values, curvature change rules, corner values, salient values, discrete values, or stepping values as well [22]. As shown in Figure 5, when we regarded the turbulence kinetic energy value of one objective runner position point as a vertical coordinate of z axis which is perpendicular to its belonging section plane denoted by x and y ones, the three-dimensional visual point cloud of energy distribution can be obtained by calculating kinetic energy values, and some key position points with representative kinetic energy value (or control points in a geometrical sense) are highlighted in red, which helps to quantitatively evaluate the mutual fuzzy relation mechanism between energy distribution characteristics and surface fitting algorithms in a specific experimental condition.

As we denote a geometrical corner point as origin O, an absolute coordinate system $O(X, Y, Z)$ is established. We import the kinetic energy values computed on the base of aforementioned algorithms; the energy values can be regarded as the z-axis coordinates. Figure 6 shows the kinetic energy distribution in the objective runner obtained by experimental simulation, and Figure 7 shows the dynamic transitional vectors as well. Under the idea of guidance of this newly proposed concept, using Catia v5r19 we established the skeleton frame of kinetic energy distribution surface constructed by using key position points and the dynamic transitional vectors between them, as one skeleton patch as shown in Figure 8. Furthermore, in order to optimize the surface precision qualities, Figure 9 presents the smoothing operation of one spatial grid obtained from trigonometry Bernstein-Bezier method, which can be achieved on the base of wiping off the redundant small surface patches in

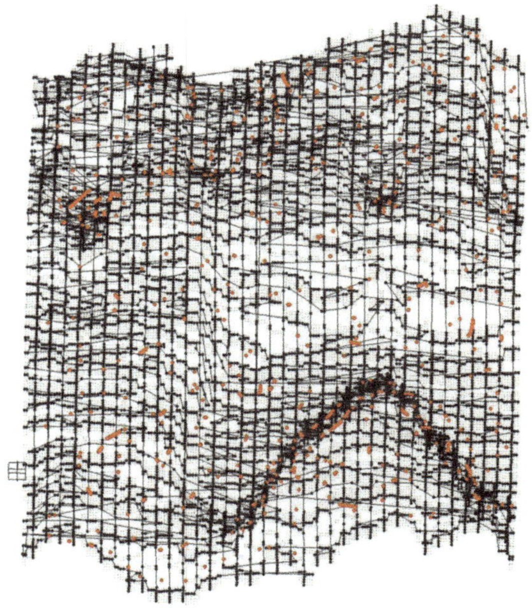

Figure 5: The point cloud of energy distribution section obtained by calculating kinetic energy values on each position point.

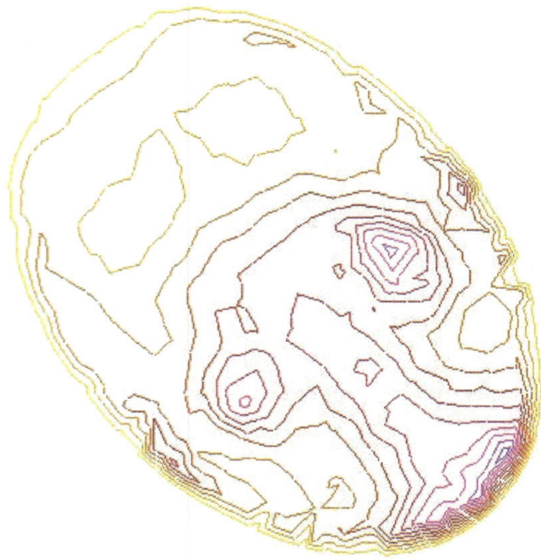

Figure 6: The kinetic energy value spatial distribution area in the objective runner.

boundary areas. Thus their respective energy distribution surface can be finally established; for example, the constructed turbulence kinetic energy distribution surface of NURBS on high-pressure reversing valve runner, with its energy values, is highlighted by different color areas in this figure, as Figure 10 shows, the same as energy optimization modeling surface (Figure 11), B-spline surface of quasiuniform bicubic (Figure 12), trigonometry Bernstein-Bezier surface (Figure 13), and scattered data interpolation surface (Figure 14); it is worth noting that the detailed surface fitting processes are abbreviated in the interest of an obvious limiting length and research focus of this paper. As the constructed surface is enclosed by the boundary control

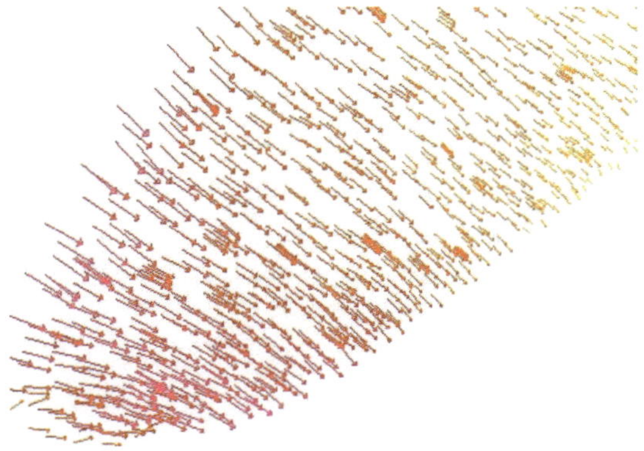

FIGURE 7: The dynamic transitional vectors of kinetic energy distribution.

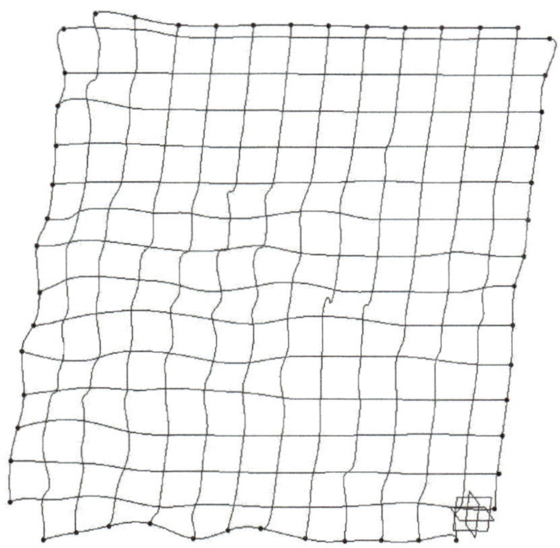

FIGURE 8: The skeleton frame of kinetic energy distribution surface.

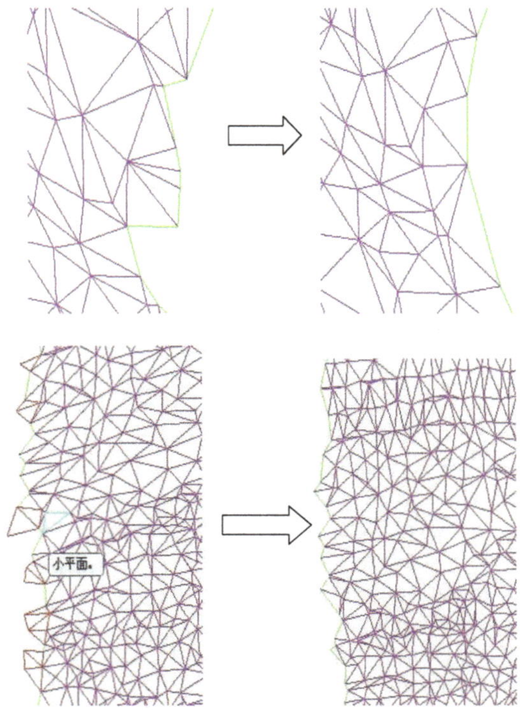

FIGURE 9: The smoothing operation of one specific spatial grid of target kinetic energy distribution surface model.

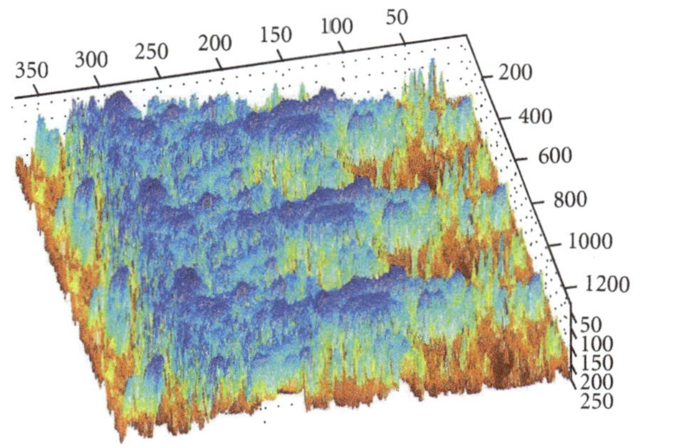

FIGURE 10: The constructed turbulence kinetic energy distribution surface of NURBS in high-pressure reversing valve runner.

curves, we use u and v axes to denote the transverse and longitudinal orientations with their value scales being $[0, 4000]$ and $[0, 5000]$, respectively, (um). Simultaneously, as the result of the data value of the vertical z axis (z direction) has a completely different meaning from that of u and v axis, therefore the value range of kinetic energy is used to clearly indicate its stereo features and spatial surface details. All surface feature blocks are highlighted by pseudocolors in order to illustrate energy distribution details as well.

In order to accurately quantify turbulence energy distribution with the help of these surfaces, by using (7)–(11) we determine the spatial mathematical features, with the mean values of these feature results being demonstrated in Tables 3, 4, 5, 6, and 7. On this basis a specific analysis and data evaluation can be successfully conducted as follows.

5. Three-Dimensional Fuzzy Performance Analysis and Evaluations

Table 2 defines the experimental parameters for different surface modeling methods. In this paper, we propose an improved three-dimensional fuzzy parameter system to establish a reliable influence evaluation mechanism as required. Different from those traditional ones, it does not require any previous information other than the three dimensional data to be disposed, but which needed by fuzzy ones [23]. Feature parameter sequence $f_i(k)$ can be determined as

$$\text{feature}_i(k) = (f_i(1), f_i(2), \ldots, f_i(n)). \quad (12)$$

Here $f_i(k)$ denotes the surface feature sequence parameters obtained from the aforementioned steps (objective sequence), $i \in [1, 2, 3, 4, 5]$ denotes the number of surface features, and k denotes the sample surface blocks with their

Table 1: Computation process and result comparison of turbulence parameter in different experimental condition by using k-ε model.

μ (cm/s)	ε (cm^2/s^3)	K_{th} (cm^2/s^2)	K (cm^2/s^2)	$K_{measured}$ (cm^2/s^2)	Deviation value (cm^2/s^2)
5	2.75×10^{-3}	3418.23	4351.19	6278.24	1927.05
10	7.43×10^{-2}	4759.12	6183.17	7187.69	1004.52
15	8.75×10^{-3}	6418.23	6351.19	6278.24	−72.95
20	9.33×10^{-2}	7359.12	8183.17	8487.69	304.52
25	0.244	10795.16	10698.33	11627.49	929.16
30	0.343	12697.26	11876.24	11394.06	−482.18
35	0.544	18795.16	17698.33	17627.49	70.84
40	0.943	37697.26	40876.24	41394.06	517.82

Table 2: External experimental parameters for different surface modeling methods.

Number of control points (u)	400	600	800	1000	1200	1400	1600	1800
Number of control points (v)	400	600	800	1000	1200	1400	1600	1800
Order of spline surface	1		2		3		4	
Order of knot vector	1		2		3		4	
Order of normal vectors	1		2		3		4	
Order of derivative vectors	1		2		3		4	
Number of constraint vectors	0	2	4	6	8	10	12	14
Order of constraint vectors	1		2		3		4	
Order of continuous level	1		2		3		4	
Kinetic energy coefficient	1.0	1.5	2.0	2.5	3.0	3.5	4.0	4.5
Rank of derivative matrix	10	20	30	40	50	60	70	80
Approximate error range	0.4	0.5	0.6	0.7	0.8	0.9	0.95	1.0

Table 3: Energy distribution surface mathematical features in NURBS surface.

No.	Feature 1	Feature 2	Feature 3	Feature 4	Feature 5
a	253.4	1250.3	896.4	2254.1	654.2
b	452.1	1147.6	887.6	2269.5	669.5
c	336.5	1584.6	992.4	2547.1	698.5
d	239.8	1325.8	865.4	2365.4	447.5
e	339.5	2014.6	857.4	3026.9	853.2
f	458.6	1854.3	887.4	3365.1	602.5
g	447.2	2231.0	836.5	3022.8	598.7
h	369.5	2014.8	884.9	3369.4	558.4
i	147.8	1578.9	759.5	3157.4	569.5
j	664.6	1148.9	771.4	4012.6	559.4
k	669.8	1574.8	725.6	3995.5	554.7
l	558.7	1369.5	771.2	3624.5	563.9

Table 4: Energy distribution surface features in energy optimization modeling surface.

No.	Feature 1	Feature 2	Feature 3	Feature 4	Feature 5
a	336.5	1458.7	889.7	2014.5	447.8
b	395.8	1169.5	895.4	2236.5	456.9
c	402.5	1785.5	887.5	2541.8	552.4
d	448.6	1126.6	886.9	2214.7	568.4
e	451.2	1255.4	892.6	2365.9	556.3
f	395.8	2014.4	902.5	2548.5	529.6
g	336.5	2230.1	911.4	2547.8	547.8
h	365.7	1254.6	875.6	2364.5	551.2
i	602.5	1138.6	884.5	3014.5	536.9
j	605.4	1204.5	902.6	2893.2	547.8
k	663.1	1169.5	923.5	2965.4	548.9
l	625.4	1247.5	933.6	2268.2	557.4

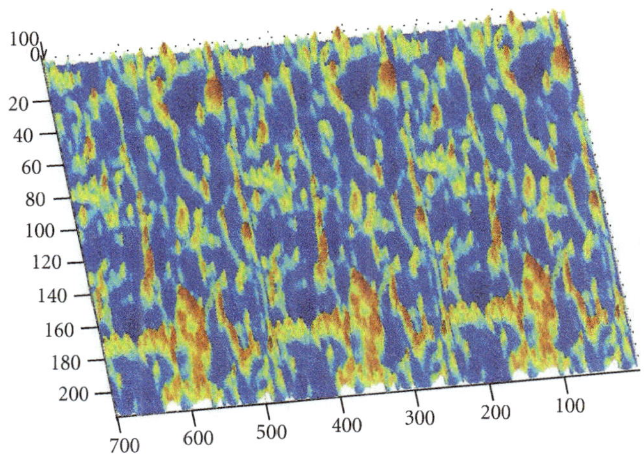

Figure 11: The constructed turbulence kinetic energy distribution surface of energy optimization modeling in high-pressure reversing valve runner.

Table 5: Energy distribution surface features in B-spline quasiuniform bicubic.

	Feature 1	Feature 2	Feature 3	Feature 4	Feature 5
a	625.4	1478.6	1024.5	2014.7	698.5
b	557.8	1159.6	1152.3	2233.6	815.2
c	512.6	1203.6	1045.6	2014.5	478.6
d	654.3	1475.2	1102.3	2159.8	556.9
e	478.5	1149.8	1069.8	2305.5	547.8
f	551.2	1523.6	1024.5	2147.8	558.6
g	526.3	1402.5	1147.6	2236.5	526.3
h	554.7	1163.2	1036.5	2105.6	547.2
i	625.4	1299.8	1029.6	2452.3	551.2
j	663.2	1475.6	1302.5	2144.5	529.8
k	602.5	1254.6	1336.5	2036.5	547.8
l	663.1	1178.9	1025.4	2045.6	553.6

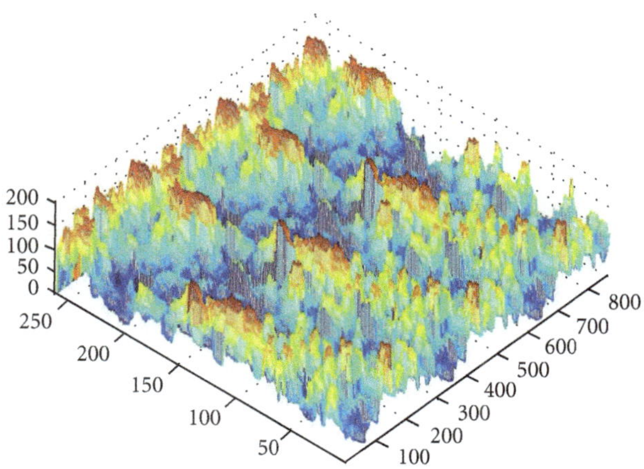

Figure 12: The constructed turbulence kinetic energy distribution B-spline surface of quasiuniform bicubic in high-pressure reversing valve runner.

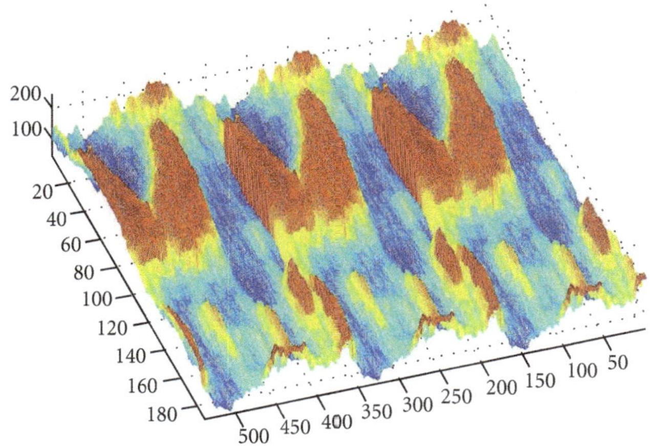

Figure 13: The constructed turbulence kinetic energy distribution surface of trigonometry Bernstein-Bezier in high-pressure reversing valve.

total number being n. On the other hand, the parameter sequence of modeling condition is illustrated as:

$$\text{parameter}_i = (tp_i(1), tp_i(2), \ldots, tp_i(n)). \quad (13)$$

Here $tp_i(k)$ denotes the condition feature sequence parameters (objective sequence), and $i \in [1,2,3,4,5,6,7,8,9,10,11,12]$ denotes the specific feature numbers [24–26].

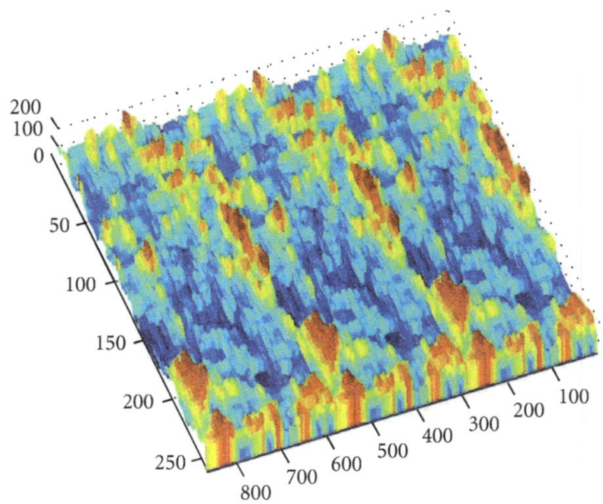

FIGURE 14: The constructed turbulence kinetic energy distribution surface of scattered data interpolation in high-pressure reversing valve runner.

TABLE 6: Energy distribution surface features in trigonometry Bernstein-Bezier surface.

	Feature 1	Feature 2	Feature 3	Feature 4	Feature 5
a	712.5	1214.5	1025.6	2147.5	558.6
b	658.6	1147.8	998.5	2236.5	893.6
c	669.5	1326.5	952.6	2014.6	602.5
d	625.4	1145.6	869.5	2514.6	621.5
e	615.4	1026.3	884.5	2014.5	548.6
f	639.5	1058.4	872.6	2231.5	554.7
g	702.5	1024.5	893.6	1987.5	523.6
h	742.3	1136.5	902.6	2036.5	547.8
i	605.4	1024.5	993.6	2245.6	523.2
j	663.5	1203.6	924.5	2011.4	514.8
k	685.2	1147.5	923.6	2036.5	556.3
l	635.4	1258.6	914.5	1877.5	529.6

TABLE 7: Energy distribution surface features in scattered data interpolation surface.

	Feature 1	Feature 2	Feature 3	Feature 4	Feature 5
a	478.9	998.5	1125.6	1995.6	635.8
b	448.5	895.6	1136.2	1897.5	665.4
c	524.6	885.4	1025.4	1785.6	639.5
d	485.6	875.6	995.6	1887.9	624.7
e	485.2	932.5	984.5	1987.6	702.5
f	402.6	902.6	975.2	1988.6	711.6
g	446.2	887.5	996.5	2015.6	711.3
h	475.8	893.6	924.5	2114.5	589.6
i	425.6	845.2	968.5	1899.6	588.9
j	485.2	869.3	889.6	1955.6	602.3
k	441.5	887.5	1021.3	2036.5	633.1
l	442.6	879.6	1125.4	1988.6	654.9

The sequence of surface fitting methods is described as

$$\text{method}_i = (m_i(1), m_i(2), \ldots, m_i(n)). \quad (14)$$

Then we compute the fuzzy relation operator $\text{fuzzy}_i{}'(k, i)$ as follows, with which an integrated fuzzy relation matrix can be established:

$$\text{fuzzy}_i{}'(k,i) = \frac{mn\sum_{k=j}^{m}\sum_{r=i}^{n}\left[\text{feature}_i(k)-\overline{\text{feature}_i(k)}\right]*\mathcal{A}*\left[m_i(k)-\overline{m_i(k)}\right]}{\sqrt{\sum_{r=j}^{m}\sum_{k=i}^{n}\left[\text{feature}_i(k)-\overline{\text{feature}_i(k)}\right]\sum_{r=1}^{m}\sum_{k=1}^{n}}*\sqrt{\sum_{r=j}^{m-j+1}\sum_{k=i}^{n-i+1}\left[m_i(k)-\overline{m_i(k)}\right]}*\sqrt{\sum_{r=j}^{m-j+1}\sum_{k=i}^{n-i+1}\mathcal{A}}}, \quad (15)$$

where $\mathcal{A}$ denotes $[\text{technical\_parameter}_i(k) - \overline{\text{technical\_parameter}_i(k)}]$. Here $k = 1, 2, \ldots, n$, $i = 1, 2, \ldots, m$, and $\overline{\text{feature}_i(k)}$, $\overline{\text{technical\_parameter}_i(k)}$, $\overline{m_i(k)}$ are the average function vectors of $\text{feature}_i(k)$ and $\text{technical\_parameter}_i(k)$, $m_i(k)$, respectively. $m_i(k)$ denotes the surface fitting methods (reference sequence), and $i \in [1, 2, 3, 4, 5]$ denotes the number of fitting methods.

The fuzzy relation coefficient $\lambda_i(k, r)$ between the approximate target and the practical surface can be calculated as follows:

$$\lambda_i(k,r) = \frac{\sum_{i-1}^{r}\left(\min_{i\in I}\min_k\left|\begin{matrix}\text{fuzzy}_0{}^*(k,r)\\-\text{fuzzy}_i{}^*(k,r)\end{matrix}\right|\right)+\beta\sum_{i=1}^{r}\left(\max_{i\in I}\max_k\left|\begin{matrix}\text{fuzzy}_0{}^*(k,r)\\-\text{fuzzy}_i{}^*(k,r)\end{matrix}\right|\right)}{\sum_{i=1}^{r}\left(\left|\begin{matrix}\text{fuzzy}_0{}^*(k,r)\\-\text{fuzzy}_i{}^*(k,r)\end{matrix}\right|\right)+\beta\sum_{i=1}^{r}\left(\max_{i\in I}\max_k\left|\begin{matrix}\text{fuzzy}_0{}^*(k,r)\\-\text{fuzzy}_i{}^*(k,r)\end{matrix}\right|\right)}. \quad (16)$$

Here β is the distinguishing parameter set as 0.5~0.7. The details are shown in Tables 8–12, which illustrates the fuzzy performance mechanism between them in different experimental parameter conditions.

With Table 8 it can be observed that NURBS fitting method exerts an obvious fuzzy relation influence on amendment quantity of external load and fairing error. And it is also highly impacted by the number of control points in u and v domain, number of boundary constraint vectors, and rank range of derivative coefficient matrix, and so forth. Energy optimization surface of turbulence kinetic energy distribution, as Table 9 shows, obviously keeps a rather close fuzzy relation with elasticity variance ratio and Zernike moment, and so forth. It is highly impacted by the number of boundary constrain vectors, order of normal vectors, and kinetic energy coefficient of external loading. Quasiuniform bicubic B-spline surface of turbulence kinetic energy (Table 10), markedly keeps close fuzzy relation with energy dispersive-ratio or faring error in the proposed experimental parameter conditions. It can be affected by the number of boundary constrain vectors, order of knot vector, and number of boundary constrain vectors, and so forth. The Bernstein-Bezier surface used for fitting turbulence kinetic energy distribution, as Table 11 demonstrates, obviously exerts a fuzzy influence on elasticity variance ratio and amendment quantity of external load. Scattered data interpolation used for turbulence kinetic energy distribution models, as shown by Table 12, keeps a close fuzzy relation with energy-dispersive ratio and amendment quantity.

Table 13 shows the performance comparisons of these proposed surface fitting algorithms in the whole experimental process. With its detailed demonstration we conclude that trigonometry bicubic B-spline fitting method has a wide application in a characteristic surface fitting condition that puts greater emphasis on the accuracy rate and shape precision of energy distribution; with the similar evaluation approach we observe that B-spline surface of quasiuniform bicubic will be more suitable for constructing a simpler approximate fluid energy distribution model; trigonometry Bernstein-Bezier ensures the high fitting accuracy and control precision of knot vectors and surface models; finally we achieve a good experimental process in computation time, computation storage, and approximate error when using scattered data interpolation. After data comparison and detailed analysis we found that energy optimization surface modeling gets an optimum integrated performance capability in practice and therefore becomes our preferred choice.

The following suggestions are proposed for developing turbulence energy distribution modeling. (1) In the interest of obtaining accurate coordinate information it is suggested that the distribution intensity when selecting position points with representative kinetic energy values (or being called the control points in geometrical modeling sense) should be in a medium-low level from 20 points per mm^2 to 40 points per mm^2, the function order of knot vectors or boundary constrain vectors be kept in a relatively low state from three to four, and the signal probing frequency of coordinate acquirement be lower than 300 times/minute, which will strongly support for a high fidelity reflection of energy distribution characteristics in a detailed pattern. (2) Fluid kinetic energy distribution properties keep a close fuzzy relation with their inherent surface fitting qualities and practical measuring parameters, with the analysis details being clearly shown in the previously mentioned paragraphs. (3) It is proposed that the three-dimensional fuzzy performance mechanism possessed by these typical energy distribution surface fitting methods be quantized by an explicit mathematical expression through detailed experimental evaluations. (4) We can make a final assessment on the specific applications of typical surface fitting algorithms when dealing with turbulence kinetic energy distribution problems: NURBS can be widely used in a preliminary characteristic classification of

TABLE 8: Fuzzy relation degrees between experimental condition parameters and kinetic energy distribution surface features in the form of NURBS surface.

Condition	Feature 1	Feature 2	Feature 3	Feature 4	Feature 5
Number of control points (u)	0.8956	0.9254	0.4478	0.5849	0.6589
Number of control points (v)	0.6254	0.9125	0.5963	0.8965	0.5542
Order of spline surface	0.2546	0.9632	0.9254	0.8475	0.6235
Order of knot vector	0.3365	0.8654	0.9125	0.4521	0.8452
Order of normal vectors	0.2547	0.7748	0.6589	0.9214	0.5742
Order of derivative vectors	0.4156	0.7639	0.4785	0.8523	0.6214
Number of constrain vectors	0.6215	0.5482	0.9921	0.6214	0.8523
Order of constrain vectors	0.2558	0.1254	0.8848	0.6144	0.9214
Order of continuous level	0.5478	0.3654	0.6215	0.5248	0.5426
Kinetic energy coefficient	0.3654	0.8457	0.5144	0.6395	0.3654
Rank of derivative matrix	0.5523	0.3369	0.4478	0.6235	0.5846
Approximate error range	0.6245	0.8452	0.9542	0.5145	0.6695

TABLE 9: Fuzzy relation degrees between experimental condition parameters and kinetic energy distribution surface features in the form of energy optimization modeling surface.

Condition	Feature 1	Feature 2	Feature 3	Feature 4	Feature 5
Number of control points (u)	0.1542	0.5547	0.6354	0.2254	0.3654
Number of control points (v)	0.6635	0.6235	0.5524	0.1687	0.5547
Order of spline surface	0.2587	0.5842	0.4852	0.6245	0.6325
Order of knot vector	0.9856	0.3369	0.2653	0.6952	0.9236
Order of normal vectors	0.8854	0.3158	0.4875	0.9025	0.2254
Order of derivative vectors	0.7536	0.4758	0.2654	0.9125	0.5784
Number of constraint vectors	0.6658	0.6125	0.5524	0.2458	0.2654
Order of constraint vectors	0.6235	0.4857	0.6932	0.6214	0.2036
Order of continuous level	0.4859	0.6235	0.9245	0.6254	0.5147
Kinetic energy coefficient	0.3698	0.5547	0.8954	0.3652	0.2658
Rank of derivative matrix	0.9325	0.2584	0.8214	0.3924	0.4852
Approximate error range	0.9254	0.2415	0.6352	0.9226	0.6214

TABLE 10: Fuzzy relation degrees between experimental condition parameters and kinetic energy distribution surface features in B-spline surface of quasiuniform bicubic.

Condition	Feature 1	Feature 2	Feature 3	Feature 4	Feature 5
Number of control points (u)	0.5524	0.3958	0.2658	0.5578	0.5547
Number of control points (v)	0.1025	0.6589	0.6354	0.6247	0.6524
Order of spline surface	0.2514	0.9254	0.8547	0.5478	0.6635
Order of knot vector	0.4152	0.9254	0.2458	0.6692	0.5874
Order of normal vectors	0.5598	0.6354	0.3654	0.5896	0.4025
Order of derivative vectors	0.9025	0.4258	0.9221	0.6254	0.2214
Number of constraint vectors	0.9254	0.4475	0.9223	0.6698	0.6354
Order of constraint vectors	0.9125	0.5263	0.6245	0.8547	0.2954
Order of continuous level	0.2635	0.3958	0.5587	0.9365	0.6245
Kinetic energy coefficient	0.2875	0.6322	0.8475	0.2584	0.2258
Rank of derivative matrix	0.6395	0.1475	0.6354	0.6635	0.1587
Approximate error range	0.9254	0.5874	0.6698	0.1547	0.6325

turbulence kinetic energy distribution; energy optimization can be used to promulgate or improve the distribute rationality of modeling information and the strain intensity of kinetic energy signal illustration, which becomes our top-preferred selection item in this experiment; quasiuniform bicubic B-spline can be conveniently used for optimizing the energy details of those newly constructed turbulence energy distribution; Bernstein-Bezier has been frequently used in the shape optimization of energy distribution microsurface when an external dynamic energy loading is exerted; we can also use scattered data interpolation to produce a more robust surface model of energy distribution when facing with some noise point coordinates (position points with useless or misleading kinetic energy information or things like that)

TABLE 11: Fuzzy relation degrees between experimental condition parameters and kinetic energy distribution surface features trigonometry Bernstein-Bezier surface.

Condition	Feature 1	Feature 2	Feature 3	Feature 4	Feature 5
Number of control points (u)	0.8831	0.5521	0.6311	0.5538	0.9831
Number of control points (v)	0.8836	0.7732	0.7713	0.8831	0.9042
Order of spline surface	0.6637	0.5572	0.6637	0.6618	0.7748
Order of knot vector	0.6648	0.6618	0.9057	0.5173	0.1623
Order of normal vectors	0.6627	0.1649	0.1724	0.8946	0.0641
Order of derivative vectors	0.3326	0.4621	0.5731	0.6681	0.8862
Number of constraint vectors	0.8845	0.1713	0.5538	0.7852	0.5824
Order of constraint vectors	0.6411	0.4813	0.5294	0.7524	0.7711
Order of continuous level	0.6411	0.6582	0.9057	0.7742	0.6630
Kinetic energy coefficient	0.7842	0.8952	0.9083	0.6638	0.6491
Rank of derivative matrix	0.6381	0.6749	0.7482	0.8849	0.7381
Approximate error range	0.6481	0.7749	0.8940	0.7759	0.7391

TABLE 12: Fuzzy relation degrees between experimental condition parameters and kinetic energy distribution surface features in scattered data interpolation surface.

Condition	Feature 1	Feature 2	Feature 3	Feature 4	Feature 5
Number of control points (u)	0.2145	0.5874	0.5596	0.5587	0.5526
Number of control points (v)	0.6332	0.8214	0.8254	0.7154	0.4475
Order of spline surface	0.6254	0.9025	0.6325	0.4896	0.8215
Order of knot vector	0.2014	0.6684	0.7152	0.8854	0.6635
Order of normal vectors	0.6005	0.7025	0.9025	0.8256	0.5547
Order of derivative vectors	0.3958	0.1547	0.8869	0.7742	0.5596
Number of constraint vectors	0.5874	0.3654	0.1452	0.9025	0.1475
Order of constraint vectors	0.3369	0.5586	0.6054	0.4856	0.6524
Order of continuous level	0.9021	0.4852	0.6625	0.5514	0.5589
Kinetic energy coefficient	0.3654	0.6635	0.7012	0.6325	0.9325
Rank of derivative matrix	0.8878	0.9201	0.5584	0.4475	0.6635
Approximate error range	0.9356	0.5547	0.9245	0.5589	0.5548

TABLE 13: Performance comparison of different surface fitting algorithms during the modeling processes of turbulence kinetic energy distribution.

Algorithms	Surface distortion rate	Computation time	Approximate error	Computation storage
NURBS surface	3.254%	44.5 s	6.55%	4475.6 kb
Energy optimization modelling	6.225%	36.5 s	4.15%	33025.4 kb
B-spline quasi-uniform bicubic	4.115%	39.5 s	6.99%	11475.2 kb
Trigonometry Bernstein-Bezier	5.114%	62.5 s	3.58%	8896.5 kb
Scattered data interpolation	6.226%	44.5 s	4.77%	9924.5 kb

and truncation errors during data acquisition as well; all these suggestions provide a valuable reference and improve evaluation practices for solving all similar problems in the future.

6. Conclusions

This paper sought to investigate the fuzzy performance mechanism exerted by surface fitting algorithms on the constructed turbulence kinetic energy distribution models in different experimental parameter conditions. With a newly proposed three dimensional fuzzy relation evaluation method, we verified a series of quantified turbulence energy distribution surface features to analyze the complicated fuzzy relation mechanism between them.

This investigation has the following theoretical superiorities over other traditional researches For the traditional methods simply focused on establishing a turbulence energy distribution model without any further considerations about its spatial distribution surface, the surface fitting algorithm, and its consequent impact on turbulence energy modeling results, we are concerned with the mutual-performance mechanism and uncertainty principle from miscellaneous data analysis; different from other traditional ones in concluding turbulence energy distribution properties on one given high-pressure fluid field from macroscale dimensional analysis, we proposed a new three dimensional fuzzy performance mechanism of surface fitting and realized its resulting quantization by discussing the microturbulence characteristic details in an experimental condition; considering the absence of fuzzy relation calibration between turbulence energy distribution and surface fitting in a traditional research, we investigated their internal mutual-performance mechanism and then assessed the respective fuzzy influence factors and inherent mathematical principles as respected.

The following major contributions are included in our work. As the traditional method has not touched upon turbulence kinetic energy distribution surface on one reversing valve's high-pressure runner, we proposed several new mathematical features to accurately show the objective surface and quantitatively evaluated their inherent features in geometrical domain; through using surface fitting for modeling turbulence kinetic energy distribution in a geometrical domain, we analyzed and quantified the fuzzy influences of surface fitting on the constructed energy distribution surface models in different experimental conditions, with their inherent change rules also being clearly indicated; we proposed an improved three dimensional fuzzy relation evaluation system to establish reliable performance mechanism which does not require any previous information other than the experimental data to be disposed, and thereafter an in-depth discussion about fuzzy performance has been made. And finally, several original suggestions concerning the specific surface fitting processes and their fuzzy performance in geometrical surface domain and turbulence energy distribution sense have been presented as well.

All the above-mentioned processes could be successive stages of computation and analysis, with the second operating on the output of the first. It solves the difficulties and dilemma in quantitatively assessing an optimum fuzzy performance evaluation method or surface fitting algorithm when researching turbulence energy distribution characteristics on one high-pressure reversing valve runner. For the experimental process has a deliberate theoretical foundation, thus the mathematical analysis process can be founded and simplified, and this research also provides a new idea for following turbulence characteristic quantitative evaluation.

Acknowledgments

The author acknowledges the funding of following science foundation: National Natural Science Foundation of China (No. 51205073, No. 51275100), China Postdoctoral Science Foundation funded project (No. 2012M510197), Foundation for Distinguished Young Talents in Higher Education of Guangdong, China (No. LYM09110), the science and technology project of Guangzhou city (No. 2012J4100053, 12C42011566), the Research Project of Guangzhou Municipal University (No. 10A068), and Foundation Project of National Engineering Research Center of Near-Net-Shape Forming for Metallic Materials, South China University of Technology (No. 2012007) is also appreciated for supporting this work. We also want to thank the editors for their hard work and the referees for their comments and suggestions to improve this paper.

References

[1] C. V. Tran, "Nonlinear transfer and spectral distribution of energy in α turbulence," *Physica D*, vol. 191, no. 1-2, pp. 137–155, 2004.

[2] A. Mokrani, C. Castelain, and H. Peerhossaini, "Experimental study of the influence of the rows of vortex generators on turbulence structure in a tube," *Chemical Engineering and Processing*, vol. 48, no. 2, pp. 659–671, 2009.

[3] C. V. Tran and T. G. Shepherd, "Constraints on the spectral distribution of energy and enstrophy dissipation in forced two-dimensional turbulence," *Physica D*, vol. 165, no. 3-4, pp. 199–212, 2002.

[4] I. Roghair, J. M. Mercado, M. V. Sint Annaland, H. Kuipers, C. Sun, and D. Lohse, "Energy spectra and bubble velocity distributions in pseudo-turbulence: numerical simulations versus experiments," *International Journal of Multiphase Flow*, vol. 37, no. 9, pp. 1093–1098, 2011.

[5] C. Liu, W. X. Zhou, and W. K. Yuan, "Statistical properties of visibility graph of energy dissipation rates in three-dimensional fully developed turbulence," *Physica A*, vol. 389, no. 13, pp. 2675–2681, 2010.

[6] T. Panidis, "The development of the structure of water—air bubble grid turbulence," *International Journal of Multiphase Flow*, vol. 37, no. 6, pp. 565–575, 2011.

[7] K. Ghorbanian, M. R. Soltani, and M. D. Manshadi, "Experimental investigation on turbulence intensity reduction in subsonic wind tunnels," *Aerospace Science and Technology*, vol. 15, no. 2, pp. 137–147, 2011.

[8] M. Kashyap, B. Chalermsinsuwan, and D. Gidaspow, "Measuring turbulence in a circulating fluidized bed using PIV techniques," *Particuology*, vol. 9, no. 6, pp. 572–588, 2011.

[9] T. S. Cheng and W. J. Yang, "Numerical simulation of three-dimensional turbulent separated and reattaching flows using a modified turbulence model," *Computers and Fluids*, vol. 37, no. 3, pp. 194–206, 2008.

[10] S. A. Ahmed, "Velocity measurements and turbulence statistics of a confined isothermal swirling flow," *Experimental Thermal and Fluid Science*, vol. 17, no. 3, pp. 256–264, 1998.

[11] A. Fujiwara, D. Minato, and K. Hishida, "Effect of bubble diameter on modification of turbulence in an upward pipe flow," *International Journal of Heat and Fluid Flow*, vol. 25, no. 3, pp. 481–488, 2004.

[12] M. Tanaka and N. Yokoyama, "A numerical study on statistical characteristics of amplitude fluctuations in a wave turbulence," *Physica D*, vol. 240, no. 14-15, pp. 1145–1155, 2011.

[13] Y. Wang, Z. Lian, and L. Lan, "The effect of turbulence intensity on local skin temperature and subjective responses to draft," *Energy and Buildings*, vol. 43, no. 10, pp. 2678–2683, 2011.

[14] M. Shahraeeni and M. Raisee, "Investigation of turbulent mixed convection of air flow in vertical tubes using a zonal turbulence model," *International Journal of Heat and Fluid Flow*, vol. 31, no. 2, pp. 179–190, 2010.

[15] C. M. B. Nobre, R. A. Braga Jr., A. G. Costa, R. R. Cardoso, W. S. da Silva, and T. Sáfadi, "Biospeckle laser spectral analysis under inertia moment, entropy and cross-spectrum methods," *Optics Communications*, vol. 282, no. 11, pp. 2236–2242, 2009.

[16] D. Erdogmus, E. G. Larsson, R. Yan, J. C. Principe, and J. R. Fitzsimmons, "Measuring the signal-to-noise ratio in

magnetic resonance imaging: a caveat," *Signal Processing*, vol. 84, no. 6, pp. 1035–1040, 2004.

[17] J. Hua, L. Meng, Z. Xu, and G. Li, "An adaptive signal-to-noise ratio estimator in mobile communication channels," *Digital Signal Processing*, vol. 20, no. 3, pp. 692–698, 2010.

[18] Y. Z. Chen, "Complex potentials and singular integral equation for curve crack problem in antiplane elasticity," *International Journal of Engineering Science*, vol. 38, no. 5, pp. 565–574, 2000.

[19] M. Griebel and G. Zumbusch, "Parallel adaptive subspace correction schemes with applications to elasticity," *Computer Methods in Applied Mechanics and Engineering*, vol. 184, no. 2– 4, pp. 303–332, 2000.

[20] J. L. C. Mission and H. J. Kim, "Design charts for elastic pile shortening in the equivalent top-down load-settlement curve from a bidirectional load test," *Computers and Geotechnics*, vol. 38, no. 2, pp. 167–177, 2011.

[21] M. Manera and A. Marzullo, "Modelling the load curve of aggregate electricity consumption using principal components," *Environmental Modelling and Software*, vol. 20, no. 11, pp. 1389–1400, 2005.

[22] T. P. Vo and J. Lee, "Interaction curves for vibration and buckling of thin-walled composite box beams under axial loads and end moments," *Applied Mathematical Modelling*, vol. 34, no. 10, pp. 3142–3157, 2010.

[23] G. A. Papakostas, Y. S. Boutalis, D. A. Karras, and B. G. Mertzios, "Pattern classification by using improved wavelet compressed Zernike moments," *Applied Mathematics and Computation*, vol. 212, no. 1, pp. 162–176, 2009.

[24] K. N. Ngan and S. B. Kang, "Geometric modelling of IC die bonds for inspection," *Pattern Recognition Letters*, vol. 10, no. 1, pp. 47–52, 1989.

[25] C. S. Lin, J. T. Huang, T. C. Wei, M. S. Yeh, and D. C. Chen, "High speed and high accuracy inspection of in-tray laser IC marking using line scan CCD with a new calibration model," *Optics and Laser Technology*, vol. 43, no. 1, pp. 218–225, 2011.

[26] K. C. Stevens and T. J. Wilson, "Locating IC defects in process monitors and test structures using optical beam induced current," *Microelectronic Engineering*, vol. 12, no. 1– 4, pp. 397–404, 1990.

3

Design and Experimental Evaluation of a Robust Position Controller for an Electrohydrostatic Actuator Using Adaptive Antiwindup Sliding Mode Scheme

Ji Min Lee, Sung Hwan Park, and Jong Shik Kim

School of Mechanical Engineering, Pusan National University, Jangjeon-dong, Geumjeong-gu, Busan 609-732, Republic of Korea

Correspondence should be addressed to Jong Shik Kim; jskim@pnu.edu

Academic Editors: U. Lee and S. Liu

A robust control scheme is proposed for the position control of the electrohydrostatic actuator (EHA) when considering hardware saturation, load disturbance, and lumped system uncertainties and nonlinearities. To reduce overshoot due to a saturation of electric motor and to realize robustness against load disturbance and lumped system uncertainties such as varying parameters and modeling error, this paper proposes an adaptive antiwindup PID sliding mode scheme as a robust position controller for the EHA system. An optimal PID controller and an optimal anti-windup PID controller are also designed to compare control performance. An EHA prototype is developed, carrying out system modeling and parameter identification in designing the position controller. The simply identified linear model serves as the basis for the design of the position controllers, while the robustness of the control systems is compared by experiments. The adaptive anti-windup PID sliding mode controller has been found to have the desired performance and become robust against hardware saturation, load disturbance, and lumped system uncertainties and nonlinearities.

1. Introduction

Recently, lots of hydraulic servo systems have been replaced by electric motor-driven systems to overcome the drawbacks of traditional hydraulic servo systems such as oil leakage, maintenance, and laying complex pipe. In spite of these drawbacks, hydraulic systems are still used in fields that require large force and high power. In order to solve these issues, electrohydrostatic actuator (EHA) systems are in the process of being developed for industrial use. The EHA systems can also be referred to as a direct drive volume controlled (DDVC) actuator [1–3] or valveless hydraulic servo (VHS) actuator [4].

The EHA systems generally consist of an electric motor, hydraulic pump, hydraulic cylinder, accumulator, check valves, and relief valves for safety. An electric motor that is directly tied to the hydraulic pump controls the cylinder, whose directional change depends upon the electric motor's rotation direction. Also, an electric motor installed in the EHA system controls the flow rate and pressure of the working fluid by regulating the velocity and torque of the motor.

The EHA systems are kind of compact high-power actuator modules that integrate the electric motor, hydraulic pump, hydraulic cylinder, accumulator, check valves, and relief valves into one component by using the manifold block. Compared to conventional hydraulic systems, EHAs are very small and do not need complex pipe lines. The EHA systems are seen as environment-friendly hydraulic servo system because it does not leak oil.

The main advantage of using EHA systems on the whole comes from its increased energy efficiency [5–10]. The EHA systems designed in this research do not use any control valves such as a servo valve and can also supply the minimum energy needed for the hydraulics system by controlling the servo motor, meaning that the EHA systems are extremely energy efficient. Also, by using an electric servo motor as a control device, EHA systems can quickly achieve fast response and high accuracy. However, since the EHA systems are a kind of hydraulic servo system, it has nonlinearity related to the friction force mainly located in the hydraulic cylinder. The EHA system also has the closed hydraulic circuit which can use relatively little working fluid. The performance of the EHA system can also be sensitive to variation within the system parameters of the working fluid such as variety of effective bulk modulus and the leakage coefficient due to changes in the working environment [5, 7, 9, 11–13]. Those nonlinearity and uncertainty may cause the position control system of the EHA to be unstable or have a much degraded performance.

In order to obtain the desirable performance of an EHA position control system, a number of control strategies have been applied and investigated by many researchers. Some of the work has used linear control theory for the position control of EHA [14–16]. The nonlinear robust control schemes such as sliding mode control, back-stepping control, and Fuzzy logic control have been proposed to achieve robustness of position control system for EHA by compensating system nonlinearity and uncertainties and to compensate for the nonlinear dynamics of EHA system such as friction [3, 5, 6, 9–11, 14, 17, 18]. Adaptive control approaches have been also proposed to make up for parameter disturbances such as varying the viscous friction coefficient and the effective bulk modulus due to change of working condition [1–3, 9, 11, 19, 20].

However, a fast response with small overshoot should be the objective for the practical use of the EHA position control systems in industrial machine such as plastic injection molding machine, aircraft actuation system, mobile machinery, CNC pipe bending machines, and hydraulic servo control of steering system for ship and hydraulic press [1, 2] even though there are lots of uncertainties and nonlinearity in position control system of EHA. The problem concerning large overshoot from the saturation in the electric motor has been considered in recent studies [2]. In [2], the saturation of voltage and current of the PWM driver and DC power is taken into account for position control system of EHA. The researchers of [2] also consider dead zone torque due to the static friction of motor pump and variable load torque. The PD-PI hybrid control scheme is proposed to control the BLDC motor in order to achieve small overshoot and to compensate for the load disturbances. However, only a computer simulation can verify the proposed linear control scheme although there are lots of uncertainty and nonlinearity besides the dead zone and motor saturation. In [2, 19], they proposed a self-turning Fuzzy PID control scheme to compensate for the dead zone, saturation of motor angular velocity, parameter uncertainties, and load disturbance with experimental verification. However, improved accuracy and fast response appear necessary to apply the position control results to servo press. Also, the design of fuzzy rules depends largely on the experience of experts or input-output data.

In this research, an adaptive antiwindup PID sliding mode control scheme is proposed to solve problems of overshoot in position control caused by saturation of angular velocity to the electric motor. The previous researches considered an antiwindup scheme based on the conventional sliding mode controller [21, 22]. However, the proposed controller consists of an adaptive control and a conventional sliding mode control with antiwindup scheme. Therefore, the proposed control scheme looks to implement robustness to parameter variations in a variety of working condition, nonlinearity, and load disturbance. The validation of the proposed control scheme is verified by experiment.

This paper will proceed as follows: Section 2 reveals the composition and details of the position control test rig of an EHA prototype, followed by system modeling and parameter identification of the EHA systems in Section 3. In Section 4, position controllers for the EHA system using traditional PID, antiwindup PID, and adaptive antiwindup PID sliding mode control schemes are presented. The comparative results of the experimental tests executed on the position control test rig of EHA applying PID, antiwindup PID, and adaptive antiwindup PID sliding mode control schemes are revealed in Section 5, followed by conclusions in Section 6.

2. Experimental Setup of the Electrohydraulic Actuator Prototype

In this study, the EHA system consists of a bidirectional brushless direct current (BLDC) motor, axial displacement hydraulic piston pump, acting doubly as a double rod hydraulic cylinder, accumulator, check valves, and relief valves. The accumulator serves as a hydraulic reservoir, pressurizing to obtain improved response speed by increasing the suction ability of the hydraulic pump and to prevent cavitations of the EHA system. The electric motor driving the hydraulic pump directly controls the hydraulic cylinder. The direction change of the hydraulic cylinder depends on the change of rotation direction of the electric motor. Also, an electric motor installed in the EHA system controls the flow rate and pressure of the working fluid by controlling the velocity and torque of the electric motor. The DC power supply, which has a maximum DC power supply capacity of 27 kW, supplies electric power to the electric motor. The detailed specifications of the components are shown in Table 1.

In this paper, we developed an EHA prototype which can be used for various purposes in industries such as an aircraft actuation system, mobile construction machinery, servo press, steering gear, and injection molding machine, with the overall specifications shown in Table 2. Figure 1 is a photo of the position control test rig of the EHA system, followed by the schematic diagram of hydraulic position control test rig for the developed EHA prototype in Figure 2.

The test rig for the developed EHA prototype consists of two hydraulic cylinders: a main cylinder which is a part of the EHA and a load cylinder. Two cylinders are directly coupled by a shaft, and the load mass is installed between the two cylinders. The electric motor controls the main cylinder of EHA, while the load cylinder is controlled by the servo valve.

The hydraulic flow to the main cylinder is controlled by the electric BLDC servo motor operated through a driver directly connected to the proposed controller. The position of the main cylinder gets measured by a linear variable displacement transducer (LVDT) installed on the bracket attached to the load mass, and another one is also installed in the hollow of the main cylinder rod. The signal from the LVDT is used for the closed-loop control of the main cylinder via proposed controllers. The output force gets measured by the load cell installed at the coupling between two cylinders.

Table 1: Specification of components for the EHA prototype.

Specification	Value
Pump	
Type	Fixed displacement axial piston pump (bidirectional)
Number of pistons	7
Displacement	5×10^{-6} m^3
Maximum pressure	20.6×10^6 Pa
Motor	
Type	Brushless DC motor
DC link voltage	270 V
Rated current	52.4 Arms
Rated torque	13.4 N·m
Max speed	10,000 rpm
Power	9 kW
Hydraulic cylinder	
Type	Double rod, double acting
Piston diameter	0.108 m
Diameter of rod	0.044 m
Length of stroke	0.0650 m
Maximum pressure	20.6×10^6 Pa
Accumulator	
Acc. Type	Bladder
Gas precharge pressure	0.49×10^6 Pa
Accumulator volume	0.7 L
Load mass	
Weight	50 kg

Table 2: Specifications of the EHA prototype.

Specifications	Values
Stall force	Max. 1,541 N
Stroke	Max. ±33.7 mm
Velocity	123.0~142.5 mm/s
Power	Max. 9.0 kW

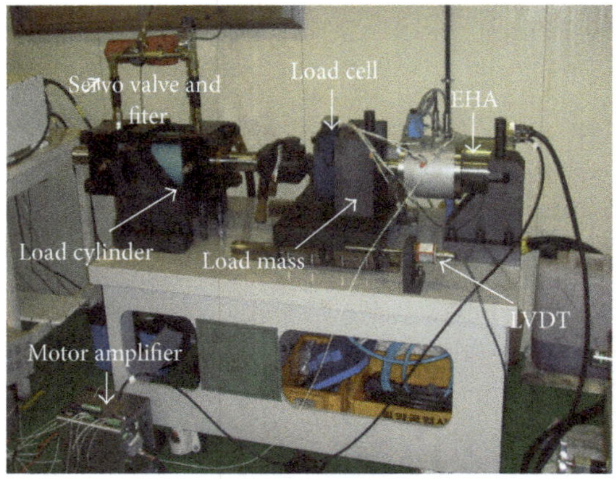

Figure 1: Photo of the position control test rig of the EHA.

DS1104 dSPACE board is used as a main control device to implement proposed position controllers: optimal PID, optimal antiwindup PID controller, and an adaptive antiwindup PID sliding mode controller. The real-time control system is programmed by MATLAB/Simulink and gets transferred to the dSPACE board through the Real-Time Workshop. The parameters of the controller implemented on the board can be changed and monitored through the Control-Desk software in real time.

3. Modeling of Position Control System of EHA

In this section, we derive a mathematical model of EHA systems, identifying system parameters by the signal compression method (SCM) in order to design the position controller of EHA.

3.1. Mathematical Model of EHA System. Figure 3 shows the schematic diagram of EHA system. As shown in Figure 3, the pump in/outflows Q_a, Q_b can be represented with the pump inlet/outlet chamber pressures p_a, p_b and the cylinder displacement x as follows:

$$\begin{aligned} Q_a &= D_p\omega_p - \xi(p_a - p_b) - \frac{V_a}{\beta_e}\frac{dp_a}{dt} - C_{ep}(p_a - p_r), \\ Q_b &= D_p\omega_p - \xi(p_a - p_b) + \frac{V_b}{\beta_e}\frac{dp_b}{dt} + C_{ep}(p_b - p_r), \end{aligned} \tag{1}$$

where D_p is displacement of the pump, ξ is the pump cross port leakage coefficient, C_{ep} is the pump leakage coefficient, p_r is the accumulator pressure, β_e stands for the effective bulk modulus of the working fluid, and V_a, V_b are inlet/outlet volumes of pipe and cylinder chamber, respectively.

In addition, the cylinder inlet/outlet flows Q_1, Q_2 can be represented by

$$\begin{aligned} Q_1 &= A\dot{x} + \frac{V_a}{\beta_e}\frac{dp_1}{dt} + L(p_1 - p_2), \\ Q_2 &= A\dot{x} - \frac{V_b}{\beta_e}\frac{dp_2}{dt} - L(p_2 - p_1), \end{aligned} \tag{2}$$

where A is pressurized area of hydraulic piston, L is internal leakage coefficient of the cylinder, and p_1, p_2 are the cylinder inlet/outlet chamber pressures, respectively. In this study, the external leakage of the cylinder is neglected.

In this study, V_a and V_b are assumed to be equal due to the double rod, and double acting hydraulic cylinder is used in EHA system. Also, V_a and V_b can be expressed with

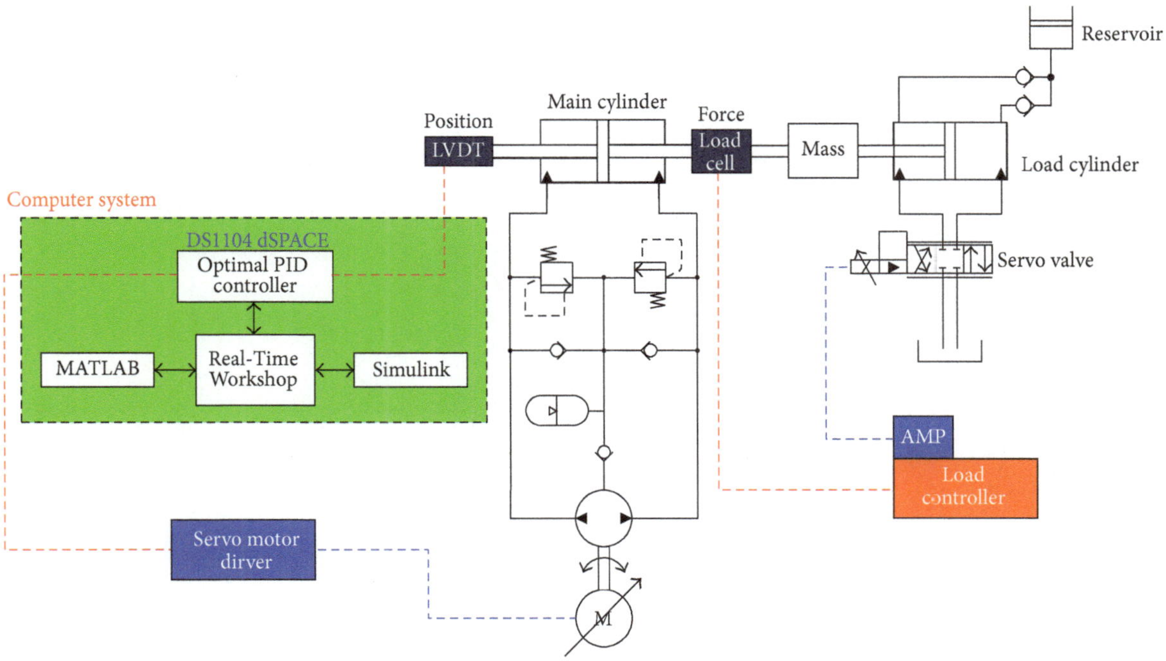

FIGURE 2: Schematic diagram for the test rig of the EHA prototype.

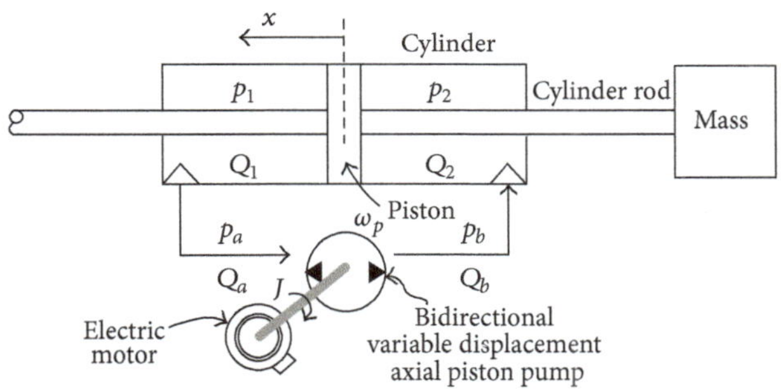

FIGURE 3: Schematic diagram of an EHA system.

the nominal volume of each EHA chamber V_0 which equals the mean volume of pipe and cylinder chamber as follows:

$$V_a = V_0 + Ax,$$
$$V_b = V_0 - Ax. \tag{3}$$

Since a solid tubing is used in the prototype, the pressure impact of pipe expansion and flexibility on the effective bulk modulus is assumed to be negligible. Therefore, the relationship pump port flows (Q_a, Q_b) and cylinder chamber flows (Q_1, Q_2) can be expressed as

$$Q_1 = Q_a,$$
$$Q_2 = Q_b. \tag{4}$$

Then, the load Q_L can be defined as follows [21]:

$$Q_L = \frac{Q_1 + Q_2}{2} = \frac{Q_a + Q_b}{2}. \tag{5}$$

The relationship between the pump port pressures (p_a, p_b) and cylinder chamber pressures (p_1, p_2) can then be expressed as

$$p_a = p_1 + p_{\text{pipe}},$$
$$p_b = p_2 - p_{\text{pipe}}, \tag{6}$$

where p_{pipe} is pressure drop.

Due to the short length of tubing, pump port pressures are assumed to be equal; that is, the pressure drop can be neglected, to the actuator inlet and outlet pressures such that [7, 8, 11, 17, 22]

$$p_a = p_1,$$
$$p_b = p_2. \tag{7}$$

Since the symmetrical double-rod hydraulic cylinder is used in this EHA system, the following equation for the cylinder chamber pressures can be established:

$$\frac{dp_1}{dt} = -\frac{dp_2}{dt}. \tag{8}$$

By combining (1), (2), (5), (7), and (8), a simplified pump/cylinder model equation can be expressed as

$$D_p\omega_p = A\dot{x} + \frac{V_0}{\beta_e}\dot{p}_L + C_T p_L + f_1, \tag{9}$$

where $p_L = p_1 - p_2$, $C_T = \xi + L + (C_p/2)$ is the equivalent leakage coefficient and f_1 is unmodeled dynamics of hydraulic part of the EHA system.

The actuator force and the displacement of the load can be represented as

$$F = p_L A = M\ddot{x} + B\dot{x} + f_2, \tag{10}$$

where A is pressurized area of hydraulic cylinder, M is mass of load, B is viscous friction coefficient and f_2 is lumped uncertain nonlinearities due to external disturbance, the unmodeled friction forces, and other hard-to-model terms of mechanical part of the EHA system.

By combining (9) and (10) and neglecting f_1 and f_2, the linear transfer function between the angular velocity of the pump and the displacement of the main cylinder can be expressed as

$$G_h(s) = \frac{x(s)}{\omega_p(s)} = \left(\frac{AD_p\beta_e}{MV_0}\right)\left(s^3 + \left(\frac{B}{M} + \frac{C_T D_p \beta_e}{V_0}\right)s^2 + \frac{\beta_e\left(BC_T D_p + A^2\right)}{MV_0}s\right)^{-1}, \tag{11}$$

where s is the Laplace operator.

Based on (11), the variation of the effective bulk modulus of the working fluid can increase or decrease the compressibility of the working fluid, and then the natural frequency of the EHA system can be adjusted, meaning that the variation of the system parameters can affect the performance of the EHA position control system.

On the other hand, the dynamics of the BLDC motor with controller, that is, electric part of the EHA system, is considered as [14, 23]

$$G_m(s) = \frac{\omega_p(s)}{v_r(s)} = \frac{K_m(T_1 s + 1)}{T_{m_1}s^2 + T_{m_2}s + 1}, \tag{12}$$

where K_m, T_1, T_{m_1}, and T_{m_2} are parameters of the BLDC motor with controller.

3.2. Parameter Identification. In general, the unknown system parameters of the linear elements in a nonlinear system can be identified by the signal compression method (SCM) [24]. In this study, SCM is applied to identify the hydraulic part of the EHA prototype.

However, hydraulic servo systems engage in significant nonlinear behavior due to system uncertainties such as pipe losses, leakages, and parameter variations of the working fluid [9–11, 16, 17, 19]. Some of those uncertainties are sometimes neglected because the SCM can only identify the linear part of the system in the identification process. This study then compensates for the neglected nonlinear behavior of the EHA system by using a robust controller, that is adaptive PID sliding mode control scheme.

As shown in Figure 4, the test signal that has the same amplitude up to 4 Hz in the frequency domain is applied to the position control system of EHA by constructing a close-loop proportional control system to obtain a more accurate equivalent impulse response for the precise identification of unknown parameters. Figure 5 shows the comparison of the frequency response of the position control system between the closed-loop nominal model and the closed-loop actual system with a proportional controller, whose gain is 1,000.

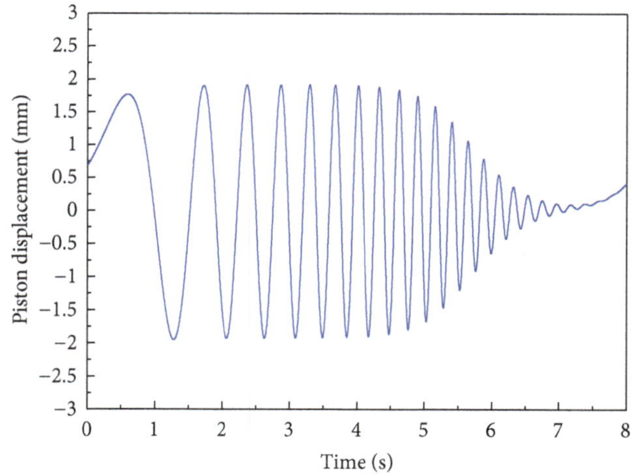

FIGURE 4: Test signal.

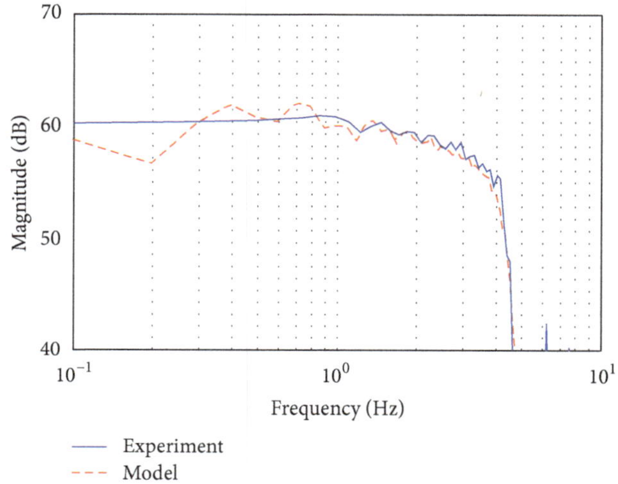

FIGURE 5: Bode plots of the closed loop nominal model and the closed loop actual system with a proportional controller.

The identified transfer function of a closed-loop actual system with a proportional controller for EHA position control system is as follows:

$$\frac{KG(s)}{1 + KG(s)} = \frac{11.8 \times 10^7}{s^3 + 2231s^2 + 71.8 \times 10^5 s + 11.8 \times 10^7}, \tag{13}$$

where K is proportional gain of closed-loop actual system.

From the identified nominal model for the closed-loop position control system for EHA with the proportional controller, the parameters of the position control system can be acquired by eliminating the effect of the proportional controller mathematically.

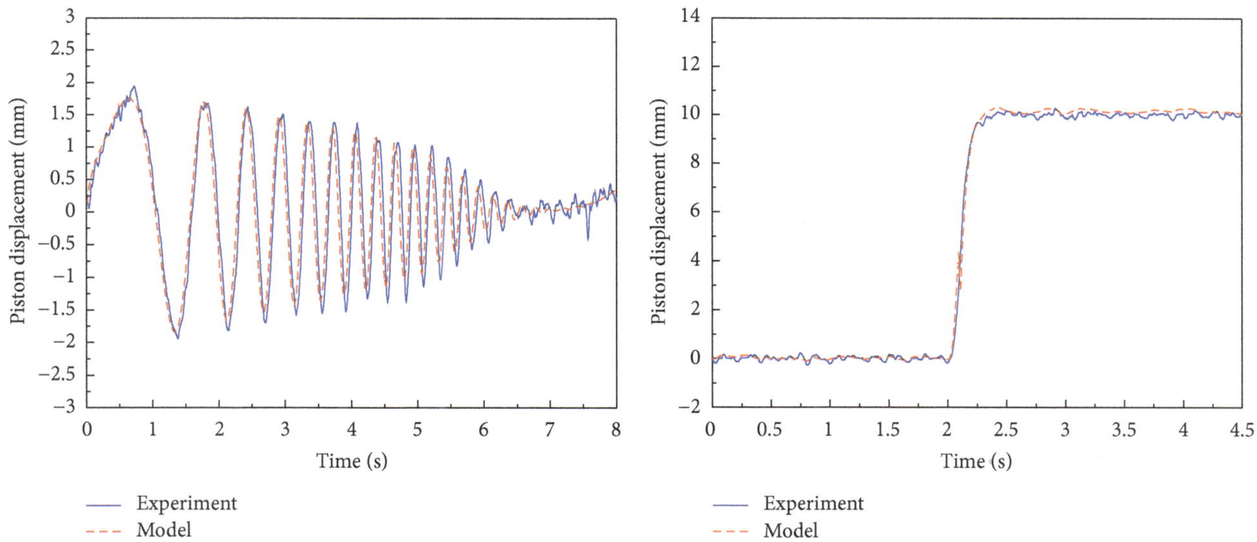

(a) Response for the test signal

(b) Step response

FIGURE 6: Time responses of the position control system obtained by the experiment and nominal model.

After eliminating the effect of the proportional controller, the identified transfer function of the hydraulic part of the EHA prototype is as follows:

$$G(s) = \frac{11.8 \times 10^4}{s^3 + 2231s^2 + 71.8 \times 10^5 s}. \quad (14)$$

The identified system parameters were verified on the time domain. Figures 6(a) and 6(b) show the response against test signal and the step responses of the nominal model and actual system, respectively. The validations of the identified parameters are estimated quantitatively by the cross-correlation as shown in [24]

$$\text{Corr} = \frac{\sum_{k=1}^{N}\left(Y_p(k) - \overline{Y}_p\right)\left(Y_m(k) - \overline{Y}_m\right)}{\sqrt{\sum_{k=1}^{N}\left(Y_p(k) - \overline{Y}_p\right)}\sqrt{\sum_{k=1}^{N}\left(Y_m(k) - \overline{Y}_m\right)^2}}, \quad (15)$$

where $Y_p(k)$ and $Y_m(k)$ are the responses of the real and nominal model in time train k, respectively. $\overline{Y}_p$ and $\overline{Y}_m$ are the mean values of $Y_p(k)$ and $Y_m(k)$. N is the acquired data number. The value of the cross-correlation is 0.904 which can be calculated from the identification results. The identified system parameters are then available to controller design. The identified transfer function in (14) is used to obtain the optimal control gains for designed position controllers by using MATLAB/Simulink.

In addition, the dynamics characteristic of the BLDC motor with controller is also investigated by using SCM. The transfer function of the BLDC motor with controller can be simplified to a second-order linear system as described in (12). Figure 7 shows bode plot of the actual BLDC motor with controller. The bandwidth of the actual open loop BLDC motor with controller is about 400 rad/s. The bandwidth of the BLDC motor with controller is 20 times larger than that of hydraulic part of the EHA system. Therefore, in this study,

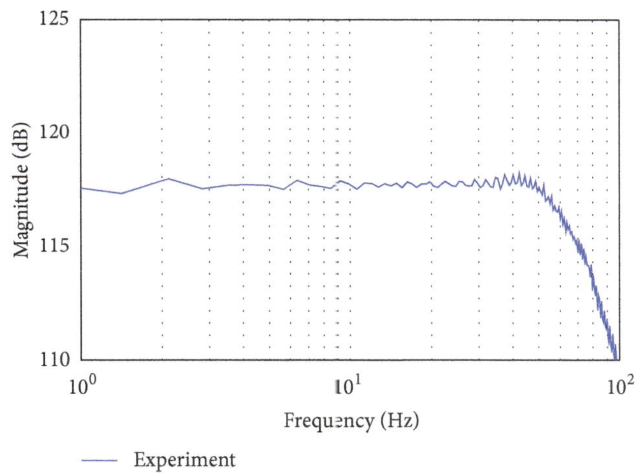

FIGURE 7: Bode plots of the open-loop actual BLDC motor with controller system.

we consider that the dynamics of BLDC motor with controller is negligible because the motor dynamic is sufficiently faster than hydraulic part of the EHA. Then, the electric part of the EHA system is taking account of the motor gain K_v, and current saturation is reflected in the controller design and the computer simulations stage.

4. Position Controller Design for a Prototype EHA

In this study, an EHA prototype is developed for application in the industrial fields. As mentioned in Section 2, performance characteristics such as no overshoot and rapid response time in position control are the typical demands of the EHA for industrial use. Therefore, appropriate control

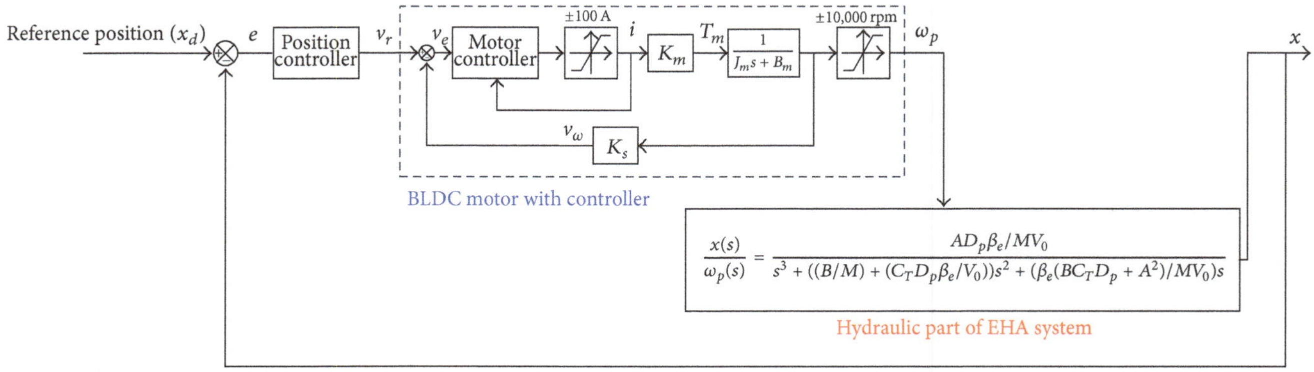

FIGURE 8: Block diagram for the position control system of the EHA.

strategy should be applied to EHA position control system in order to achieve these outcomes.

Figure 8 shows the block diagram of position control system of the EHA. As mentioned in the previous section, in Figure 8, the dynamics of the BLDC motor with controller is considered as just the motor gain K_m, and saturation of current of the BLDC motor is considered as an actuator constraint.

In order to compare and show the desirable performance characteristics for the proposed EHA control system, three kinds of position controllers are designed based on identified simple transfer function. First, an optimal PID controller is designed and applied to the system in order to understand the basic characteristics of EHA position control system. The optimal antiwindup PID controller is then designed to reduce overshoot due to saturation of current in the electric motor. Finally, an adaptive antiwindup PID sliding mode controller is designed to obtain the desirable performance and robustness against the effect of hardware saturation, load disturbance, and parameter variation.

4.1. Optimal PID and Optimal Antiwindup PID Control Systems. An optimal PID position control system is designed to understand basic control performance of EHA prototype. Also, an optimal antiwindup PID control system for the EHA is designed to obtain desirable performance characteristic considering saturation in electric motor.

The cost function is considered during the design of the optimal PID and optimal antiwindup PID controllers as follows [25]:

$$J\left(K_p, K_i, K_d\right) = \sum_{t=0}^{\infty}\left(y_{\text{step}}(t) - y_{\text{step}}^{d}(t)\right), \tag{16}$$

where $y_{\text{step}}^{d}(t)$ is the desired step response of the optimal PID and optimal antiwindup PID control systems and $y_{\text{step}}(t)$ is the step response by the identified transfer function in (14). The optimal PID and optimal antiwindup PID controller designs can be stated as

$$\min_{K_p, K_i, K_d} J\left(K_p, K_i, K_d\right), \tag{17}$$

where K_p, K_i, and K_d are the proportional, integral, and differential control gains, respectively.

There are many optimization algorithms in the optimization toolbox of MATLAB/Simulink. The cost function is given by (K_p, K_i, K_d). And the gains of the optimal PID and optimal antiwindup PID controllers K_p, K_i and K_d can be found by using the optimization toolbox of MATLAB/Simulink. To find optimal control gains, the reference step input is applied in optimization process, and the response performance is set as follows. The amplitude of reference is 20 mm, the rising time is 0.4 seconds, settling time is 0.8 seconds, percent overshoot is under 5%, and steady-state error is under 0.1 mm.

In the optimization process of optimal PID and optimal antiwindup PID controllers, system parameter uncertainties due to the modeling error are considered. Specifically, ±10% of modeling error is considered in terms of identified transfer function associate with variable system parameters such as effective bulk modulus of the working fluid and total leakage coefficient.

With respect to optimal antiwindup PID and antiwindup PID sliding mode controllers, the antiwindup algorithm shown in Figure 9 is used to consider saturation in an electric motor [2, 26–28].

Figure 10 shows the results of a computer simulation with optimal antiwindup PID control gains using MATLAB/Simulink optimization toolbox based on identified transfer function considering model uncertainty. The dotted line there shows upper bound and lower bound considering the ±10% of modeling error, and straight line shows nominal simulation result with optimal antiwindup PID control gains. As shown in Figure 10, the simulation result with optimal gains satisfies set response performance.

4.2. Adaptive Antiwindup Sliding Mode Controller Considering the PID Sliding Surface. In this study, the adaptive antiwindup PID sliding mode control scheme is also designed to achieve desirable performance and robustness of EHA position control systems. Throughout the controller design process, the effective bulk modulus and leakage coefficients are considered as variable system parameters. In order to consider saturation in electric motor, the antiwindup algorithm presented in Figure 8 also applies to the adaptive antiwindup PID sliding mode control scheme.

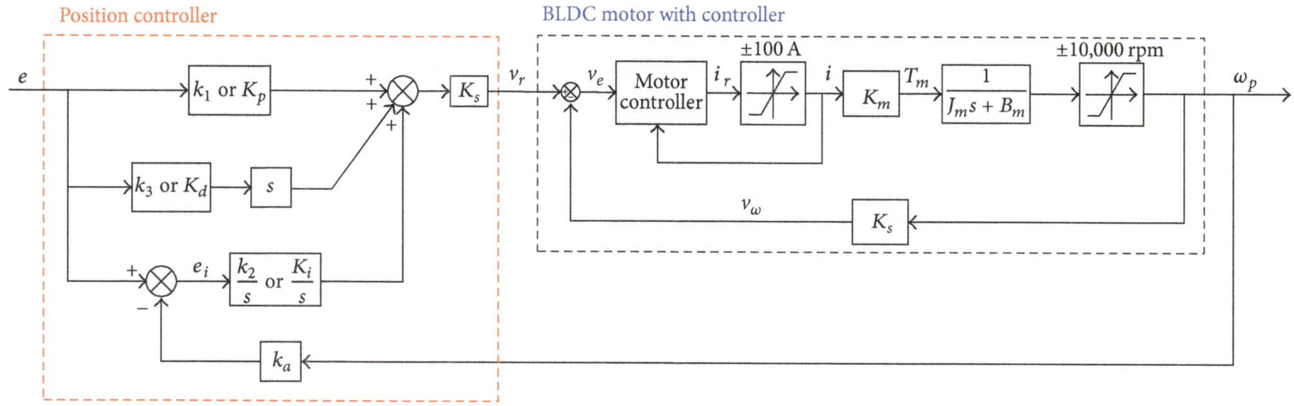

FIGURE 9: Block diagram for adopted antiwindup algorithm in controller design.

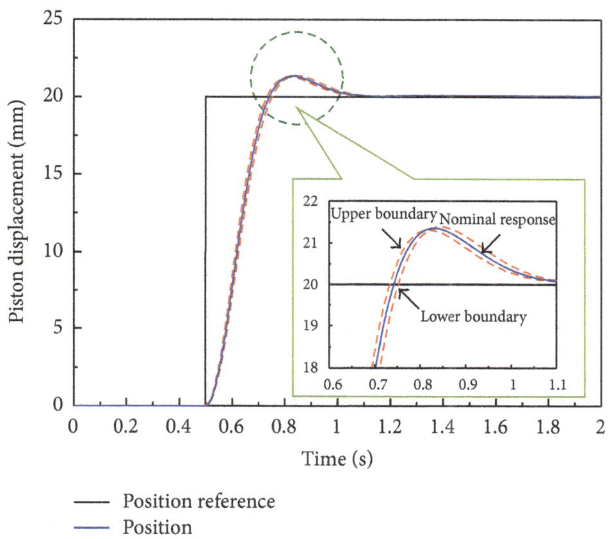

FIGURE 10: Step response of the optimal antiwindup PID control system in computer simulation based on identified transfer function of EHA system.

The PID sliding surface ζ for the design of the adaptive sliding mode control system is defined as [29–32]

$$\zeta = k_1 e + k_2 \int e\, dt + k_3 \dot{e}, \tag{18}$$

where the tracking error of the position $e = x_d - x$, x is the displacement of hydraulic cylinder and k_1, k_2, and k_3 are positive design parameters.

The sliding mode control law consists of equivalent and robust control term; that is,

$$\omega_p = \omega_{p,eq} + \omega_{p,\text{robust}}. \tag{19}$$

By combining (11) and (18) with consideration of the noise term in the jerk $\dddot{x} = \dddot{x}_m + \dddot{x}_n$, the derivative of the sliding surface $\dot{\zeta}$ can be written as

$$\begin{aligned}\dot{\zeta} = {} & -k_3 \frac{AD_p\beta_e}{BV_0 + C_T D_p M\beta_e}\omega_p \\ & + k_3 \frac{MV_0}{BV_0 + C_T D_p M\beta_e}\left(\dddot{x}_m + \dddot{x}_n\right) \\ & + k_3 \frac{\left(A^2 + BC_T D_p\right)\beta_e}{BV_0 + C_T D_p M\beta_e}\dot{x} + k_3 \ddot{x}_d + k_1 \dot{e} + k_2 e \\ & + \frac{k_3}{BV_0 + C_T D_p M\beta_e} f,\end{aligned} \tag{20}$$

where $f = f_1 + (C_T/A) f_2 + (V_0/AD_p\beta_e)\dot{f}_2$ is lumped uncertain nonlinearities of the EHA system and the subscripts m and n denote the nominal and noisy values, respectively.

To determine the equivalent control term $\omega_{p,eq}$, the noise term in the jerk is neglected, and it is assumed that the sliding surface ζ is at steady-state, that is, $\dot{\zeta} = 0$, and then the equivalent control law can be determined as

$$\begin{aligned}\omega_{p,eq} = {} & \frac{BV_0 + C_T D_p M\beta_e}{AD_p\beta_e} \\ & \times \left\{ \frac{MV_0}{BV_0 + C_T D_p M\beta_e}\dddot{x}_m + \frac{\left(A^2 + BC_T D_p\right)\beta_e}{BV_0 + C_T D_p M\beta_e}\dot{x} \right. \\ & \left. + \frac{k_1}{k_3}\dot{e} + \frac{k_2}{k_3} e + \ddot{x}_d \right\}.\end{aligned} \tag{21}$$

Thus, $\dot{\zeta}$ can be rewritten as

$$\dot{\zeta} = k_3 \frac{MV_0}{BV_0 + C_T D_p M\beta_e}\dddot{x}_n + \frac{k_3}{BV_0 + C_T D_p M\beta_e} f. \tag{22}$$

In the standard sliding mode control, to satisfy the reachability condition that directs system trajectories toward a sliding surface where they remain and to attenuate chattering in the control input, the derivative of the sliding surface is selected as [33, 34]

$$\dot{\zeta} = -D\zeta - K \operatorname{sgn}(\zeta), \tag{23}$$

where D and K are positive design parameters.

To determine a robust control term $\omega_{p,\text{robust}}$, that achieves robustness to uncertainties such as noise terms in the jerk and unmodeled system dynamics as well as the disturbance force, it is assumed that

$$D_0|\zeta| + K_0 > \left|\frac{k_3 M V_0}{BV_0 + C_T D_p M\beta_e}\left(\dddot{x}_n + \frac{1}{MV_0} f\right)\right| + \eta, \tag{24}$$

where $K_0 = (AD_p\beta_e/(BV_0 + C_T D_p M\beta_e))K$ and $D_0 = (AD_p\beta_e/(BV_0 + C_T D_p M\beta_e))D$ are positive design parameters and η is sufficiently small positive parameter. On the assumption of (24), the robust control law is determined as

$$\omega_{p,\text{robust}} = D\zeta + K\,\text{sgn}(\zeta). \tag{25}$$

Finally, the sliding mode control law is selected as

$$\omega_p = \frac{BV_0 + C_T D_p M\beta_e}{AD_p\beta_e} \times \left\{\dddot{x}_d + \frac{MV_0}{BV_0 + C_T D_p M\beta_e}\dddot{x}_m + \frac{(A^2 + BC_T D_p)\beta_e}{BV_0 + C_T D_p M\beta_e}\dot{x} + \frac{k_1}{k_3}\dot{e} + \frac{k_2}{k_3}e\right\} + D\zeta + K\,\text{sgn}(\zeta). \tag{26}$$

If the total leakage coefficient C_T and the effective bulk modulus of EHA systems β_e are considered as variable parameters, that is, $C_T = C_{T,n} + C_{T,p}$ and $\beta_e = \beta_{e,n} + \beta_{e,p}$, then the uncertainty ψ is defined as

$$\psi = \theta^T\varphi = \left(\frac{MV_0}{BV_0 + C_{T,p}D_p M\beta_{e,p}}\right)\dddot{x}_m + \left(\frac{(A^2 + BC_{T,p}D_p)\beta_{e,p}}{BV_0 + C_{T,p}D_p M\beta_{e,p}}\right)\dot{x}, \tag{27}$$

where $\theta^T = [(MV_0/(BV_0 + C_{T,p}D_pM\beta_{e,p}))\ ((A^2 + BC_{T,p}D_p)\beta_{e,p}/(BV_0 + C_{T,p}D_pM\beta_{e,p}))]$ and $\varphi = \left[\begin{smallmatrix}\dddot{x}_m \\ \dot{x}\end{smallmatrix}\right]$. The parameter vector θ is considered as an unknown parameter vector, and it can be estimated by using the update law. From (26) and estimated unknown parameter vector $\widehat{\theta}$, the estimated sliding mode control law can be selected as

$$\widehat{\omega}_p = \frac{BV_0 + C_{T,n}D_p M\beta_{e,n}}{AD_p\beta_{e,n}} \times \left\{\dddot{x}_d + \widehat{\theta}^T\varphi + \frac{MV_0}{BV_0 + C_{T,n}D_p M\beta_{e,n}}\dddot{x}_m + \frac{(A^2 + BC_{T,n}D_p)\beta_{e,n}}{BV_0 + C_{T,n}D_p M\beta_{e,n}}\dot{x} + \frac{k_1}{k_3}\dot{e} + \frac{k_2}{k_3}e\right\} + D\zeta + K\,\text{sgn}(\zeta). \tag{28}$$

In order to obtain the update law for the unknown parameters, the Lyapunov candidate is defined as

$$V = \frac{1}{2}\zeta^2 + \frac{1}{2\gamma}\widetilde{\theta}^T\widetilde{\theta}, \tag{29}$$

where $\widetilde{\theta} = \theta - \widehat{\theta}$, γ is a positive parameter. Also, $\widetilde{\theta}$ and $\widehat{\theta}$ are the nominal and estimated parameter vectors, respectively. The derivative of the Lyapunov candidate including sliding dynamics is expressed as

$$\dot{V} = \zeta\left[\dddot{x}_d + \widetilde{\theta}^T\varphi + \frac{MV_0}{BV_0 + C_{T,n}D_p M\beta_{e,n}}\dddot{x}_m + \frac{(A^2 + BC_{T,n}D_p)\beta_{e,n}}{BV_0 + C_{T,n}D_p M\beta_{e,n}}\dot{x} + \frac{k_1}{k_3}\dot{e} + \frac{k_2}{k_3}e - k_3\frac{AD_p\beta_e}{BV_0 + C_T D_p M\beta_e}\widehat{\omega}_p\right] - \frac{1}{\gamma}\widetilde{\theta}^T\dot{\widehat{\theta}}. \tag{30}$$

Substituting the estimated angular velocity of the pump $\widehat{\omega}_p$ given by (28) into (30) yields

$$\dot{V} = -D\zeta^2 - K\zeta\,\text{sgn}(\zeta) + \widetilde{\theta}^T\left(\zeta\varphi - \frac{1}{\gamma}\dot{\widehat{\theta}}\right). \tag{31}$$

From the derivative of the Lyapunov included sliding dynamics, the update law for the unknown parameters is selected as

$$\dot{\widehat{\theta}} = \gamma\zeta\varphi. \tag{32}$$

Therefore, the derivative of the Lyapunov is given as

$$\dot{V} = -D\zeta^2 - K\zeta\,\text{sgn}(\zeta) \le 0. \tag{33}$$

By applying invariant set theorem for (33), the asymptotical stability of the EHA position control system is guaranteed [33, 34].

Finally, the sliding mode control law can be selected as

$$\widehat{\omega}_p = \frac{BV_0 + C_T D_p M\widehat{\beta}_e}{AD_p\widehat{\beta}_e} \times \left(\widehat{\theta}^T\varphi + \frac{MV_0}{BV_0 + C_{T,n}D_p M\beta_{e,n}}\dddot{x}_m + \frac{(A^2 + BC_{T,n}D_p)\beta_{e,n}}{BV_0 + C_{T,n}D_p M\beta_{e,n}}\dot{x} + \frac{k_1}{k_3}\dot{e} + \frac{k_2}{k_3}e\right) + D\zeta + K\,\text{sgn}(\zeta), \tag{34}$$

where $\widehat{\beta}_e(t + T_s) = (d\widehat{\beta}_e(t)/dt)T_s + \widehat{\beta}_e(t)$, $\widehat{C}_T(t + T_s) = (d\widehat{C}_T(t)/dt)T_s + \widehat{C}_T(t)$ and T_s is the sampling time.

Table 3: Optimal gains of optimal PID, antiwindup PID, and adaptive anti-windup PID sliding mode control systems in experiments.

Gain	Optimal PID	Optimal antiwindup PID	Adaptive antiwindup PID sliding mode
Proportional (K_p or k_1)	1034.3	1694.7	95.0
Integral (K_i or k_2)	18.4	1087.7	0.001
Derivative (K_d or k_3)	4.7	416.0	14.5
Antiwindup (K_a)	—	0.001	0.001

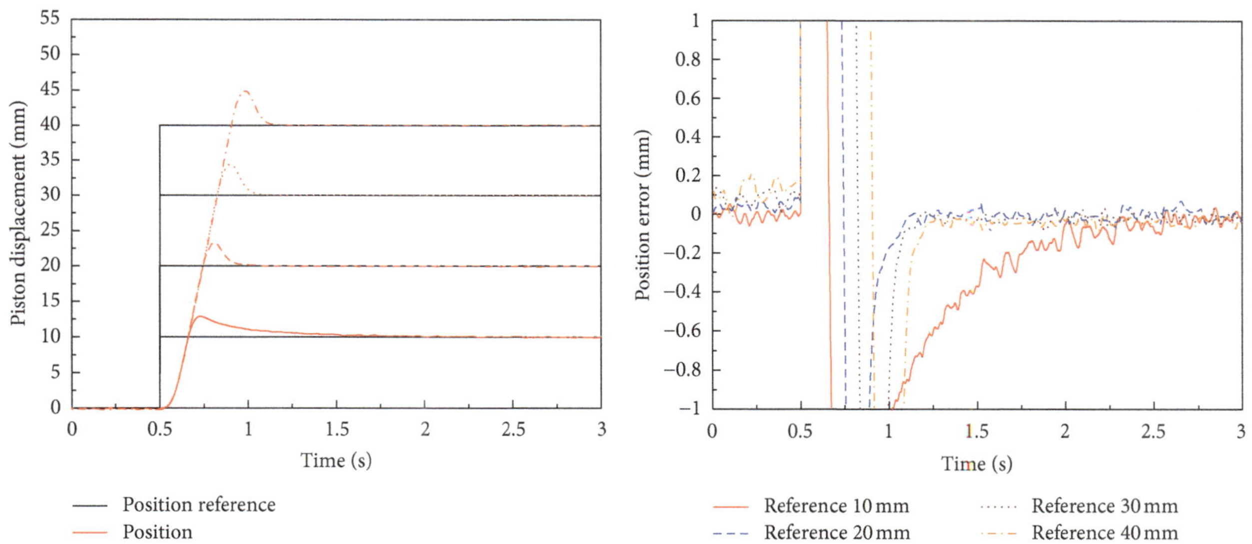

(a) Step responses

(b) Position errors

Figure 11: Performances of optimal PID position control system for EHA without load disturbance (only mass load is applied).

5. Experimental Results and Discussion

This section looks to the experimental results in order to compare the performance of a position control systems. Three kinds of controllers are compared in order to show improved performance of overshoot reduction and robustness to the parameter change and load disturbance of the proposed controller for EHA position control system.

Firstly, optimal PID controller is designed and implemented by experiments. Figure 11 shows experimental results of the PID position controller with step references of four different amplitudes which are 10, 20, 30, and 40 mm without load disturbance; that is, only mass load is applied as load to the position control system.

The gains of the three-position controllers are optimally selected by using MATLAB/Simulink optimization toolbox based on identified system dynamic model of (14). In the process of optimal gain selection, the amplitude of position reference is set to 20 mm and target performance is 0.4 seconds of the rising time, 0.8 seconds of settling time, under 5% of percent overshoot, and under 0.1 mm of steady-state error. In experiments, the optimal gains derived from MATLAB/Simulink optimization toolbox are tuned to obtain best responses. The finally tuned optimal gains of the three controllers are represented in Table 3.

As shown in Figure 11, large overshoot is observed in position control result for using the optimal PID position controller. The performance of optimal PID position control system is represented in Table 4. The performance of optimal PID controller in experiment is not satisfactory the aimed value. In detail, large overshoot is observed in experiments. The current and angular velocity saturations of BLDC motor in position experiment result of optimal PID controller are shown in Figure 12. The large overshoot in position controller can be caused due to saturation of BLDC motor, as shown in Figure 12.

To reduce such a large overshoot, an antiwindup algorithm which is described in Figure 9 is applied to the position control system of the EHA. The performance characteristics of an antiwindup PID position controller are depicted in Figure 13, and also the values of performance indices are shown in Table 4. As shown in Figure 13, the overshoot of the optimal antiwindup PID controller is more diminished than that of the optimal PID controller. However, as shown in Table 4, the antiwindup PID controller still has a performance of relatively long settling times which do not meet the aimed performance. Moreover, as the amplitude of the reference input increases, the steady-state error is increasing as shown in Figure 13.

Figure 14 shows the experimental results of optimal adaptive antiwindup sliding mode position control system for the EHA system. As shown in Figure 14, the optimal antiwindup PID sliding mode position control system has a better performance than the system with optimal PID and antiwindup PID controllers such that relatively small overshoot and short settling time are obtained by the adaptive antiwindup PID

Table 4: Comparison of the performance of optimal PID, optimal antiwindup PID, and adaptive antiwindup PID sliding mode control systems without load disturbance.

Performance	Reference	Optimal PID	Optimal antiwindup PID	Adaptive antiwindup PID sliding mode
Overshoot	10 mm	38.9%	20.8%	1.0%
	20 mm	16.1%	13.3%	0.5%
	30 mm	14.7%	11.9%	0.2%
	40 mm	12.2%	8.2%	0.1%
Settling time	10 mm	1.78 sec	2.2 sec	0.6 sec
	20 mm	0.56 sec	2.2 sec	0.6 sec
	30 mm	0.58 sec	0.63 sec	0.7 sec
	40 mm	0.69 sec	2.6 sec	0.7 sec

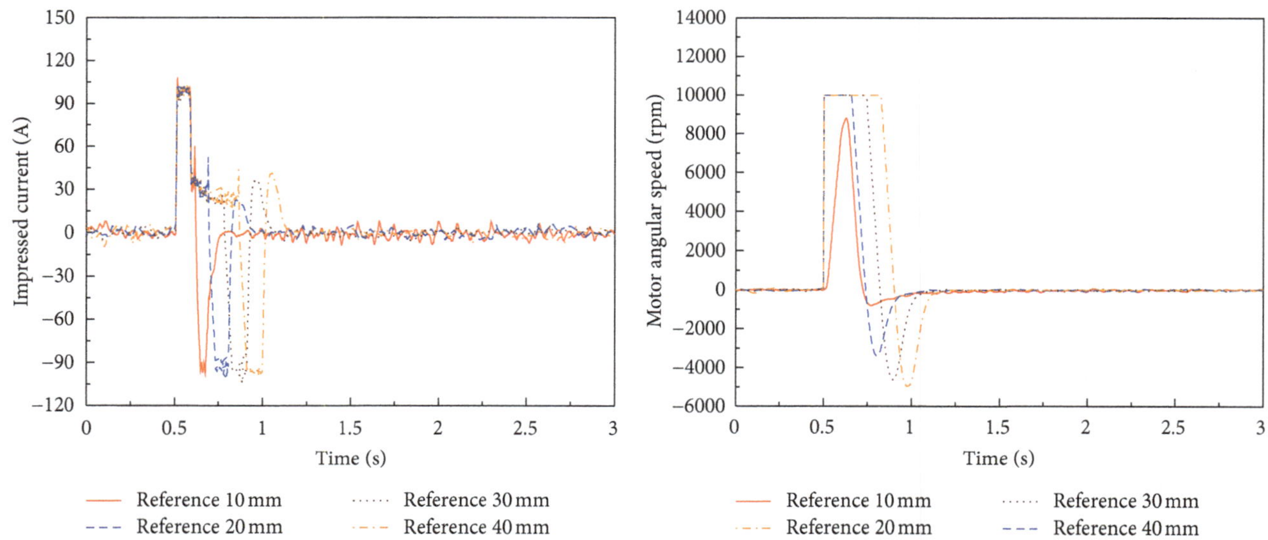

(a) Saturation of impressed current (b) Saturation of angular velocity

Figure 12: Saturation of BLDC motor for optimal PID position control system for EHA without load disturbance (only mass load is applied).

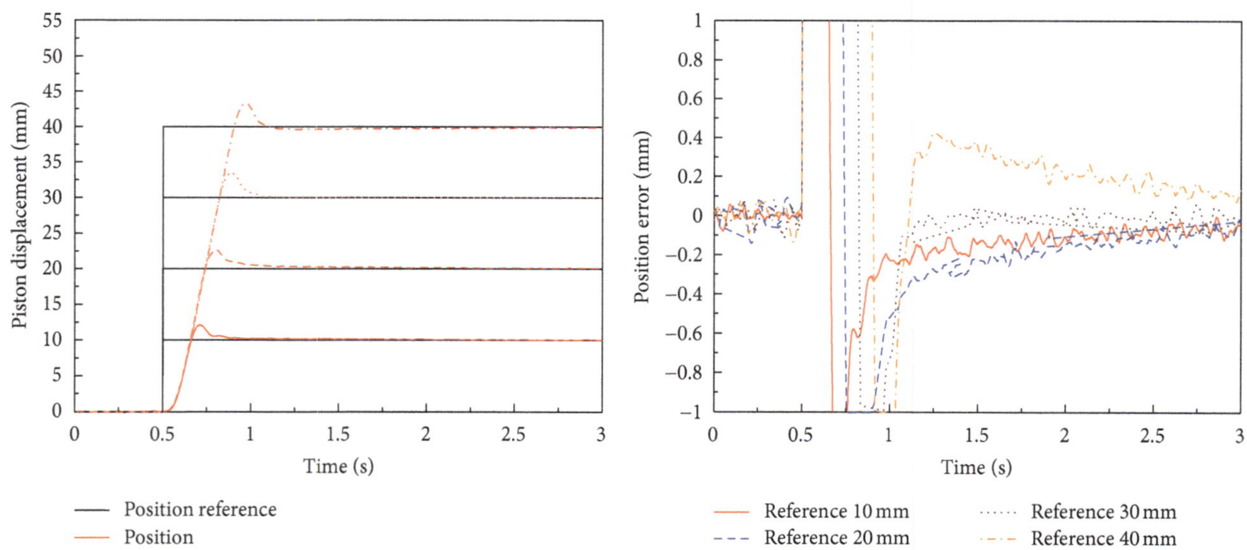

(a) Step responses (b) Position errors

Figure 13: Performances of optimal antiwindup PID position control system for EHA without load disturbance (only mass load is applied).

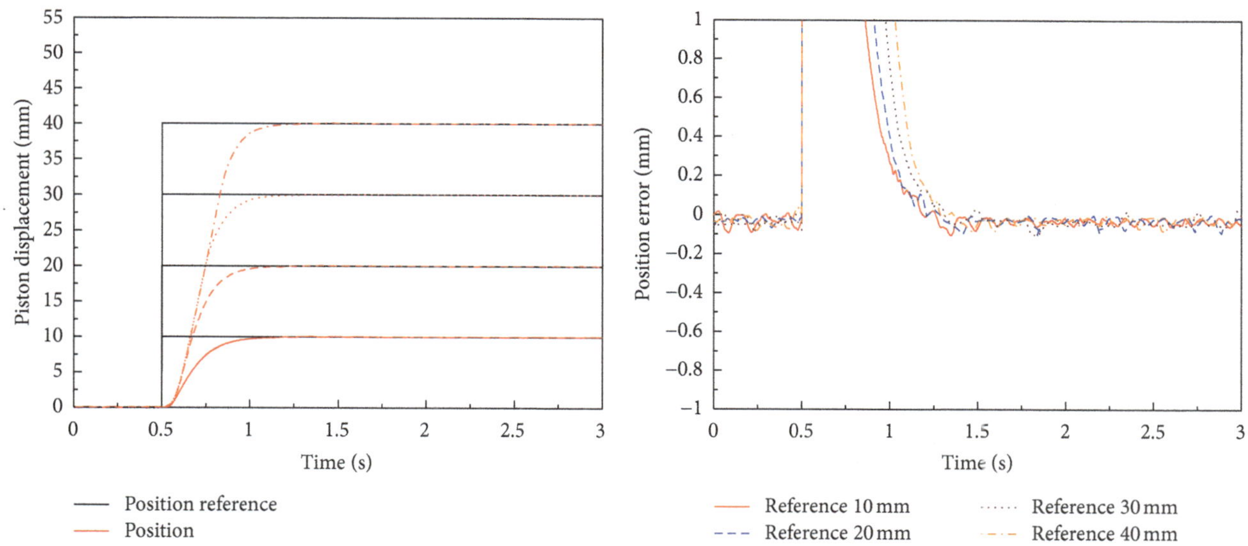

(a) Step responses

(b) Position errors

FIGURE 14: Performances of optimal adaptive antiwindup PID sliding mode position control system for EHA without load disturbance (only mass load is applied).

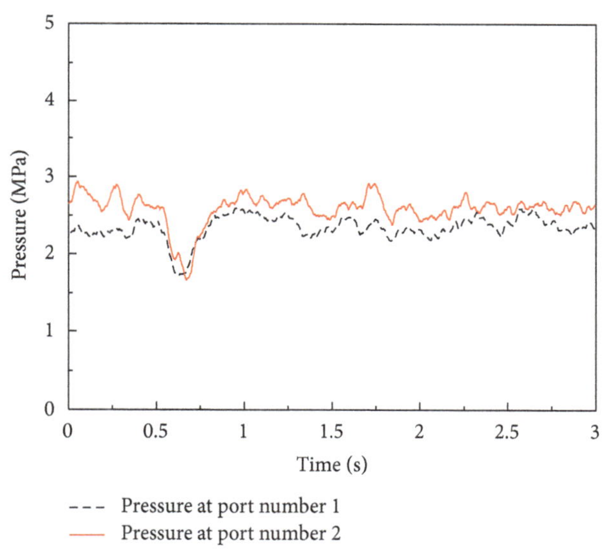

FIGURE 15: Pressure variations of optimal PID position control system for EHA without load disturbance (only mass load is applied).

sliding mode control system. As shown in Figure 14 and Table 4, the consistent performance is realized by the optimal adaptive antiwindup PID sliding mode control system. The time-domain quantitative performance indicators are also summarized in Table 4 for optimal adaptive antiwindup PID sliding mode control system without load.

On the other hand, Figure 15 depicts pressure variations of the optimal PID EHA position control system without load disturbance. As shown in Figure 15, the range of operating pressure of the EHA system is under 3 MPa in which the effective bulk modulus can be significantly changed according to contained air ratio of working fluid [5, 7, 11–13] as shown in Figure 16. The range of operating pressure for antiwindup PID and antiwindup PID sliding mode control system is also under 3 MPa.

Figure 17 depicts pressures of the EHA system when the position control systems are operating with load disturbance. As shown in Figure 17, the maximum operating pressures of the EHA system is about 10 MPa when the position control systems are working under the condition of adding disturbance load. The maximum operating pressure for antiwindup PID and antiwindup PID sliding mode control system is also up to 10 MPa. Thus, the effective bulk modulus can be changed greatly according to variation of operating pressure of the EHA systems. In addition, total leakage coefficient also can be changed in response to the level of working pressure.

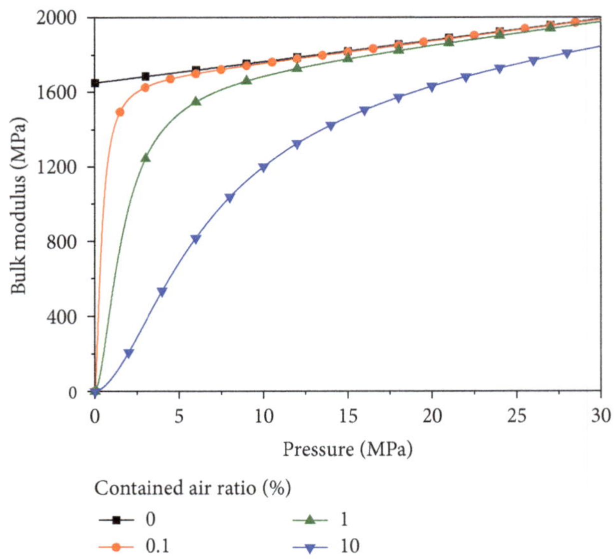

FIGURE 16: Example of the relationship between pressure and effective bulk modulus [12, 13].

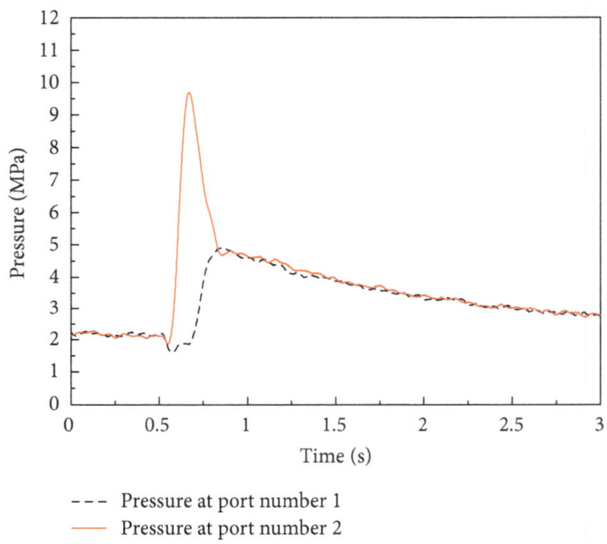

FIGURE 17: Pressure variations of optimal PID position control system for EHA with load disturbance.

In order to investigate robustness to the lumped system uncertainties such as varying parameters and modeling error and load disturbance, the performances of the three-position controllers of EHA system are compared in experiments by applying load disturbance. The experiments are conducted with 20 mm, position reference, and the load disturbance is applied by controlling the opening area of servo valve which is installed in a meter outline of load cylinder.

As shown in Figure 18, the optimal PID and optimal antiwindup PID control systems do not achieve robustness to the parameter change and load disturbance. In spite of the load disturbance and saturation of the BLDC motor in the position control system for EHA, the adaptive antiwindup PID sliding mode control system has a better performance and robustness than the optimal PID and optimal antiwindup PID control systems such that the adaptive antiwindup PID sliding mode control system recovers the parameter change and external disturbance in about 0.9 sec with a small overshoot. The time-domain quantitative performance indicators are summarized in Table 5 for optimal PID, antiwindup PID, and adaptive antiwindup PID sliding mode control systems with load disturbance. The list of terms are represented in Table 6.

6. Conclusion

This paper discussed the application of the optimal PID controller with optimal control gains using the optimization toolbox of MATLAB/Simulink to the position control system of EHA in order to understand the basic performance characteristics of the system. In experiment with optimal PID controller, large overshoot due to saturation of electric motor is observed.

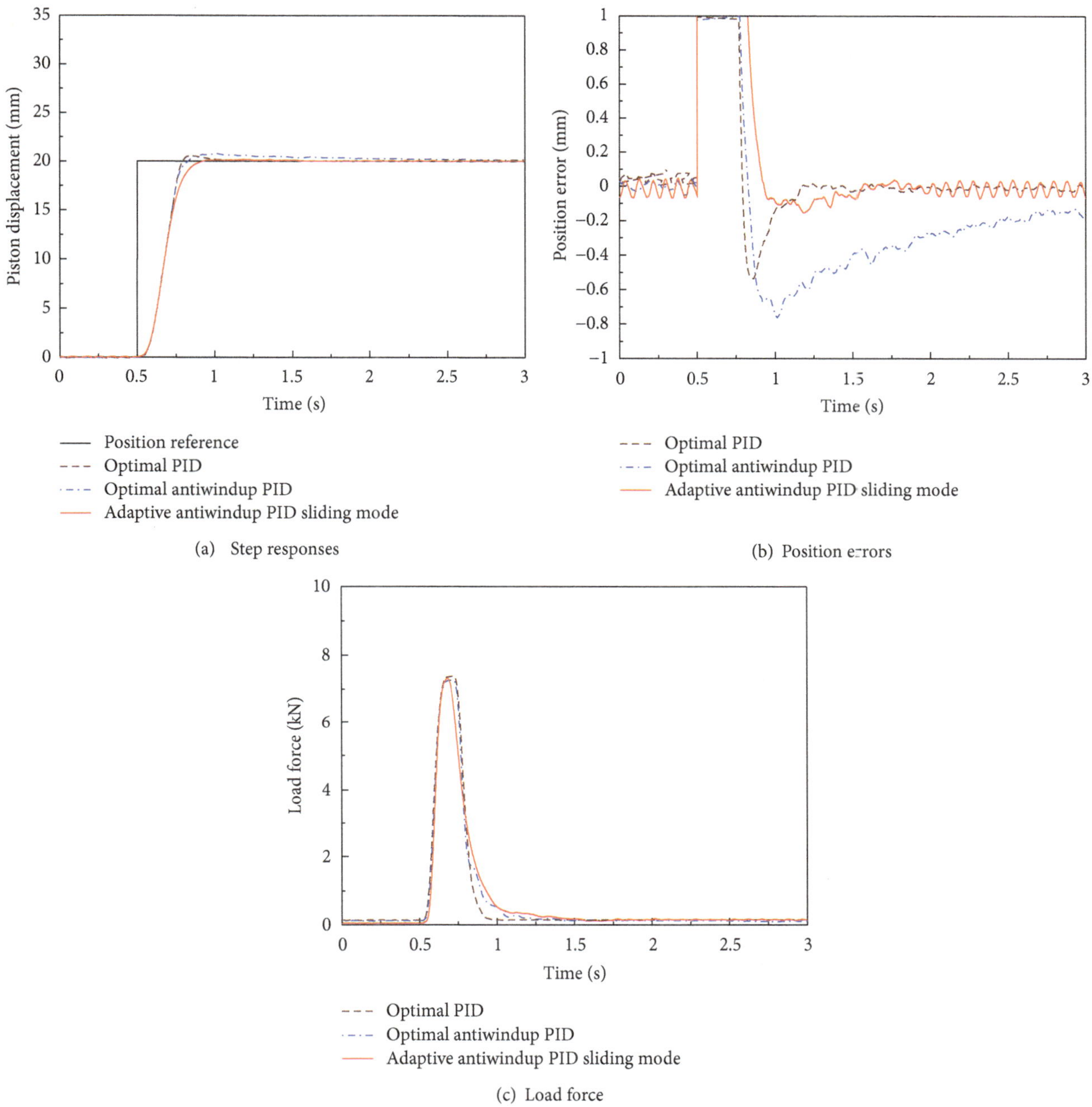

FIGURE 18: Performances of optimal PID, antiwindup PID, and adaptive antiwindup PID sliding mode control systems for EHA with load disturbance.

TABLE 5: Comparison of the performance of optimal PID, optimal antiwindup PID, and adaptive antiwindup PID sliding mode control systems with load disturbance.

Performance	Optimal PID	Optimal antiwindup PID	Adaptive antiwindup PID sliding mode
Overshoot	4.5%	3.9%	1.2%
Settling time	3.8 sec	1.8 sec	0.8 sec

In order to achieve the desirable performance of EHA position control system, optimal antiwindup PID controller and adaptive antiwindup PID sliding mode controller are also applied to control the position of EHA systems. In the process of robust control system design, mathematical modeling was carried out based on theoretical analysis, and the simple third-order nominal model was derived. The system parameters of the simple second-order nominal model were identified by the signal compression method.

Based on the experimental results and the time-domain quantitative performance indicators, the performance of the EHA position control system can be improved and robust to the load disturbance, parameter variation, and system

TABLE 6: List of terms.

Notation	Description	Unit
A	Pressurized area of hydraulic cylinder of EHA	m^2
B	Viscous friction coefficient	N/m/sec
C_T	Equivalent leakage coefficient	m^3/sec/Pa
D_p	Pump displacement	m^3/rad
e	Position error	mm
i	Current	Arms
K_{pipe}	Pipe coefficient relating pressure drop to flow	Pa/m^6/sec^2
K_s	Angular velocity to voltage gain	—
K_p	Proportional control gain	—
K_i	Integral control gain	—
K_d	Differential control gain	—
k_a	Antiwindup gain	—
$k_1, k_2, k_3, D, K, \gamma$	Positive design parameters of adaptive PID sliding mode controller	—
$K_m, T_1, T_{m_1}, T_{m_2}$	Parameters of dynamic model for the BLDC motor with controller	—
L	Internal leakage coefficient of the cylinder	m^3/Pa/sec
M	Mass load of position control test rig of EHA	N
p_1, p_2	Cylinder inlet/outlet chamber pressure	Pa
p_{pipe}	Pressure drop of pipe between the pump and the cylinder	Pa
Q_1, Q_2	Cylinder inlet/outlet flow	m^3/sec
s	Laplace operator	—
T_m	Torque of BLDC motor	Nm/(rad/sec)
T_s	Sampling time	sec
V_a, V_b	Volumes of inlet/outlet pipe and cylinder chamber	m^3
V_0	Mean volume of pipe and cylinder chamber.	m^3
v_e	Voltage signal of error for angular velocity of BLDC motor	V
v_r	Voltage signal of reference angular velocity of BLDC motor	V
v_ω	Voltage signal of angular velocity	V
x	Cylinder displacement	m
$y^d_{\text{step}}(t)$	Desired step response of the optimal PID and antiwindup PID control system	—
$y_{\text{step}}(t)$	Step response by the identified transfer function	—
β_e	Effective bulk modulus	Pa
ω_p	Pump angular velocity	rad/sec
ζ	Sliding surface	—
ξ	Pump cross port leakage coefficient	m^3/Pa/sec

uncertainty such as saturation of electric motor by using the adaptive antiwindup PID sliding mode controller.

References

[1] J.-M. Zheng, S.-D. Zhao, and S.-G. Wei, "Fuzzy iterative learning control of electro-hydraulic servo system for SRM direct-drive volume control hydraulic press," *Journal of Central South University of Technology*, vol. 17, no. 2, pp. 316–322, 2010.

[2] J.-M. Zheng, S.-D. Zhao, and S.-G. Wei, "Application of self-tuning fuzzy PID controller for a SRM direct drive volume control hydraulic press," *Control Engineering Practice*, vol. 17, no. 12, pp. 1398–1404, 2009.

[3] D. K. Pitipongsa Guansak and W. Po-Ngaen, "Two layer fuzzy generalized Maxwell-slip compensator in direct drive servo hydraulic system," in *Applied Mechanics and Materials*, pp. 420–427, 2012.

[4] Y. Cao, X. Y. Tian, G. Q. Wu, Q. M. Li, and Y. H. Wu, "Simulation experiment of roll bending machine in valveless servo system," *Advanced Materials Research*, vol. 619, pp. 410–414, 2012.

[5] Y. Chinniah, R. Burton, and S. Habibi, "Failure monitoring in a high performance hydrostatic actuation system using the extended Kalman filter," *Mechatronics*, vol. 16, no. 10, pp. 643–653, 2006.

[6] Y. Chinniah, R. Burton, S. Habibi, and E. Sampson, "Identification of the nonlinear friction characteristics in a hydraulic actuator using the extended Kalman filter," *Transactions of the Canadian Society for Mechanical Engineering*, vol. 32, no. 2, pp. 121–136, 2008.

[7] Y. Chinniah, S. Habibi, R. Burton, and E. Sampson, "Parameter identification in a high performance hydrostatic actuation system using the Unscented Kalman filter," *Transactions of the Canadian Society for Mechanical Engineering*, vol. 30, no. 3, pp. 375–389, 2006.

[8] K. Rongjie, J. Zongxia, W. Shaoping, and C. Lisha, "Design and simulation of electro-hydrostatic actuator with a built-in power regulator," *Chinese Journal of Aeronautics*, vol. 22, no. 6, pp. 700–706, 2009.

[9] H. M. Kim, S. H. Park, J. H. Song, and J. S. Kim, "Robust position control of electro-hydraulic actuator systems using the adaptive back-stepping control scheme," *Journal of Systems and Control Engineering*, vol. 224, no. 6, pp. 737–746, 2010.

[10] Y. Lin, Y. Shi, and R. Burton, "Modeling and robust discrete-time sliding-mode control design for a fluid power electrohydraulic actuator (EHA) system," *IEEE/ASME Transactions on Mechatronics*, vol. 18, no. 1, pp. 1–10, 2013.

[11] H. M. Kim, S. H. Park, J. M. Lee, and J. S. Kim, "A robust control of electro hydrostatic actuator using the adaptive back-stepping scheme and fuzzy neural networks," *International Journal of Precision Engineering and Manufacturing*, vol. 11, no. 2, pp. 227–236, 2010.

[12] J. Kajaste, H. Kauranne, A. Ellman, and M. Pietola, "Experimental validation of different models for effective bulk modulus of hydraulic fluid," in *Proceedings of the 9th Scandinavian International Conference on Fluid Power (SICFP '05)*, p. 16, Linköping, Sweden, 2005.

[13] K. Sunghun and M. Hubertus, *Measurement of Effective Bulk Modulus for Hydraulic Oil at Low Pressure*, vol. 134, ETATS-UNIS: American Society of Mechanical Engineers, New York, NY, USA, 2012.

[14] M. A. El Sayed and S. Habibi, "Inner-loop control for electro-hydraulic actuation systems," *Journal of Dynamic Systems, Measurement and Control*, vol. 134, no. 1, 2012.

[15] S. Habibi and A. Goldenberg, "Design of a new high performance electrohydraulic actuator," *IEEE/ASME Transactions on Mechatronics*, vol. 5, no. 2, pp. 158–164, 2000.

[16] M. Pachter, C. H. Houpis, and K. Kang, "Modelling and control of an electro-hydrostatic actuator," *International Journal of Robust and Nonlinear Control*, vol. 7, no. 6, pp. 591–608, 1997.

[17] S. Wang, R. Burton, and S. Habibi, "Sliding mode controller and filter applied to an electrohydraulic actuator system," *Journal of Dynamic Systems, Measurement and Control*, vol. 133, no. 2, 2011.

[18] S. Wang, S. Habibi, and R. Burton, "Sliding mode control for an electrohydraulic actuator system with discontinuous non-linear friction," *Journal of Systems and Control Engineering*, vol. 222, no. 8, pp. 799–815, 2008.

[19] K. K. Ahn and Q. T. Dinh, "Self-tuning of quantitative feedback theory for force control of an electro-hydraulic test machine," *Control Engineering Practice*, vol. 17, no. 11, pp. 1291–1306, 2009.

[20] S. H. Cho and R. Burton, "Position control of high performance hydrostatic actuation system using a simple adaptive control (SAC) method," *Mechatronics*, vol. 21, no. 1, pp. 109–115, 2011.

[21] H. E. Merritt, *Hydrostatic Control Systems*, Wiley Interscience, New York, NY, USA, 1967.

[22] S. R. Habibi and R. Burton, "Parameter identification for a high-performance hydrostatic actuation system using the variable structure filter concept," *Journal of Dynamic Systems, Measurement and Control*, vol. 129, no. 2, pp. 229–235, 2007.

[23] J. Y. Oh, G.-H. Jung, G.-H. Lee, Y.-J. Park, and C.-S. Song, "Modeling and characteristics analysis of single-rod hydraulic system using electro-hydrostatic actuator," *International Journal of Precision Engineering and Manufacturing*, vol. 13, no. 8, pp. 1445–1451, 2012.

[24] N. Aoshima, "Measurement of nonlinear vibration by signal compression method," *Journal of the Acoustical Society of America*, vol. 76, no. 3, pp. 794–801, 1984.

[25] G. P. Liu and S. Daley, "Optimal-tuning nonlinear PID control of hydraulic systems," *Control Engineering Practice*, vol. 8, no. 9, pp. 1045–1053, 2000.

[26] S.-U. Lee and P. H. Chang, "The development of anti-windup scheme for time delay control with switching action using integral sliding surface," *Journal of Dynamic Systems, Measurement and Control*, vol. 125, no. 4, pp. 630–638, 2003.

[27] M. Yokoyama, G.-N. Kim, and M. Tsuchiya, "Integral sliding mode control with anti-windup compensation and its application to a power assist system," *Journal of Vibration and Control*, vol. 16, no. 4, pp. 503–512, 2010.

[28] M. Yokoyama, Y. Tohta, and K. Saitoh, "A design method of sliding mode controller for servo-systems subject to actuator saturation," *JSME International Journal, Series C*, vol. 46, no. 3, pp. 960–966, 2003.

[29] I. Eker, "Second-order sliding mode control with PI sliding surface and experimental application to an electromechanical plant," *Arabian Journal for Science and Engineering*, vol. 37, no. 7, pp. 1969–1986, 2012.

[30] I. Eker, "Second-order sliding mode control with experimental application," *ISA Transactions*, vol. 49, no. 3, pp. 394–405, 2010.

[31] I. Eker, "Sliding mode control with PID sliding surface and experimental application to an electromechanical plant," *ISA Transactions*, vol. 45, no. 1, pp. 109–118, 2006.

[32] I. Eker and Ş. A. Akınal, "Sliding mode control with integral augmented sliding surface: design and experimental application to an electromechanical system," *Electrical Engineering*, vol. 90, no. 3, pp. 189–197, 2008.

[33] J. J. E. Slotine and W. Li, *Applied Nonlinear Control*, Pearson Education, Upper Saddle River, NJ, USA, 1991.

[34] M. Krstic, I. Kanellakopoulos, and P. Kokotovic, *Nonlinear and Adaptive Control Design*, Wiley Interscience, New York, NY, USA, 1995.

Quantitative Analysis of Electroplated Nickel Coating on Hard Metal

Hassan A. Wahab, M. Y. Noordin, S. Izman, and Denni Kurniawan

Faculty of Mechanical Engineering, Universiti Teknologi Malaysia, 81310 Skudai, Malaysia

Correspondence should be addressed to Denni Kurniawan; denni@utm.my

Academic Editors: L. Nobile and S. Torii

Electroplated nickel coating on cemented carbide is a potential pretreatment technique for providing an interlayer prior to diamond deposition on the hard metal substrate. The electroplated nickel coating is expected to be of high quality, for example, indicated by having adequate thickness and uniformity. Electroplating parameters should be set accordingly for this purpose. In this study, the gap distances between the electrodes and duration of electroplating process are the investigated variables. Their effect on the coating thickness and uniformity was analyzed and quantified using design of experiment. The nickel deposition was carried out by electroplating in a standard Watt's solution keeping other plating parameters (current: 0.1 Amp, electric potential: 1.0 V, and pH: 3.5) constant. The gap distance between anode and cathode varied at 5, 10, and 15 mm, while the plating time was 10, 20, and 30 minutes. Coating thickness was found to be proportional to the plating time and inversely proportional to the electrode gap distance, while the uniformity tends to improve at a large electrode gap. Empirical models of both coating thickness and uniformity were developed within the ranges of the gap distance and plating time settings, and an optimized solution was determined using these models.

1. Introduction

Nickel coating is widely used for decorative and functional applications, by improving corrosion resistance, wear resistance, or by modifying other properties of the coated articles [1]. With tungsten carbide/cobalt (WC-Co) cemented carbides, nickel coating is explored as well for the purpose of jointing these hard metals with other substances, both softer and harder ones. Cemented carbides have high hardness and strength contributed to the tungsten carbide with some toughness contributed to the cobalt binder. These properties make them suitable as fortifying coating for metallic substrates for some applications. This is where brazing using electroplated nickel becomes a potential jointing technique. Chen et al. [2] jointed WC-Co to stainless steel by a brazing process using nickel interlayer between the two metals and found that fracture of the brazed joints occurred in the bulk WC-Co substrates, indicating the success of the jointing nickel interlayer.

In other applications requiring extreme wear resistance, cemented carbide properties are still insufficient, and coating it with harder and stronger substances such as diamond or diamond-like carbon is applicable. Electroplated nickel interlayer is explored for this purpose where adhesion between the two substances is the main issue. Common diamond coating technique, that is, chemical vapor deposition, requires high temperature in which the cobalt binder within cemented carbide favors reaction towards graphitic phase, an unexpected interlayer during diamond deposition process which leads to deleterious effect to the diamond coating [3, 4]. There is also problems related to residual stress at the interface caused by mismatch in thermal expansion coefficients between the two substances. These issues make pretreatments on the hard metal substrate prior to diamond deposition required. Introducing intermediate layer between the cemented carbide substrate and the diamond coating is a way for this purpose. The use of electroplated nickel as the interlayer is of interest in this study considering nickel's thermal expansion coefficient which is close to that of hard metal, and that diamond can deposit and grow on nickel substrate [3, 5, 6].

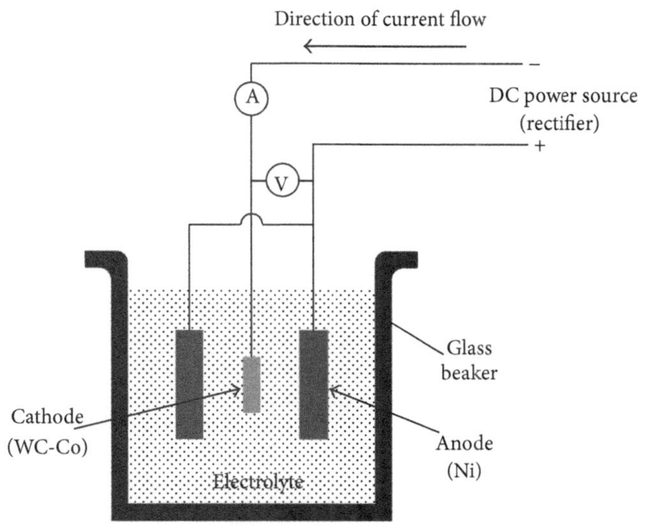

FIGURE 1: Schematic of the electroplating cell.

Electroplating for nickel deposition uses an electrolytic path. This method has advantages of having low reaction temperature which avoids residual stress caused by mismatch in thermal expansion coefficient, being economically viable, and being easy to control by manipulating the deposition parameters [7–9]. Thickness and uniformity are some of quality measures of electroplating results. This study attempts to improve the quality of electroplated nickel on hard metal substrate by selecting proper deposition parameters combination. Intensity of electric field and electrolyte resistance between anode and cathode are affected by the gap between electrodes [10]. Also, considering kinetics of reaction, electroplating is affected by the duration of reaction. Accordingly, in this study the electroplating time and the gap between electrodes varied, and their influence on the quality of deposited nickel layer is quantified by empirical models using design of experiment (DOE) so that the result could be objectively analyzed [11].

2. Experimental

2.1. Sample Preparation. A WC-6%Co rod of size 5 mm diameter × 150 mm length was cut using a precision cutter machine into moderate thin samples. The surface roughness of the samples was modified to $Ra = 0.3\,\mu$m by sand blasting using SiC with 180 μm grit size. After blasting, these samples were ultrasonically cleaned with acetone for 20 minutes to remove any residuals, followed by rinsing with deionized water. Finally, the samples were cleaned by steam jet for 30 seconds to remove any remaining residuals.

2.2. Plating Setup. For the nickel deposition, an electroplating solution containing 400 g/L $NiSO_4{\cdot}6H_2O$, 30 g/L $NiCl_2{\cdot}6H_2O$, and 30 g/L H_3BO_4 was used. The pH level of the electrolyte solution was regulated to a value of 3.5 by adding 10% of dilute sulfuric acid to the electrolyte. The experiments were carried out in a bath with a temperature of 55°C and 1.0 V electric potential to produce 0.1 Amp of current under continuous agitation by a magnetic stirrer. Two pure nickel round plates of 20 mm diameter × 3 mm thickness were used as anodes. The cathode was the thin sample previously cut (5 mm diameter × 2.05 mm thick) WC-6%Co substrate, which was fixed at distinct distances from each anode. Figure 1 illustrates the electroplating configuration and setup. The electrodeposition of nickel onto the substrate surface was performed at different holding times.

TABLE 1: Design layout of experiment.

Standard	Distance (mm)	Time (min)	Coded form	
			x_1	x_2
1	15	20	1	0
2	10	20	0	0
3	15	10	1	−1
4	5	10	−1	−1
5	5	20	−1	0
6	10	10	0	−1
7	15	30	1	1
8	5	30	−1	1
9	10	30	0	−1

2.3. Characterization. Electroplated samples were mounted with resin and then sectioned by precision cutter for metallographic tests. The nickel coating thickness and interface were characterized by scanning electron microscope (SEM). Value of coating thickness is obtained from the average of three measurements on each SEM image.

The layer thickness variation (k) between the substrate face and its circumference was calculated using (1) as follows:

$$k = \Big[\big(\text{coating thickness on surface} - \text{coating thickness on circumference}\big) \times \left(\text{coating thickness on surface}\right)^{-1}\Big] \times 100\%. \quad (1)$$

2.4. Deposition Parameters. Electrode gap and plating time were the input variables, termed in coded form as x_1 and x_2, respectively. The upper and lower limit values of the input variables were chosen in order to maintain interlayer thickness greater than 2 μm with adequate quality. The selected upper and lower limit values for the electrode gap were 5 mm and 15 mm, respectively, and for the plating time the values were 10 min and 30 min, respectively. The middle limit values were 10 mm for the electrodes gap and 20 min for the plating time.

Design of experiments and results analysis were performed using commercial statistical analysis software (Design Expert 6.0 of Stat-Ease Inc.). A three-level factorial design with two variables as the input was employed [11]. The total treatment combinations required for the design were 9 combinations (Table 1).

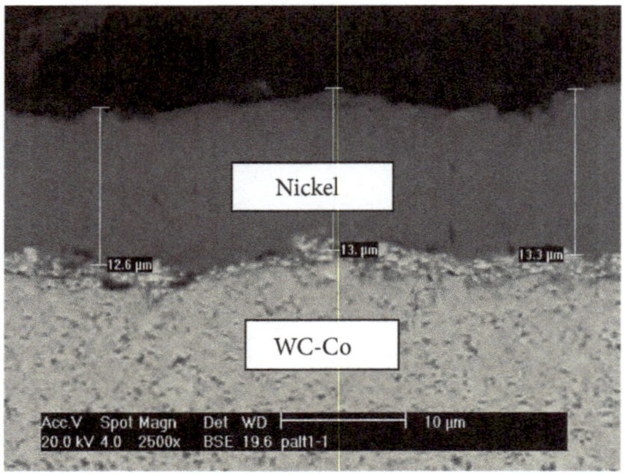

FIGURE 2: SEM image of electroplated nickel coating on hard metal surface when plating time is 10 min and anode/cathode gap distance is 5 mm.

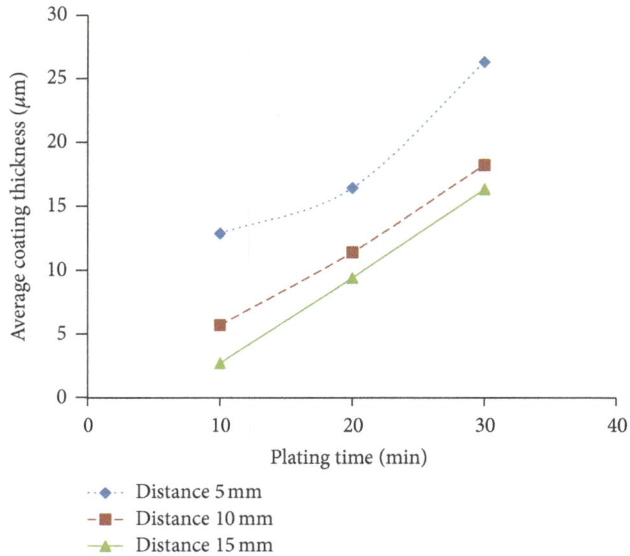

FIGURE 3: Average coating thickness versus plating time for three different anode/cathode gaps (pH: 3.5, current: 0.1 A, and voltage: 1.0 V).

For determining the relationship between the input and response, the collected data were analyzed using regression. In regression model, the empirical variable (response) is approximated based on a functional relationship between the estimated variable, y, and one or more input variables, x_1 and/or x_2. The model equation containing the input variables was fitted by using least square technique. Accordingly, the residual error measured by the sum of square deviations between the actual and the estimated responses was minimized. This involved the calculation of estimates for the regression coefficients, that is, the coefficients of the model variables including the intercept or constant term for statistical significance test.

3. Result and Discussion

Typical result of electroplated nickel coating on hard metal surface is as shown in Figure 2. The thickness and uniformity of the electroplated nickel coating were indeed affected by the plating time and electrode gap distance as presented in following subsections.

3.1. Coating Thickness. The experimental results for coating thickness are shown in Figure 3. The electroplated nickel coating thickness is inversely proportional to the gap between the electrodes and is proportional to the deposition time. The highest coating thickness of 26.3 μm was obtained at electrode gap 5 mm and deposition time of 30 minutes, while the lowest coating thickness of 2.7 μm was resulted when the electrode gap is 15 mm and deposition time is 10 minutes.

To develop the empirical model of the coating thickness, a fit summary output to determine the most suitable regression model was evaluated. Linear model was selected for having the least probabilistic value, Prob $> F$, for representing coating thickness results. Analysis of variance was performed to test the significance of the regression model and its coefficients, with probabilistic value of less than 0.05 indicating significance. The linear model and its coefficients have a probabilistic value of well below 0.05 (Table 2). Furthermore, its coefficient of determination, R^2, of 0.96 is very close to unity, also indicating that the model closely approximates the coating thickness data. The model's adequate precision ratio of 22.6, which compares the range of the predicted values at the design points to the average prediction error, is well beyond the minimum adequacy limit of 4.

The empirical model equation which relates between resulted coating thickness and the setting of gap distance and plating time is

$$T = 9.16 - 0.91d + 0.66t, \quad (2)$$

where T is coating thickness, d is gap distance between electrodes (in mm), and t is plating time (in minutes). The model equation can be depicted in a contour graph (Figure 4). The developed model's predicted coating thickness for any combination of gap distance and plating time was confirmed within the range of 95% confidence interval. This verifies that the model is sufficient to represent the coating thickness data for this particular electroplated nickel deposition.

The finding that electrode gap distance is inversely proportional to coating thickness is in line with previous reports which found that when gap between electrodes is narrow the intensity of electric field increases and electrolyte resistance between anode and cathode decrease [10, 12]. These result in increasing metal deposition rate or thicker coating within a certain period of time. This study adds that the relation in particular settings is linear. Regarding the linear effect of deposition time to coating thickness, it is in agreement with Faraday's law in which deposition rate (or in this case the coating thickness resulted within a certain period) is proportional to the plating current and time [13, 14].

TABLE 2: Analysis of variance of the coating thickness model.

Source	Sum of square	Degree of freedom	Mean square	F value	Prob > F
Model	383.35	2	191.67	66.27	<0.05
Gap distance	123.10	1	123.10	42.63	<0.05
Plating time	260.04	1	260.04	89.91	<0.05
Residual	17.35	6	2.89		
Pure error		0			
Cor total	400.70	8			

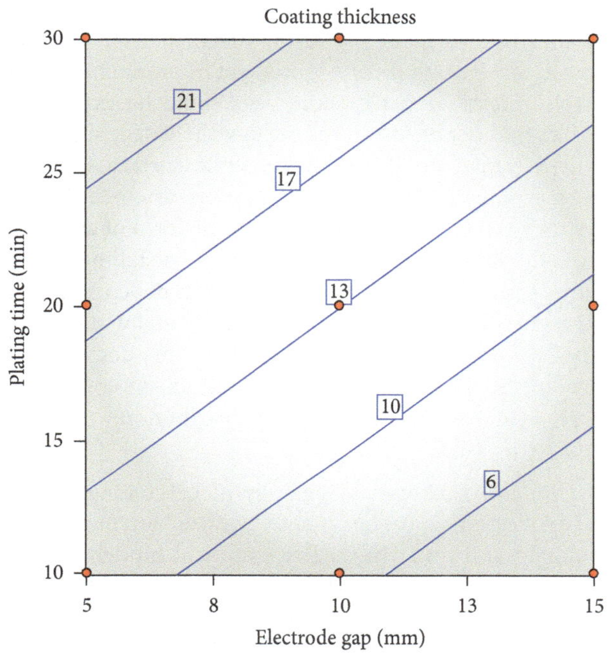

FIGURE 4: Response surface graph of contours for coating thickness.

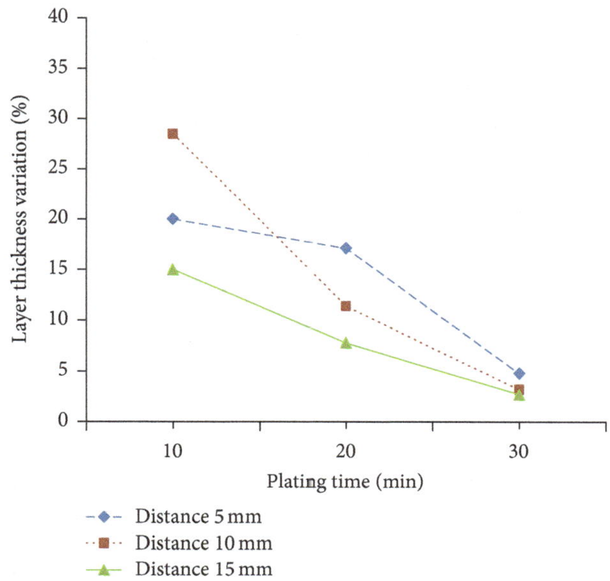

FIGURE 5: Thickness uniformity versus plating time at various electrode gaps.

3.2. Layer Uniformity. The layer thickness variations between the substrate face and its circumference are shown in Figure 5. The combination between the electrode gaps and plating time used in this study can generate coating layer with up to a maximum of 26.5% thickness variation. The increase of both electrode gap and plating time tend to produce more uniform nickel coating on the hard metal substrate.

In developing the empirical model for coating thickness uniformity as function of gap distance and plating time, similar steps used to develop model for coating thickness was employed. A preliminary diagnosis was conducted to determine the appropriate power transformation for the model. This was assessed by diagnosing Box-Cox plot of coating thickness uniformity. The lowest point of the plot that might result in the minimum residual sum of square in the transformed model (λ) was 0.61. Accordingly, a logarithmic transformation ($\lambda = 0$) was selected as the closest value to the actual one and as the most appropriate power transformation. Next step was determining the suitable regression model. The probabilistic value, Prob > F, was calculated, and linear model was selected for having the least probabilistic value. Then, analysis of the variance was performed in order to test the significance of the selected regression model and its coefficients (Table 3). In the same way as before, the maximum probabilistic value of 5% was set for the model, and its coefficients were to be considered significant.

The empirical model equation which relates coating thickness uniformity (k) to gap distance (d) and plating time (t) was found to be

$$\ln k = 4.48 - 0.05d - 0.08t. \quad (3)$$

This model can be represented by the graph of contours (Figure 6). The model's high coefficient of determination ($R^2 = 0.92$) and adequate precision (of 14.56) indicated that the model is adequately representative. At any combination of gap distance and plating time, the model's predicted coating thickness uniformity is within the range of 95% confidence interval. This verifies the adequacy represented by the model.

Due to the nature of the process, obtaining high degree of coating uniformity via electroplating technique is practically difficult. The ability of producing uniform coating thickness over a surface by electroplating process is associated with current distribution on that surface [1]. Current distribution is usually determined by the shape of the surface and its relative position with respect to the anode [15]. Edges and

TABLE 3: Analysis of variance for the coating thickness uniformity.

Source	Sum of square	Degree of freedom	Mean square	F value	Prob > F
Model	4.53	2	2.27	33.37	<0.05
Gap distance	0.43	1	0.43	6.36	0.04
Plating time	4.10	1	4.10	60.38	<0.05
Residual	0.41	6	0.07		
Pure error		0			
Cor total	4.94	8			

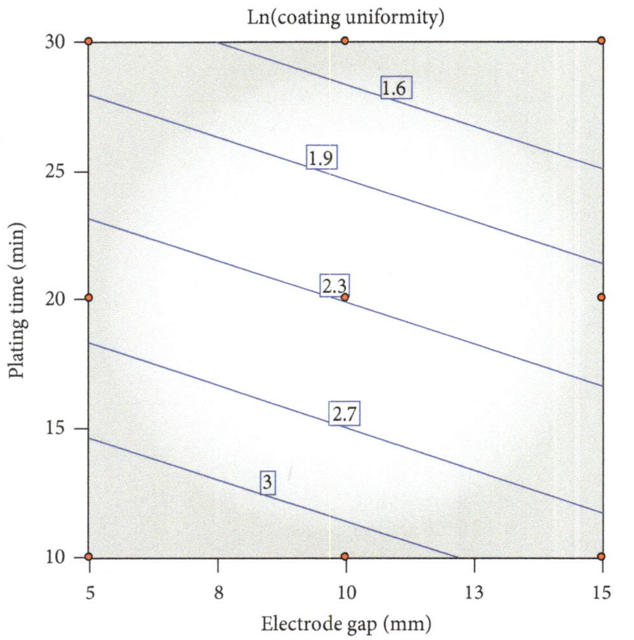

FIGURE 6: Response surface graph of contours for surface roughness.

recesses on a surface often receive more current and this consequently results in thicker deposits on them [16]. Placing the electrodes at an adequate distance relative to one another might reduce the thickness variation of an electroplated coating by making the current distribution on the surface to be nearly uniform [10].

This study, in which 3% to 26.5% layer thickness variations were obtained, found that the nickel coating was more uniform when the hard metal substrate is placed at longer distance from the anodes. The result is in good agreement with a previous report that found that short electrode gap produces high offset voltage which causes larger current to be concentrated at the edges and the corner of the electrode [10]. Under this condition, reduction of reactant on the electrode surface is faster, and variation in reactants diffusion between the electrode edges and center region surface could increase [10]. When the distance between electrodes is increased the offset set voltage decreases and plating current lowers. Consequently, the current distribution at the surface area is improved [10].

Electroplating duration is also believed to have a positive effect in acquiring uniform distribution of the coating over the substrate surface. The result obtained from this study showed that plating time is significant in producing uniform coating thickness distribution. This could be explained by the fact that deposited atoms occupy the defect sites (edges, corners, steps, and kinks) on substrate surface after either direct or lateral diffusion. When atoms reach the defected sites they start to back themselves in the form of atoms lines across the substrate surface forming the crystalline structure of the deposit [17]. This action mainly depends on local lattice energy (incorporate more atoms to metal matrix at the cathode) and the plating time, as short time does not allow the completion of this deposition mechanism and could be the reason for nonuniform coating distribution.

3.3. Optimum Condition. Generally, nickel coating for industrial applications must be obtained with an acceptable degree of quality and uniformity. The empirical model developed here revealed the possibility of producing electroplated layer thickness with minimum degree of nonuniformity along the substrate surfaces by setting the electrode gap and plating time. In practice, it is difficult to obtain 100% uniform electroplated coating over a surface. However, low coating thickness variation across the surface is desired.

This study, where the electroplated nickel is intended to be the interlayer between hard metal substrate and diamond coating, expects the nickel coating not to be excessive (low thickness) but with high uniformity. This can be achieved by selecting a suitable combination of electrodes gap and plating time. An optimized solution of these two variables should be determined. With the help of the empirical models developed previously, this optimized solution can be determined. The effect of each input variable (gap distance and plating time) on the electroplating responses (coating thickness and uniformity) is quantifiable using the models. It is presumed that coating thickness of 5 μm is adequate to withstand cracks or delamination when the hard metal is subjected to heat treatment and diamond deposition. Also the deposited coating should have thickness uniformity of no less than 15% because below that point narrow or nonsolution domains might establish. Optimization of the responses is when the electrode gap and plating time combination is within the grey area of the overlay plot (Figure 7). The optimized solution is represented by the intersection between the solutions of coating thickness and the solutions for layer thickness variation.

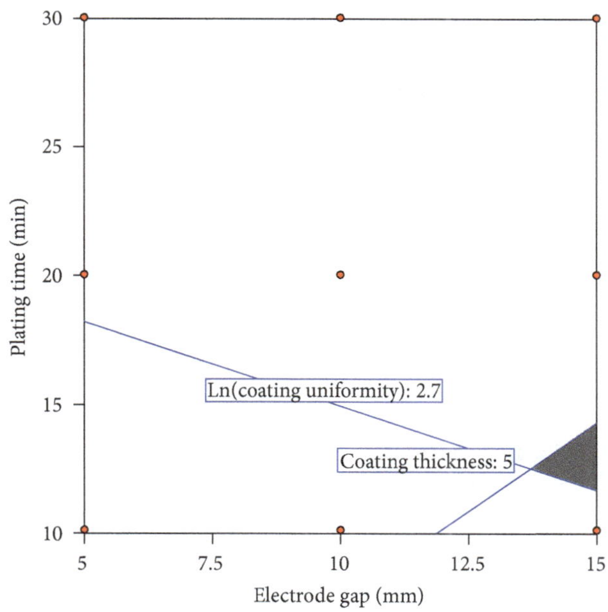

FIGURE 7: Overlay plot of the input variables for the predetermined output criteria of maximum of 5 μm coating thickness and minimum of 15% coating uniformity. (The grey area is the optimized solution.)

4. Conclusion

The attempt to obtain electroplated nickel coating on WC-6%Co substrate was performed by varying electrode gap distance and plating time, with coating thickness and uniformity as the responses. Coating thickness is proportional to the plating time and inversely proportional to the electrode gap distance. The coating thickness uniformity tends to be better at a large electrode gap. Empirical models of both electroplating responses were developed within the ranges of the input variables. The models can determine the optimized solutions of gap distance and plating time combination for obtaining adequate coating thickness with high uniformity.

Conflict of Interests

The authors declare that they have no conflict of interests. The authors gain no secondary interest from or have any direct financial relation with the commercial identities mentioned in this paper that will anyhow alter the validity of the research.

Acknowledgments

Financial supports from the Ministry of Higher Education, Malaysia, and Universiti Teknologi Malaysia through Fundamental Research Grant Scheme (no. 78572) and Research University Grant (no. 05H27) and scholarship from the Ministry of Higher Education, Sudan, are gratefully acknowledged.

References

[1] J. A. Di Bari, "Electrodeposition of nickel," in *Modern Electroplating*, M. Schlesinger and M. Paunovic, Eds., John Wiley and Sons, 2000.

[2] H. Chen, K. Feng, S. Wei, J. Xiong, Z. Guo, and H. Wang, "Microstructure and properties of WC-Co/3Cr13 joints brazed using Ni electroplated interlayer," *International Journal of Refractory Metals and Hard Materials*, vol. 33, pp. 70–74, 2012.

[3] W. Lu, D. Zuo, M. Wang, and F. Xu, "Analysis of interlayer between WC-Co and CVD diamond film," *Key Engineering Materials*, vol. 375-376, pp. 92–96, 2008.

[4] R. Polini, M. Barletta, G. Rubino, and S. Vesco, "Recent advances in the deposition of diamond coatings on Co-cemented tungsten carbides," *Advances in Materials Science and Engineering*, vol. 2012, Article ID 151629, 14 pages, 2012.

[5] H. Zhu, T. Kuang, B. Zhu, S. Lei, Z. Liu, and S. P. Ringer, "Flame synthesis of carbon nanostructures on Ni-plated hardmetal substrates," *Nanoscale Research Letters*, vol. 6, article 331, 2011.

[6] J. M. Huh and D. Y. Yoon, "Enhanced nucleation of diamond on polycrystalline Ni by d.c. glow discharge in hot filament CVD," *Diamond and Related Materials*, vol. 9, no. 8, pp. 1475–1479, 2000.

[7] W. Bacher, K. Bade, B. Matthis, M. Saumer, and R. Schwarz, "Fabrication of LIGA mold inserts," *Microsystem Technologies*, vol. 4, no. 3, pp. 117–119, 1998.

[8] A. M. Rashidi and A. Amadeh, "Effect of electroplating parameters on microstructure of nanocrystalline nickel coatings," *Journal of Materials Science and Technology*, vol. 26, no. 1, pp. 82–86, 2010.

[9] V. Vasilache, S. Gutt, G. Gutt, T. Vasilache, C. Filote, and I. Sandu, "Studies of hardness for the electrodeposited nickel from watts baths with addition of polyvinyl pyrrolidone (PVP)," *Revue Roumaine de Chimie*, vol. 54, no. 3, pp. 243–246, 2009.

[10] Y.-J. Tan and K. Y. Lim, "Understanding and improving the uniformity of electrodeposition," *Surface and Coatings Technology*, vol. 167, no. 2-3, pp. 255–262, 2003.

[11] M. Y. Noordin, D. Kurniawan, Y. C. Tang, and K. Muniswaran, "Feasibility of mild hard turning of stainless steel using coated carbide tool," *International Journal of Advanced Manufacturing Technology*, vol. 60, pp. 853–863, 2012.

[12] Y. S. Dong, P. H. Lin, and H. X. Wang, "Electroplating preparation of Ni-Al_2O_3 graded composite coatings using a rotating cathode," *Surface and Coatings Technology*, vol. 200, no. 11, pp. 3633–3636, 2006.

[13] E. S. Masuku, A. R. Mileham, H. Hardisty, A. N. Bramley, C. Johal, and P. Detassis, "A finite element simulation of the electroplating process," *CIRP Annals*, vol. 51, no. 1, pp. 169–172, 2002.

[14] M. De Vogelaere, V. Sommer, H. Springborn, and U. Michelsen-Mohammadein, "High-speed plating for electronic applications," *Electrochimica Acta*, vol. 47, no. 1, pp. 109–116, 2001.

[15] O. Sadiku-Agboola, E. R. Sadiku, O. I. Ojo, O. L. Akanji, and O. F. Biotidara, "Influence of operation parameters on metal deposition in bright nickel-plating process," *Portugaliae Electrochimica Acta*, vol. 29, no. 2, pp. 91–100, 2011.

[16] F. A. Lowenhiem, *Modern Electroplating*, Wiley, 1974.

[17] N. Kanani, *Electroplating Basic Principles, Processes and Practice*, Elsevier, 2004.

5

Construction of a WMR for Trajectory Tracking Control: Experimental Results

R. Silva-Ortigoza,[1] C. Márquez-Sánchez,[1] M. Marcelino-Aranda,[2] M. Marciano-Melchor,[1] G. Silva-Ortigoza,[3] R. Bautista-Quintero,[4] E. R. Ramos-Silvestre,[5] J. C. Rivera-Díaz,[6] and D. Muñoz-Carrillo[1]

[1] *Instituto Politécnico Nacional, CIDETEC, Área de Mecatrónica, Unidad Profesional Adolfo López Mateos, 07700 México, DF, Mexico*
[2] *Instituto Politécnico Nacional, UPIICSA, Sección de Estudios de Posgrado e Investigación, 08400 México, DF, Mexico*
[3] *Benemérita Universidad Autónoma de Puebla, Facultad de Ciencias Físico Matemáticas, 72001 Puebla, PUE, Mexico*
[4] *Instituto Tecnológico de Culiacán, Departamento de Metal-Mecánica, 80220 Culiacán, SIN, Mexico*
[5] *Universidad Privada del Valle, Facultad de Informática y Electrónica, Tiquipaya, CBBA, Bolivia*
[6] *Centro Nacional de Actualización Docente, Área de Máquinas, 13420 México, DF, Mexico*

Correspondence should be addressed to R. Silva-Ortigoza; rsilvao@ipn.mx

Academic Editors: W. Chen and D. K. Liang Ou

This paper reports a solution for trajectory tracking control of a differential drive wheeled mobile robot (WMR) based on a hierarchical approach. The general design and construction of the WMR are described. The hierarchical controller proposed has two components: a high-level control and a low-level control. The high-level control law is based on an input-output linearization scheme for the robot kinematic model, which provides the desired angular velocity profiles that the WMR has to track in order to achieve the desired position (x^*, y^*) and orientation (φ^*). Then, a low-level control law, based on a proportional integral (PI) approach, is designed to control the velocity of the WMR wheels to ensure those tracking features. Regarding the trajectories, this paper provides the solution or the following cases: (1) time-varying parametric trajectories such as straight lines and parabolas and (2) smooth curves fitted by cubic splines which are generated by the desired data points $\{(x_1^*, y_1^*), \ldots, (x_n^*, y_n^*)\}$. A straightforward algorithm is developed for constructing the cubic splines. Finally, this paper includes an experimental validation of the proposed technique by employing a DS1104 dSPACE electronic board along with MATLAB/Simulink software.

1. Introduction

In the last decades, the control of wheeled mobile robots (WMRs) has been an interesting topic for research [1]. The differential robot configuration studied in this paper has nonholonomic constraints [2]. In order to improve the autonomy of the mobile robots, the literature in this field has generally focused on solving the following problems: (1) mobile robot positioning, (2) stabilization, (3) trajectory tracking control, (4) planning the trajectories, and (5) obstacle avoidance.

In the robot stabilization problem, according to [3], it is known that a nonholonomic system cannot be asymptotically stabilized at an equilibrium point using a differentiable control law, despite the system's being completely controllable. Accordingly, the stabilization of nonholonomic systems can only be achieved by nondifferentiable control laws [4] or time-dependent ones [5–9].

On the other hand, the trajectory tracking task in nonholonomic systems can be performed through differentiable control laws. In [10], a hierarchical control scheme based on two levels (high-level and low-level) is presented for the trajectory tracking control of a car-trailer system. The high-level control is based on a time-varying linear quadratic regulator, which provides the desired angular velocity profiles that the system has to track in order to achieve the desired trajectory.

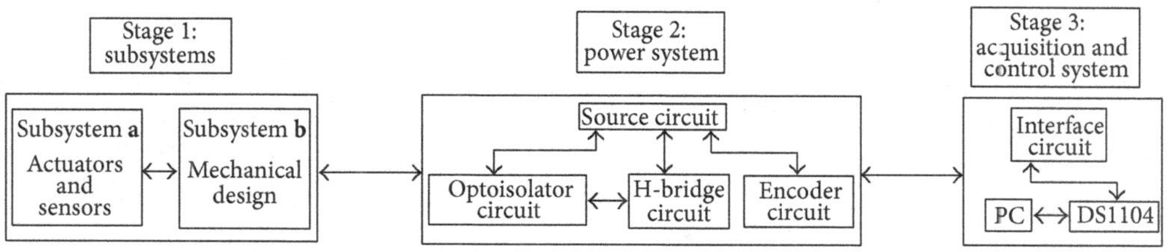

FIGURE 1: General block diagram of the WMR prototype.

Then, a low-level control is designed for controlling the traction and the steering motors by using a proportional integral derivative (PID) control. Experimental results from the application of this tracking control scheme are presented. Similarly, in [11], the control of a differential WMR is proposed via a variable structure control scheme, based on external and internal control strategies. The external control is associated to the kinematic model of the mobile robot, which is responsible for generating the desired angular velocity profiles for the motors, via an input-output linearization scheme. Meanwhile, the internal control is meant to track the velocities imposed by the external control via a PID control. In [12], via the differential flatness property of the WMR kinematic model, the dynamic controller design for trajectory tracking tasks was presented. The basis of flatness-based control can be found in the work by Fliess et al. [13]. Likewise, in [14], a sliding mode control in combination with differential flatness was presented for different types of WMRs for tracking problems. These controllers were implemented through computer simulations. In [15], a control algorithm to achieve trajectory tracking of a WMR with nonholonomic constraints was proposed by using the computed torque method and the theory of sliding mode control. The proposed algorithm was implemented on a vision-based mobile robot system. In [16], an input-output linearization design was presented in which the actuators are controlled by a sensorless scheme. More recently, in [17], a formal justification was presented for use of velocity and electric current inner loops, driven by proportional controllers, when a trajectory tracking task is to be solved. In this work, the dynamics of the brushed DC motors used as actuators is taken into account to perform the stability analysis.

Different approaches to the trajectory generation problem have been widely studied (see, e.g., [18]) where the "path planning" problem is related to generating a sequence of points through which the robot must pass (either an end-effector of an industrial robot or a mobile trajectory); see [19]. Another approach is known as "trajectory planning and optimization," which includes a structure that integrates the dynamics with the path planning. Multiobjective optimization can be addressed, such as the cost efficient use of energy, the task time, and other similar constraints; see [20]. Although the trajectory generation can be solved regardless of dynamic consideration, real constraint issues make this approach impractical for many applications. An ideal trajectory implementation with smooth transitions (human-like) as well as being robust against external disturbances must be designed under multiobjective contradictory goals. A well known implementation for this problem is based on splines (see [21–23]) which can be designed in such a way that both the velocity and acceleration transitions follow the dynamic requirements. For instance, in [24, 25], an algorithm for path planning for mobile robots used in football competitions was presented. In [26], a technique was shown for path planning based on splines, which was used to simulate differential mobile robots.

The contribution of this research is to present an integration of both theoretical and practical knowledge in order to close the gap between the two. The relevant and modern open-architecture testbed described in this paper deals with a synergetic combination of an effective control approach and a state-of-the-art technology for rapid prototyping that combines a hard real-time control implementation with a software (MATLAB/Simulink) widely used for these applications. In this context, the present paper has three aims: (1) to describe the construction of a differential drive WMR, (2) to show the implementation of a real-time hierarchical control strategy in order to carry out a trajectory tracking control task, and (3) to present a methodology for generating the WMR trajectories (based on cubic splines) which are constructed from the desired data points $\{(x_1^*, y_1^*), \ldots, (x_n^*, y_n^*)\}$. Furthermore, time-varying parametric trajectories such as straight lines and parabolic curves, are also implemented.

To this end, the present paper is organized as follows. Section 2 presents the general description of the WMR construction. Section 3 describes the hierarchical controller law for the kinematic model for the trajectory tracking task. Section 4 gives an algorithm for generating smooth curves based on specific points given in the X-Y plane via cubic splines. Section 5 shows the real-time control implementation for the WMR. Finally, some conclusions and prospects for future research are presented in Section 6.

2. Construction of the WMR

A mobile robot is, in general, composed of two mechanical subsystems: (1) *actuators and sensors* and (2) *mechanical design*. The control of these subsystems requires two electronic stages: (1) the power stage and (2) the acquisition and control stage. The interaction of these stages is shown in the block diagram presented in Figure 1. The electronic power stage (stage 2) allows interaction between the electronic control interface (stage 3) and the two dynamic subsystems (stage 1). This interaction includes the communication system, the power supply, and the conditioning circuits (in order to interconnect the electronic board DS1104). The strategy used in the PC (using MATLAB/Simulink) keeps the whole system under control, taking into consideration the physical restrictions. Likewise, the control stage comprises the control strategies used to integrate the functioning of each subsystem (based on mathematic models of the plant).

2.1. Stage 1: Subsystems. This part describes the WMR dynamic subsystems **a** and **b**, which include actuators, sensors, and the mechanical structure. Subsystem **a** (motors and sensors) allows propulsion of the WMR in a specified workspace and discrete position sensing. The subsystem **b** describes the mechanical design.

2.1.1. Subsystem **a**: *Actuators and Sensors.* The prototype described in this work uses DC motors as actuators. In order to estimate the mechanical capacity of these actuators, it is proposed to use the maximum velocity, acceleration, and weight of the WMR system as

$$v = 1\,\text{m/s}, \qquad a = 1\,\text{m/s}^2, \qquad m = 46.74\,\text{kg}. \tag{1}$$

Based on the mass m and maximum acceleration a, it is possible to calculate the required force F for a given displacement as

$$F = ma = (46.74\,\text{kg})\left(1\,\text{m/s}^2\right) = 46.74\,\text{N}. \tag{2}$$

Since two motors are used, each of them has an associated force, F_{rm} for the right wheel motor and F_{lm} for the left wheel motor. One motor provides half of the total force required, that is,

$$F_{\text{rm}} = F_{\text{lm}} = \frac{F}{2} = \frac{46.74\,\text{N}}{2} = 23.37\,\text{N}. \tag{3}$$

Since the wheel diameter is 0.15 m ($r = 0.075$ m), the required torque for each motor is given by

$$\begin{aligned}\tau_{\text{rm}} &= F_{\text{rm}} r = (23.37\,\text{N})(0.075\,\text{m}) = 1.75\,\text{Nm},\\ \tau_{\text{lm}} &= F_{\text{lm}} r = (23.37\,\text{N})(0.075\,\text{m}) = 1.75\,\text{Nm}.\end{aligned} \tag{4}$$

Since $v = \omega_{\text{w}} r$, the angular velocity is related to each wheel and it is determined by

$$\omega_{\text{w}} = \frac{v}{r} = \frac{1\,\text{m/s}}{0.075\,\text{m}} \approx 13.33\,\text{rad/s} \approx 127.35\,\text{rpm}. \tag{5}$$

Hence, the power required, P_{w}, is calculated as follows:

$$P_{\text{w}} = \tau_{\text{rm}}\omega_{\text{w}} = (1.75\,\text{Nm})(13.33\,\text{rad/s}) \approx 23.37\,\text{W}. \tag{6}$$

The energy must be transmitted to each wheel of the WMR; however, the power required is higher than the calculated power since there is a loss due to the gearbox coupling. The gearbox efficiency is around 70%, so that the estimation of the power required (P_d) is given by

$$P_d = \frac{P_{\text{w}}}{0.7} = \frac{23.37\,\text{W}}{0.7} = 33.39\,\text{W}. \tag{7}$$

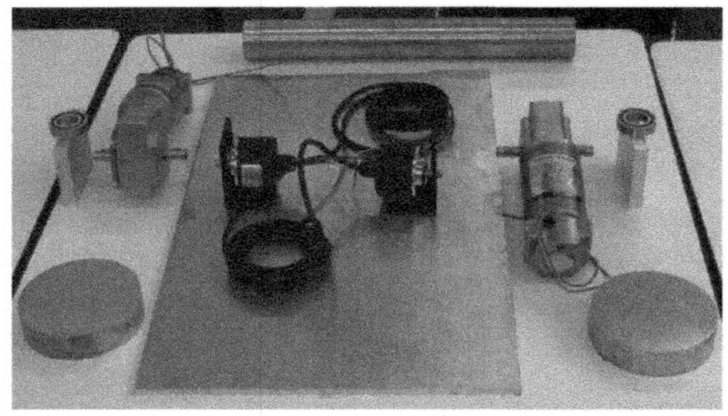

FIGURE 2: Materials used for the prototype construction.

Based on the data sheet of the motor [27] and the golden rule for selecting an appropriate motor, it is recommended to have between 1.5 and 2 times the required power P_d, and thus

$$50.09\,\text{W} < P_d < 66.78\,\text{W}, \tag{8}$$

so that the motors selected are *GNM 3150* (24 V, 55 W) and the gearbox chosen for each motor is *G 2.6*. The second golden rule given by the manufacturer shows the importance of maintaining the desired velocity ω_d between 65% and 90% of the no-load motor velocity (i.e., 3526 rpm), and thus

$$2291.9\,\text{rpm} < \omega_d < 3173.4\,\text{rpm}. \tag{9}$$

In order to satisfy this condition, the reduction ratio should be at least 20 : 1, since

$$\omega_m = 20\omega_{\text{w}} = (20)(127.35\,\text{rpm}) = 2547\,\text{rpm}. \tag{10}$$

Finally, these selected motors, including the gearboxes, are the series *GNM 3150* (24 V, 55 W) Engel brand and a gearbox *G 2.6*. The power for the motors is supplied with two YUASA batteries 12 V @ 12 Ah.

Regarding the prototype sensors, it uses two incremental optical encoders. These sensors allow estimating the angular velocity of the wheels, and consequently the robot is able to perform the trajectory tracking task. The manufacturer of the sensors is the Korean *Autonics*, and the model series is E50S8. These sensor are powered by the YUASA battery previously presented. The configuration of these sensors in the prototype can be seen in Figure 9.

2.1.2. Subsystem **b**: *Mechanical Design.* Once the components were selected (motors, encoders, and batteries), the software SolidWorks is used to integrate all the mechanical parts of the WMR into a virtual model. The 3D visual feature of this software allows the correct component distribution in the WMR in accordance with the design requirements. Also, the software can be used for specifying the material properties to the model of each part of the WMR. The parts built for this prototype, from the materials previously selected (see Figure 2, and Videos S1, S2 in Supplementary Material available online at http://dx.doi.org/10.1155/2013/723645), are as follows.

(i) The WMR base is made of aluminum sheet, which is used to support all components (see Figure 3).

(ii) The wheels used to provide traction to the WMR are constructed with an aluminum rod (see Figure 4).

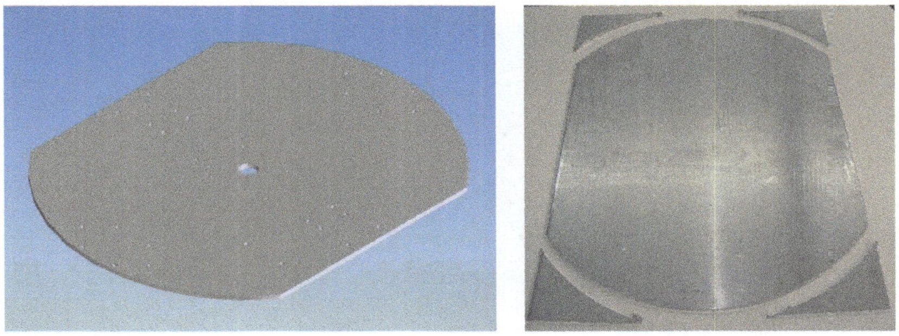

(a) Base in SolidWorks (b) Real picture of the base

FIGURE 3: Base design.

(a) Wheel in SolidWorks (b) Real wheels of the WMR

FIGURE 4: Traction wheels design.

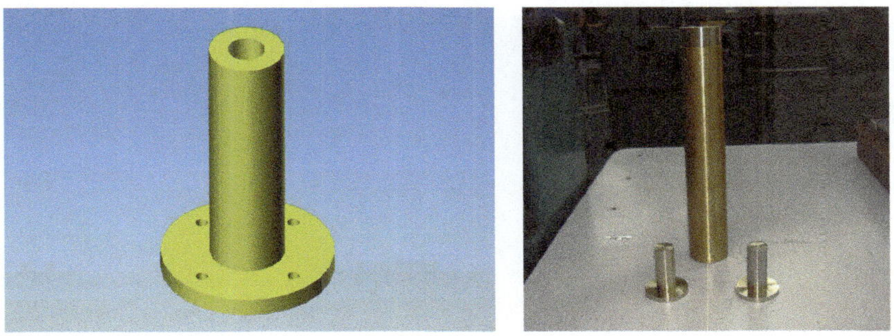

(a) SolidWorks design (b) Coupling system

FIGURE 5: Join coupling system.

(iii) The axle housings are made of yellow brass: they are used for coupling between the shaft of the motor and the wheels (see Figure 5).

(iv) Supports for the ball casters: they are two rectangular pieces made of aluminum used for assembly the WMR structure (see Figure 6).

(v) Six aluminum bars join the ball caster supports with the base of the WMR (see Figure 7).

(vi) Ball-bearings support: the ball bearings allow rotation without translation of the traction wheels due to the shaft motor rotation. Supports have two functions: keeping the ball-bearing embedded into it and joining the base with the motors. It was made of aluminum cut to the appropriate dimensions. A mechanical lathe was used for making the box in the supports in order to place the ball bearing (see Figure 8).

To conclude this subsection, Figure 9 shows a bottom view of the final mechanical design made in SolidWorks, as well as pictures of three different views of the prototype.

2.2. Stage 2: Power System. Electronic design is an essential stage for the WMR since its correct functioning depends on it. This section describes the most important features related to the interconnection between the data acquisition board (from the computer) and the conditioning signals (to the WMR).

A block diagram of stage 2 is shown in Figure 10. This block makes reference to four substages, numbered from 1 to 4: (*source circuit*, *optoisolator circuit*, *power circuit*, and *encoder circuit*). In *substage 1*, the power supply (source circuit) distributes the different voltages to the general electronic system. *Substage 2* allows electrical signal isolation between the electronic board DS1104 and *substage 3*. Also, this part of

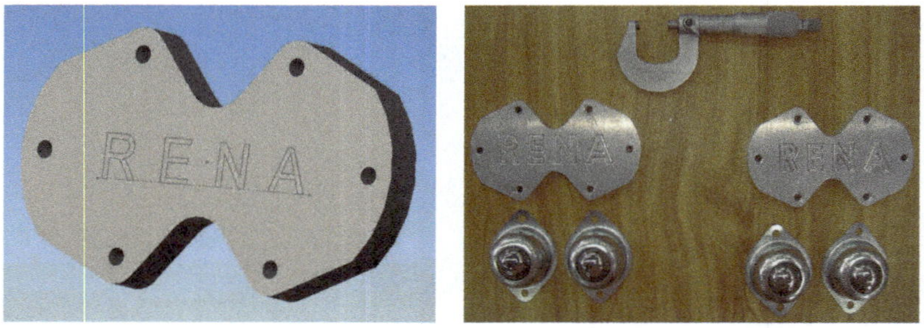

(a) SolidWorks design (b) Supports and ball casters

FIGURE 6: Ball-caster support.

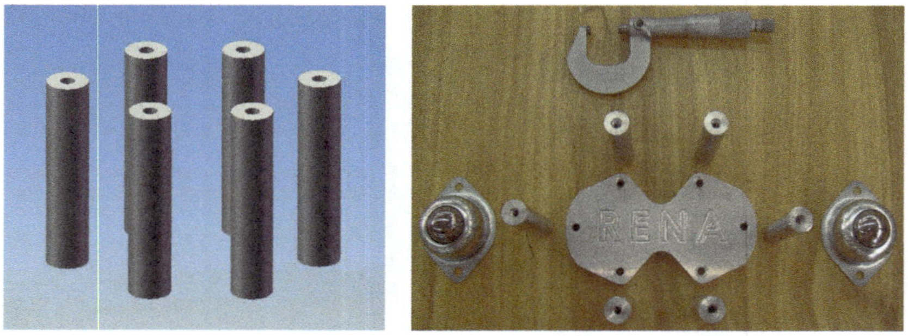

(a) SolidWorks design (b) Join bars, support, and ball casters

FIGURE 7: Holding bars design.

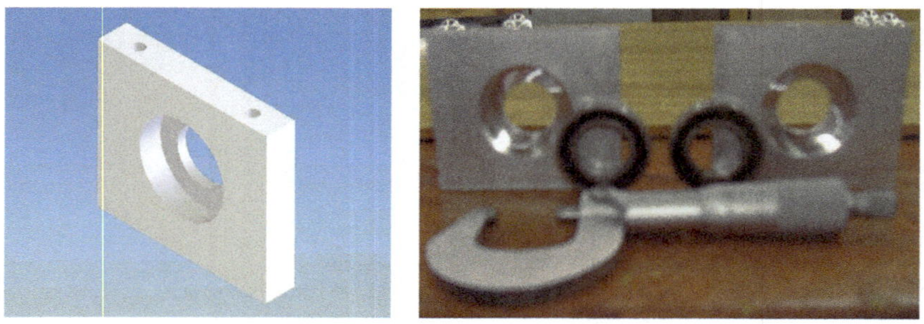

(a) SolidWorks design (b) Real holders

FIGURE 8: Ball-bearing supports.

the system allows a voltage to be applied across the motor in either direction. This circuit is based on the H-bridge LMD18200. *Substage 4* is used to acquire the encoders' signals, which can then be used to estimate the position in the workspace.

2.3. Stage 3: Data Acquisition and Control. This stage contains the last subsystem presented in the general description of the WMR (see Figure 1). As was mentioned before, the DS1104 board performs the acquisition and control of the WMR. This board was selected due to the integration software between the MATLAB and the board firmware. The high programming level of Simulink (as part of MATLAB) is a practical choice for programming complex control strategies in a graphic environment. This stage also includes the interface circuit (see Figure 1). This circuit establishes the communication between the acquisition board and the WMR. It contains an integrated circuit (74HC541) as a buffer that not only allows the signal bit interchange between the WMR and the DS1104 but also activates two signals DIR and PWM motor (LEDs indicators). Lastly, Figure 11(a) shows the final WMR mechanical design and Figure 11(b) is the real WMR with the instrumentation included (see Video S3 placed online as Supplementary Material).

3. Control for the WMR

In this section, two separate controllers are proposed, the first for the kinematic model of the WMR and the second for the DC motors (WMR actuators). The integration of both control strategies is based on the hierarchical control as in [10, 16].

3.1. Control for the WMR via Input-Output Linearization. The WMR studied in this work has two back wheels (left and right) which are identical, completely parallel, inflexible, and

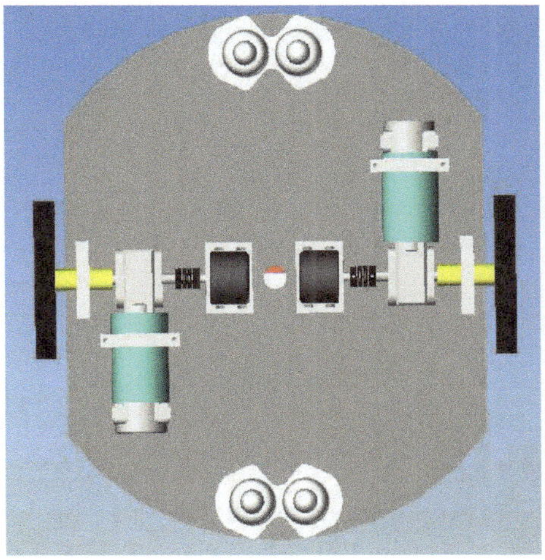

(a) SolidWorks design

(b) Bottom view of the prototype

(c) Isometric view of the prototype

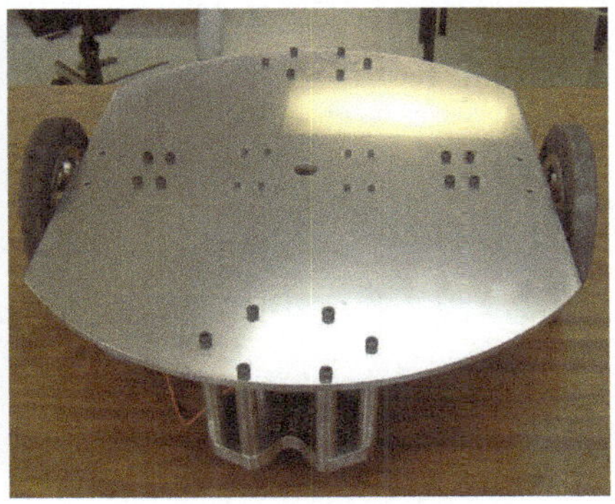

(d) Another view of the protctype

FIGURE 9: Views of both mechanical design and the real prototype.

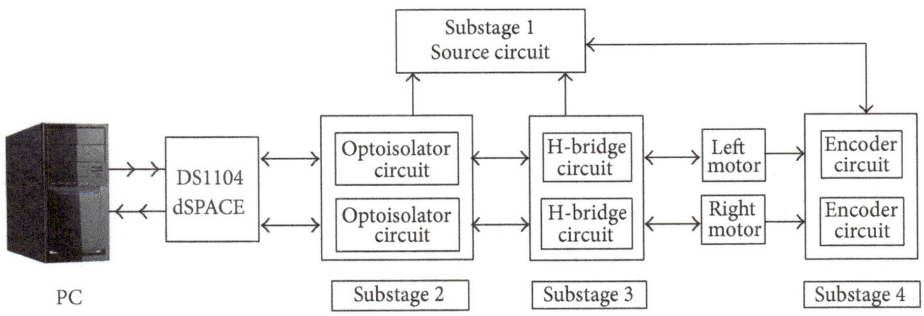

FIGURE 10: Block diagram of the power stage.

(a) Prototype 3D view elaborated in SolidWorks

(b) Real view of the WMR

FIGURE 11: The complete prototype.

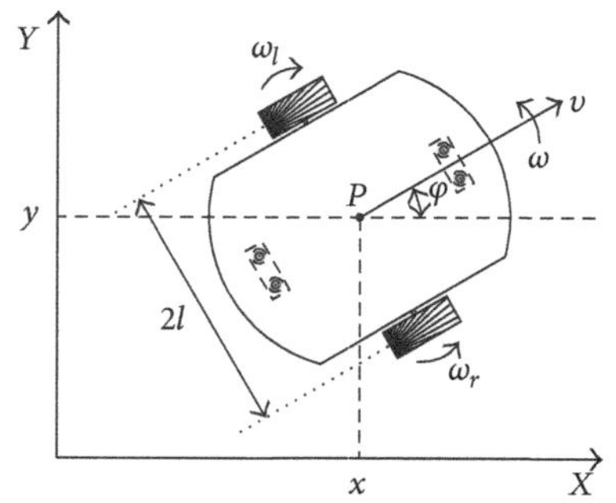

FIGURE 12: WMR diagram.

joined by a shaft. Moreover, it has front ball casters that make sure that the WMR platform moves in a plane. Assuming that the movements of the WMR are over the X-Y plane, and that there are no slippery condition, the kinematic model is given by [12]

$$\dot{x} = \frac{(\omega_r + \omega_l) r}{2} \cos\varphi,$$
$$\dot{y} = \frac{(\omega_r + \omega_l) r}{2} \sin\varphi, \quad (11)$$
$$\dot{\varphi} = \frac{(\omega_r - \omega_l) r}{2l},$$

where (x, y) is the position measured at the midpoint between the two back wheels, φ is the angle between the axis of symmetry of the WMR with respect to the positive X axis, and ω_l (ω_r) is the angular velocity of the left (right) wheel. Similarly, r is the wheel ratio and $2l$ is the gap between them (see Figure 12). In (11), the first derivative with respect to time t is denoted by a dot.

The input control is given by the angular velocity (ω_r, ω_l) and a strategy based on input-output linearization static feedback. The control can be implemented for any pair of outputs: (x, y), (x, φ), or (y, φ). Consequently, in each case there will be a remaining dynamic (dynamic zero) so that, in order to ensure a closed loop stable system, it requires an analysis of its stability.

The control design is performed over the output variables (x, φ) and the analysis of the remaining dynamic in closed loop is studied, particularly for the state variable y. First, the kinematic equations that describe the WMR are rewritten in two subsystems. The first is given by

$$\begin{pmatrix} \dot{x} \\ \dot{\varphi} \end{pmatrix} = A_1 \begin{pmatrix} \omega_r \\ \omega_l \end{pmatrix}, \quad A_1 = \begin{pmatrix} \frac{r\cos\varphi}{2} & \frac{r\cos\varphi}{2} \\ \frac{r}{2l} & -\frac{r}{2l} \end{pmatrix}. \quad (12)$$

This model only includes the state variables to be controlled (x, φ). And, the second is given by

$$\dot{y} = \frac{(\omega_r + \omega_l) r}{2} \sin\varphi. \quad (13)$$

It has a remaining dynamic associated to the state variable y, once the states (x, φ) have been controlled, so that $x \to x^*$ and $\varphi \to \varphi^*$.

Since $\det(A_1) = -r^2 \cos\varphi / 2l$, a controller can be proposed where the output is (x, φ) except when $\varphi = k\pi/2$, where $k = \pm 1, \pm 3, \pm 5, \ldots$, (i.e., A_1 must be invertible). In order to obtain the relation between the controllers (ω_r, ω_l) and the output variables (x, φ) from (12), we have

$$\begin{pmatrix} \omega_r \\ \omega_l \end{pmatrix} = \begin{pmatrix} \frac{1}{r\cos\varphi} & \frac{l}{r} \\ \frac{1}{r\cos\varphi} & -\frac{l}{r} \end{pmatrix} \begin{pmatrix} \dot{x} \\ \dot{\varphi} \end{pmatrix}. \quad (14)$$

From these input-output relations, we obtain that the controls (ω_r, ω_l) that allow the states (x, φ) to tend asymptotically to the desired trajectory (x^*, φ^*) can be defined as follows:

$$\begin{pmatrix} \omega_r \\ \omega_l \end{pmatrix} = \begin{pmatrix} \frac{1}{r\cos\varphi} & \frac{l}{r} \\ \frac{1}{r\cos\varphi} & -\frac{l}{r} \end{pmatrix} \begin{pmatrix} u_x \\ u_\varphi \end{pmatrix}, \quad (15)$$

where u_x and u_φ are the two auxiliary control variables, written as

$$u_x = \dot{x}^* - \alpha_x (x - x^*),$$
$$u_\varphi = \dot{\varphi}^* - \alpha_\varphi (\varphi - \varphi^*). \quad (16)$$

Hence, the tracking error dynamics in closed loop is determined by the following linear differential equation system:

$$\dot{e}_x + \alpha_x e_x = 0,$$
$$\dot{e}_\varphi + \alpha_\varphi e_\varphi = 0, \quad (17)$$

where $e_x = x - x^*$ and $e_\varphi = \varphi - \varphi^*$ denote the tracking errors of the variables x and φ, respectively, and α_x and α_φ are two positive constants. From (17) it is observed that $(e_x, e_\varphi) \to (0, 0)$ as $t \to \infty$, thus $(x, \varphi) \to (x^*, \varphi^*)$, which is the final control requirement.

Now, regarding the remaining dynamic associated with the state variable y, we shall analyze the behavior of this variable when $(x, \varphi) \to (x^*, \varphi^*)$. For this purpose, in (13), φ is replaced by φ^* and the controls ω_r and ω_l, which are given by (15) and (16) respectively, are substituted, obtaining

$$\dot{y} = [\dot{x}^* - \alpha_x (x - x^*)] \tan\varphi^*. \quad (18)$$

Thus, when $t \to \infty$,

$$\dot{y} = \dot{x}^* \tan\varphi^*. \quad (19)$$

If x^* and y^* satisfy

$$y^* = f(x^*), \quad (20)$$

where $f(x^*)$ is a smooth function such that $f(0) = 0$, then (11) and (20) show that

$$\varphi^* = \arctan\left(\frac{\dot{y}^*}{\dot{x}^*}\right) = \arctan\left(\frac{df(x^*)}{dx^*}\right), \quad (21)$$

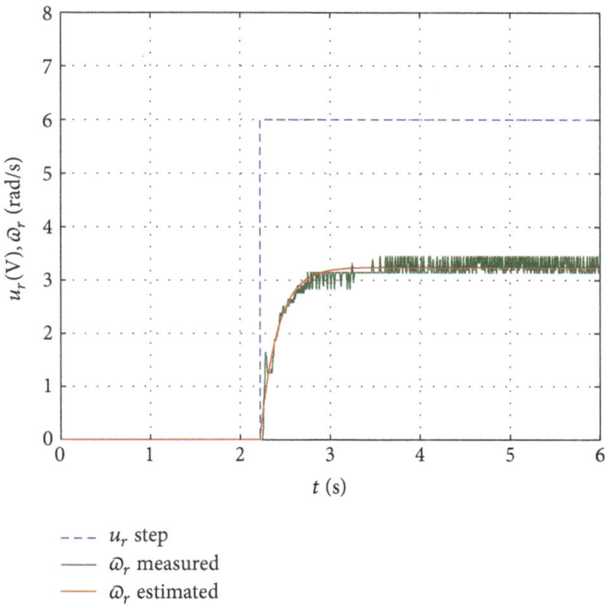

(a) Right motor

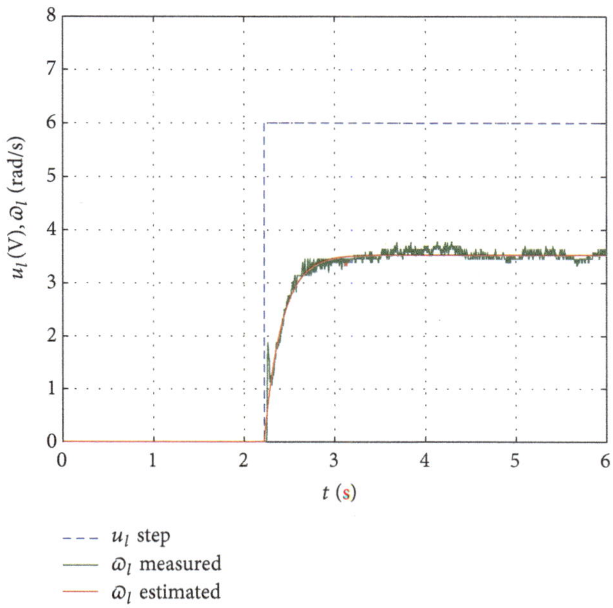

(b) Left motor

FIGURE 13: Step input response $u_r(t) = u_l(t) = 6$ V.

and (19) can then be simplified as

$$\dot{y} = \dot{x}^* \tan\varphi^* = \dot{x}^* \left(\frac{df(x^*)}{dx^*} \right) = \frac{df(x^*)}{dt}. \quad (22)$$

Now, integrating gives

$$y(t) = f(x^*(t)) = y^*, \quad (23)$$

where the constant of integration has been chosen in such a way that $y(0) = f(0) = 0$.

In conclusion, it has been shown that under the initial condition $(x, y, \varphi) = (0, 0, 0)$, the controllers (15) arrange that the state-space variables (x, y, φ) tend to (x^*, y^*, φ^*), respectively.

3.2. PI Control for the DC Motor. In general, due to the fact that motor manufacturers do not always providing the dynamic parameters of their products, there are different strategies for experimentally obtaining such parameters. This is beyond the scope of this paper; however, there is a reduction from a second-order linear system (motor model: voltage to angular position) to a first-order linear system (motor model: voltage to angular velocity). This can be obtained with the following transfer function: $G(s) = \varpi(s)/u(s)$. For that, Laplace transformation of the motor model is obtained, that is,

$$G(s) = \frac{\varpi(s)}{u(s)} = \frac{nk_m}{(Js + b)(Ls + R) + n^2 k_e k_m}. \quad (24)$$

Assuming that inductance may be neglected, $L \approx 0$, (24) simplifies

$$G(s) = \frac{\varpi(s)}{u(s)} = \frac{K}{\tau s + 1}, \quad (25)$$

where

$$K = \frac{nk_m}{(bR + n^2 k_e k_m)}, \qquad \tau = \frac{JR}{(bR + n^2 k_e k_m)}, \quad (26)$$

with u being the armature motor voltage, k_e the counter-electromotive force constant, k_m the torque constant, R the armature constant, J the rotor and inertial load, b the Coulomb friction coefficient (due to motor and load), and n represents the gearbox ratio.

In order to characterize the dynamic system (i.e., in order to obtain the values of the parameters for the DC motor and load), it is proposed to use a step input function with an amplitude of A in order to get coefficients K and τ such that

$$u(s) = \frac{A}{s}. \quad (27)$$

Substituting (27) into (25) and assuming that the system is a DC motor of which the output has the angular velocity ϖ, we obtain the following expression:

$$\varpi(s) = \frac{K}{\tau s + 1} \frac{A}{s}. \quad (28)$$

Since the coefficients K and τ are obtained experimentally using the time representation given by the inverse Laplace transformation of (28) (see (29)), we have

$$\varpi(t) = KA\left(1 - e^{-t/\tau}\right). \quad (29)$$

Once the characterization of the motor parameters K and τ is finished, from (25), with $u(t) = 6$ V, the experimental results are shown in Figure 13. The characterization of the right motor is

$$K_r = 0.54, \qquad \tau_r = 0.10. \quad (30)$$

Meanwhile for the left motor, we have

$$K_l = 0.59, \qquad \tau_l = 0.10. \tag{31}$$

Based on this, the temporal representation of the transfer function for the right and left motors is as follows:

$$\frac{d\varpi_r}{dt} = -10.20\varpi_r + 5.51u_r,$$
$$\frac{d\varpi_l}{dt} = -10.20\varpi_l + 5.99u_l. \tag{32}$$

Additionally, Figures 13(a) and 13(b) show the step input response of (29) (ϖ is labelled as the theoretical value) and the experimentally obtained values of K and τ. The validation of this characterization is carried out by comparing the theoretical and experimental responses.

Since the motors are mechanically attached to the wheels, the angular velocities ϖ_r and ϖ_l must be tracked in order to follow a desired velocity path ϖ_r^* and ϖ_l^* defined by (15). In order to attain this goal, a PI control is implemented for each motor, so that

$$u(t) = K_p e(t) + K_i \int_0^t e(t)\, dt, \tag{33}$$

$$e(t) = \overline{E}(t) - R(t) = \varpi^* - \varpi, \tag{34}$$

where $e(t)$ is the tracking error, $\overline{E}(t)$ is the desired value, $R(t)$ is the signal meant to be controlled, K_p is the proportional gain, and K_i is the integral gain.

DC motors have a nonlinear response due to friction and a dead zone in which the motor requires a certain amount of voltage before it can rotate. However, the dead-zone varies, depending on the mechanical load. In practice, the motor selected for this application requires 1.5 V. For this reason, a friction compensator has to be implemented. A friction model can use a linear gain f_v (which represents the viscous friction coefficient) and a discontinues shifting phase f_c (which represents the Coulomb friction coefficient) as

$$V_{\text{comp}}(\varpi) = [f_v \text{abs}(\varpi) + f_c] \operatorname{sign}(\varpi)$$
$$= f_v\varpi + f_c \operatorname{sign}(\varpi), \tag{35}$$

where ϖ is the angular velocity that has to be compensated and $V_{\text{comp}}(\varpi)$ is the friction compensator term. In practice, f_v represents the system viscous friction, and thus

$$f_v = b = n^2 b_m + b_L, \tag{36}$$

where b_m is the viscous friction coefficient due to the rotor and b_L is the viscous friction coefficient due to the load. In general, these two coefficients values are too small to be estimated accurately; hence, f_v can be neglected, and thus f_c is taken into account, so that,

$$V_{\text{comp}}(\varpi) = f_c \operatorname{sign}(\varpi). \tag{37}$$

Using (33), the PI controller can be redefined in order to consider the friction compensator V_{comp}, so that

$$u = K_p e(t) + K_i \int_0^t e(t)\, dt + V_{\text{comp}}(\varpi). \tag{38}$$

The control diagram for each motor can be seen in Figure 14.

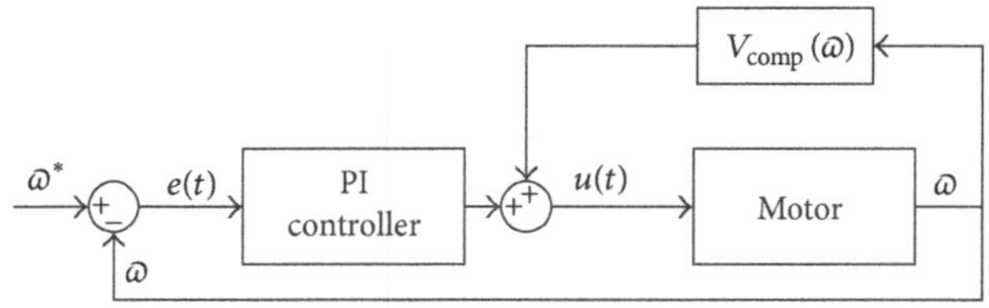

FIGURE 14: PI control strategy for a DC motor.

3.3. Hierarchical Control Integration. The control law described before has considered only the kinematic structure of the WMR, Section 3.1, and the actuator system in Section 3.2, and both systems are basically the WMR. In order to present clearly the integration of the hierarchical control, Figure 15 shows a block diagram for this purpose.

Regarding the DC motor control, the nominal output ϖ^* used in (38) is determined for the angular path required ϖ_r or ϖ_l. Since the WMR has two equal and independent motors (left and right), the dynamic models are given by

$$\frac{d\varpi_r}{dt} = -10.20\varpi_r + 5.51u_r,$$
$$y_r = \varpi_r,$$
$$\frac{d\varpi_l}{dt} = -10.20\varpi_l + 5.99u_l, \tag{39}$$
$$y_l = \varpi_l,$$

where ϖ_r and ϖ_l are the right and left angular velocity of the motors.

Thus, two controls, u_r and u_l, are required; according to Section 3.2, they are given by (38), so that

$$u_r = K_{pr} e_r(t) + K_{ir} \int_0^t e_r(t)\, dt + V_{\text{comp}}(\varpi_r),$$
$$u_l = K_{pl} e_l(t) + K_{il} \int_0^t e_l(t)\, dt + V_{\text{comp}}(\varpi_l), \tag{40}$$

with $V_{\text{comp}}(\varpi_r)$ and $V_{\text{comp}}(\varpi_l)$, defined by

$$V_{\text{comp}}(\varpi_r) = f_c \operatorname{sign}(\varpi_r),$$
$$V_{\text{comp}}(\varpi_l) = f_c \operatorname{sign}(\varpi_l), \tag{41}$$

where u_r and u_l are the control voltages for the right and left motors, respectively, K_{pr} and K_{ir}, K_{pl} and K_{il}, are the constant gains (proportional and integral) associated to each motor. Finally, $e_r(t)$ and $e_l(t)$ represent the motor tracking errors defined by

$$e_r(t) = \varpi_r^* - \varpi_r,$$
$$e_l(t) = \varpi_l^* - \varpi_l, \tag{42}$$
$$(\varpi_r^*, \varpi_l^*) = (\omega_r, \omega_l).$$

It means that the desired angular velocity profiles for the motors, (ϖ_r^*, ϖ_l^*), are determined by the (ω_r, ω_l). These angular velocity trajectories are obtained from (15) and (16).

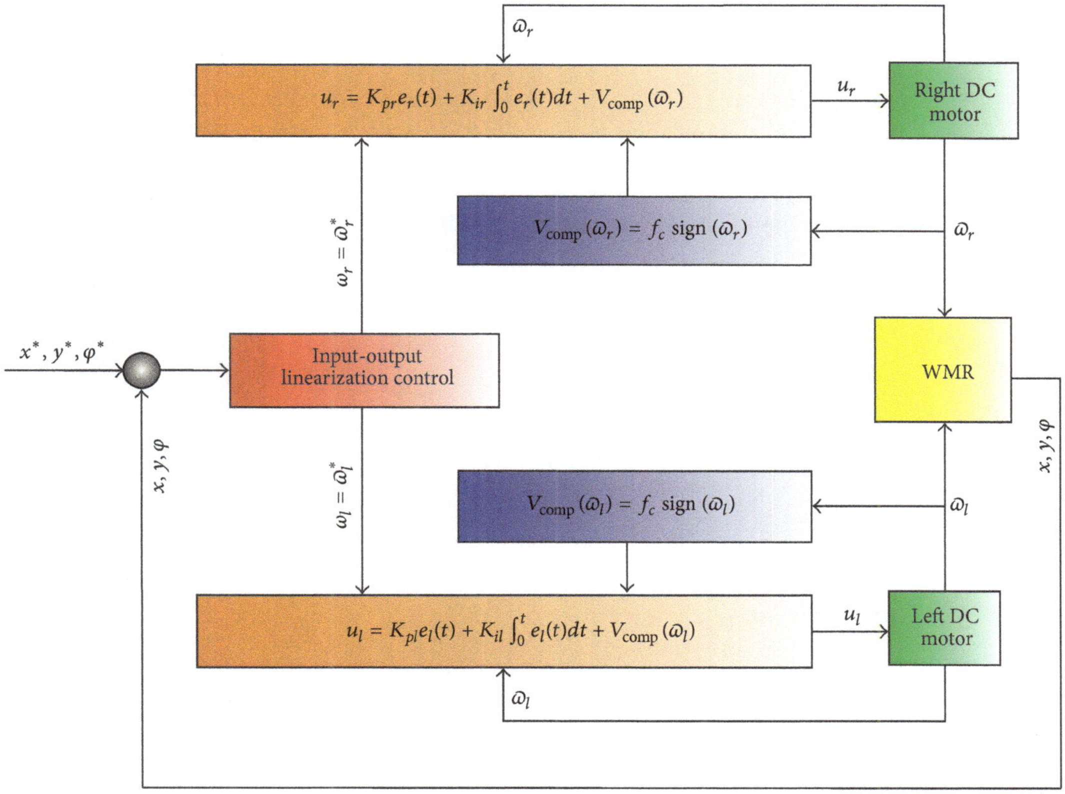

FIGURE 15: WMR hierarchical control block diagram.

4. Trajectory Generation via the Method of Interpolation of Cubic Splines

As can be seen from (15) and (16), the synthesis of these formulations requires the knowledge of the desired trajectory, i.e., (x^*, y^*, φ^*). This section presents the path generation problem solved via the cubic splines interpolation approach, in order to extend the possible trajectories to be followed (compared to the parametric approach widely reported in the literature).

4.1. Cubic Splines. Polynomial interpolation is based on a substitution of either a mathematical function or a lookup table for a polynomial. The bigger the number of points, the higher the degree of the polynomial, and consequently the greater the error between the desired trajectory and the interpolated curve due to oscillations. An alternative method is to employ several polynomials of lower degree, in subintervals along the whole trajectory. This method is the basis of the spline interpolation approach [21–23]. The method divides the path to be tracked into subintervals (given by the lookup table points) and a cubic interpolation joins such subintervals. This method is known as polynomial segmentary approximation as

$$S_j(x) = a_j + b_j(x - x_j) + c_j(x - x_j)^2 + d_j(x - x_j)^3. \tag{43}$$

For each $j = 0, 1, \dots, n-1$ and according to [21], the equation used for cubic spline interpolation is given by $Ax = b$, where

$$A = \begin{bmatrix} 1 & 0 & 0 & \cdots & 0 \\ h_0 & 2(h_0 + h_1) & h_1 & \cdots & 0 \\ 0 & h_1 & 2(h_1 + h_2) & h_2 & 0 \\ \cdots & \cdots & \cdots & \cdots & \cdots \\ 0 & \cdots & h_{n-2} & 2(h_{n-2} + h_{n-1}) & h_{n-1} \\ 0 & \cdots & 0 & 0 & 1 \end{bmatrix},$$

$$b = \begin{bmatrix} 0 \\ \frac{3}{h_1}(a_2 - a_1) - \frac{3}{h_0}(a_1 - a_0) \\ \vdots \\ \frac{3}{h_{n-1}}(a_n - a_{n-1}) - \frac{3}{h_{n-2}}(a_{n-1} - a_{n-2}) \\ 0 \end{bmatrix},$$

$$x = [c_0 \; c_1 \; c_2 \; \cdots \; c_n]^T, \tag{44}$$

where A is an $(n-1)$ by $(n-1)$ diagonal matrix, b is a constant vector, and x is an unknown vector. A numerical method is used to find $c_0, c_1, \dots, c_n$, and based on this method, the coefficients in (43), a_j, b_j, c_j, and d_j, are calculated in order to obtain a cubic polynomials for each segment.

4.2. Interpolation Algorithm. Based on the previous approach, a path planning algorithm is presented in this section. This method generates the Cartesian coordinates of the points along the path to be followed: these points are given by (x^*, y^*, φ^*), using (15) and (16) in order to allow $(x, y, \varphi) \rightarrow (x^*, y^*, \varphi^*)$.

Algorithm 1. In order to build the cubic interpolation function (43), according to [21], there must be defined the number of points n which satisfy the following condition: $x_0 < x_1 < \cdots < x_n$ with $\ddot{S}(x_0) = \ddot{S}(x_n) = 0$. The following steps are suggested to generate (x^*, y^*, φ^*).

(1) Vector coordinate points are introduced, so that, $x_0, x_1, \ldots, x_n$ and $y_0, y_1, \ldots, y_n$.

(2) Assign $a_0 = f(x_0), a_1 = f(x_1), \ldots, a_n = f(x_n)$.

(3) For $i = 0$ until $n-1$, set $h_i = x_{i+1} - x_i$.

(4) For $i = 1$ until $n-1$, set $\alpha_i = (3/h_i)(a_{i+1} - a_i) - (3/h_{i-1})(a_i - a_{i-1})$.

(5) Put $l_0 = 1$, $\mu_0 = 0$, $z_0 = 0$.

(6) For $i = 1$ until $n-1$, set

$$l_i = 2(x_{i+1} - x_{i-1}) - h_{i-1}\mu_{i-1},$$
$$\mu_i = \frac{h_i}{l_i}, \tag{45}$$
$$z_i = \frac{(\alpha_i - h_{i-1}z_{i-1})}{l_i}.$$

(7) Set $l_n = 1$, $z_n = 0$, $c_n = 0$.

(8) For $j = n-1$ until 0, set

$$c_j = z_j - \mu_j c_{j+1},$$
$$b_j = \frac{(a_{j+1} - a_j)}{h_j} - \frac{h_j(c_{j+1} + 2c_j)}{3}, \tag{46}$$
$$d_j = \frac{(c_{j+1} - c_j)}{3h_j}.$$

(9) If $x \leq x_0$, then $j = 1$; otherwise, $x \geq x_0$ and $x \leq x_1$ and then $j = 2$; otherwise, $x \geq x_{n-1}$ and $x \leq x_n$ and then $j = n-1$.

(10) If $x \leq x_n$, then

$$y_j = a_j + b_j(x - x_j) + c_j(x - x_j)^2 + d_j(x - x_j)^3. \tag{47}$$

(11) Based on the equation obtained in step 10, the following expression is obtained:

$$\dot{y}_j = b_j + 2c_j(x - x_j) + 3d_j(x - x_j)^2. \tag{48}$$

(12) φ^* is obtained as $\varphi^* = \arctan(\dot{y})$.

(13) Based on this, considering the points (1), (10), and (12), the coordinates obtained are denoted by (x^*, y^*, φ^*).

This algorithm is encoded in MATLAB in order to implement experiments using the WMR.

5. Experimental Results

This section presents the closed-loop simulation as well as experimental results in order to validate the behavior of the hierarchical approach presented in Section 3, for both parametric and generated (based on given points) curves. The simulations are implemented in MATLAB/Simulink, and the real-time experiments are based on the board DS1104 using the same software.

According to the WMR dimensions, r and l have the following values:

$$r = 0.075\,\text{m}, \qquad l = 0.22\,\text{m}. \tag{49}$$

Regarding the motors, whose characterization was presented in Section 3.2, the right motor has the following parameters:

$$K_r = 0.54, \qquad \tau_r = 0.10. \tag{50}$$

The left motor has the parameters

$$K_l = 0.59, \qquad \tau_l = 0.10. \tag{51}$$

The controller gains (15) and (16) and the gains of the controllers (40), were selected like:

$$\alpha_x = \alpha_\varphi = 2, \qquad K_{pr} = K_{pl} = 2, \qquad K_{ir} = K_{il} = 50. \tag{52}$$

5.1. Experimental Results Using the Parametric Curve Approach. General results are used when the desired path is a parabola, so that

$$y^*(t) = x^{*2}(t). \tag{53}$$

The parametrization is chosen as follows:

$$x^*(t) = A\sin\left(\frac{2\pi}{P}t\right). \tag{54}$$

It is found that $\varphi^*(t)$ is determined by the following equation:

$$\varphi^*(t) = \arctan(2x^*(t)). \tag{55}$$

This parametrization allows the displacement of the WMR from the origin to the point $(x, y) = (A, A^2)$, later to $(x, y) = (-A, A^2)$ passing through the origin and finally returning to the origin in P seconds. The parameter values A and P are associated to (54), and they are chosen to be

$$A = 1\,\text{m}, \qquad P = 15\,\text{s}. \tag{56}$$

In order to simulate the system, an initial condition chosen for (x, y, φ) is $(0, 0, 0)$, assuming $t = 30\,\text{s}$, where t is the total simulation time. The simulation results are presented in Figure 16. Similarly, the real-time experiments require the same conditions used in the simulation. The data was obtained using ControlDesk and MATLAB/Simulink along with the DS1104 board. These results are presented in Figure 17, and Video S4 in Supplementary Material.

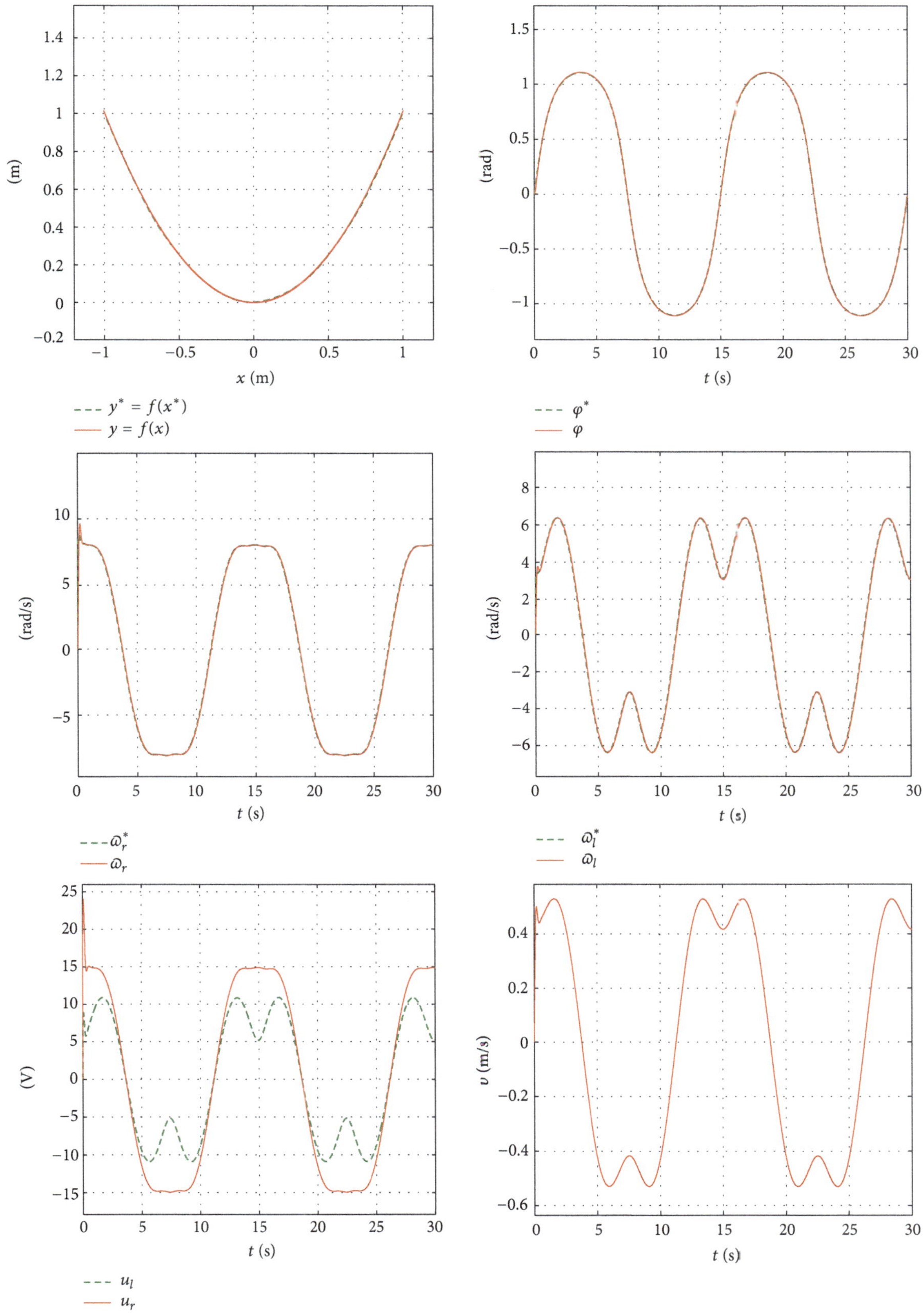

Figure 16: Simulation results.

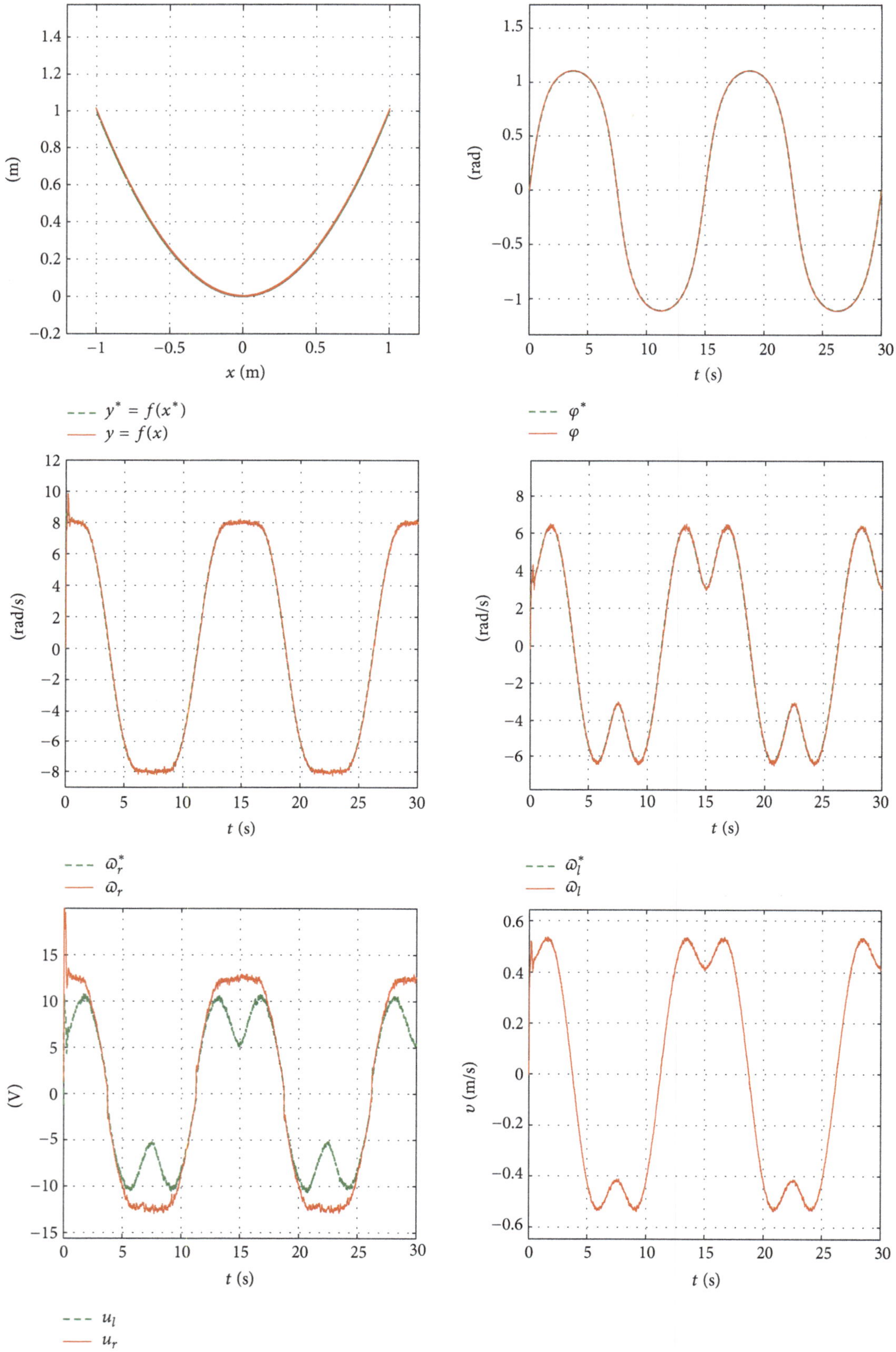

FIGURE 17: Real-time experimental results.

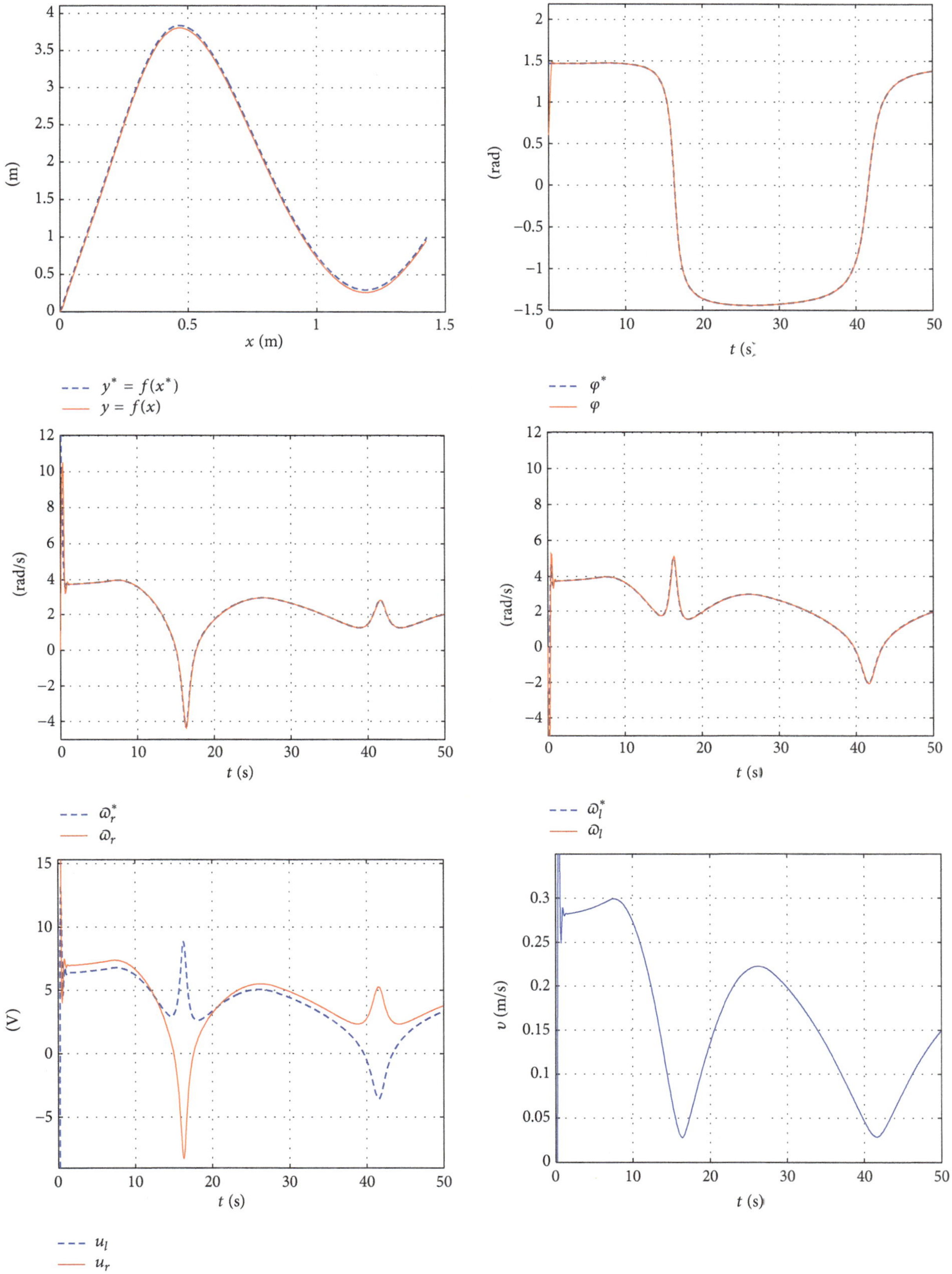

FIGURE 18: Simulation results for a curve obtained from given points.

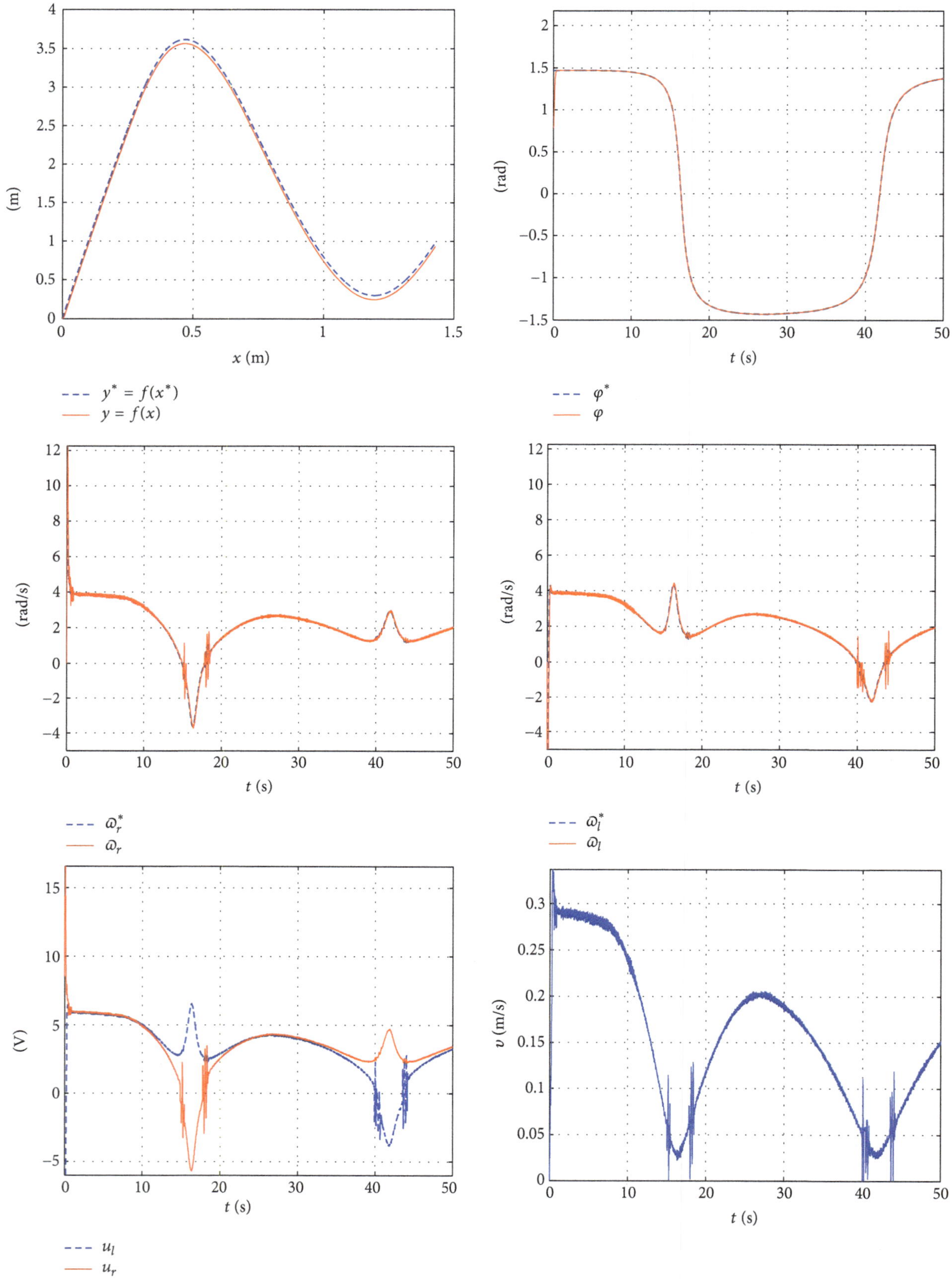

FIGURE 19: Experimental results for curve obtained by points.

TABLE 1: Points used in order to generate the desired trajectory.

x^*	y^*
$x_0 = 0.0$ m	$y_0 = 0.0$ m
$x_1 = 0.2$ m	$y_1 = 0.2$ m
$x_2 = 0.4$ m	$y_2 = 3.5$ m
$x_3 = 0.8$ m	$y_3 = 2.0$ m
$x_4 = 1.2$ m	$y_4 = 0.3$ m
$x_5 = 1.4$ m	$y_5 = 1.0$ m

5.2. Experimental Results Using Generated Curves Given by Points. This section presents both the experimental and the simulation results obtained in Section 3 where the curve generated is obtained from coordinate points. The chosen points are given by Table 1. Both the control system and the parameters used in the previous subsection are employed, (49)–(52), providing the simulation results shown in Figure 18. The experimental results are shown in Figure 19.

5.3. Discussion of the Experimental Results. Similar results are obtained in both the simulation and the experimental plots. The velocity is estimated based on the position measured directly from the encoders attached to the motor shafts. Both $\bar{\omega}_r$ and $\bar{\omega}_l$, follow the desired path angular velocity according to the kinematic structure $\bar{\omega}_r^*$ and $\bar{\omega}_l^*$ respectively. Consequently, trajectory tracking task is performed successfully, (i.e., $(x, y, \varphi) \rightarrow (x^*, y^*, \varphi^*)$). In addition, it is shown that the control voltages u_r and u_l are within the interval (−24 V, +24 V), which is convenient since the nominal motor voltages are in the range of ±24 V. Also, from the results, we find that the maximum linear value v is equal to 0.52 m/s. Based on all the previous results, the efficiency of the hierarchical control design has excellent performance.

6. Conclusions

This work shows the design and construction of a wheeled mobile robot with differential configuration architecture. In addition, a planning trajectory has been proposed and implemented with a control law that experimentally showed successful results, for two approaches: smooth parametric trajectories and interpolated trajectories determined by given points.

More specifically, this control is hierarchically distributed in two controllers: the first controller is meant to use the kinematic model of the WMR, which is based on a linearization of the input-output plant (ω_r, ω_l)-(x, φ), that generates a remaining dynamics for y that is stable. The second controller is a PI strategy that determines the output control signals in order to track the desired velocities ω_r and ω_l such that $x \rightarrow x^*$, $y \rightarrow y^*$, and $\varphi \rightarrow \varphi^*$, with the initial conditions $x = 0$, $y = 0$, and $\varphi = 0$. The main difference between this hierarchical control and the nonlinear techniques is that these last ones have to commutate between two control laws in order to keep the system stable. In this approach, there is a drawback to the use of parametric trajectories, which happens when $\varphi = k\pi/2$ for $k = \pm1, \pm3, \pm5, \ldots$. This is a significant constraint when the trajectory to be followed is a parametric curve, since the controller, in some conditions, is undetermined. For this reason, the validation of the results was carried out not only with the use of the parametric curves approach, but also with an interpolation method based on cubic splines, which enables following a wider range of trajectories than is possible with parametric curves.

The second controller is designed for the model of the motors given by (39), and this control law guarantees the trajectory tracking given for the desired trajectory established for the first controller. The so-called *hierarchical approach* was introduced in [10] and used in [16] and was graphically presented in Figure 15.

Likewise, this research found that a control based on an input-output plant linearization approach for (ω_r, ω_l)-(y, φ) is always possible in such a way that $\varphi \neq k\pi$ for $k = 0, \pm1, \pm2, \ldots$, since the matrix at these points becomes singular.

The synergetic combination of the theoretical results, implementation techniques, and construction description is the main contribution of this research. This approach has also an important impact for mathematicians willing to explore the application field as well as practical researchers who might like to know more about the theoretical results described in a real and effective prototype.

Finally, as possible direction for future research, this approach can be extended to the problem of real-time obstacle avoidance. Similarly, another possible direction for future research is the modification of the second controller (related to the actuators), where the armature voltage and current can be the feedback variables for velocity control instead of the encoders; this then will become a speed sensorless approach.

Conflict of Interests

The authors declare that the research was conducted in the absence of any commercial, financial, or personal relationships that could be construed as a potential conflict of interests.

Acknowledgments

The authors wish to thank the anonymous reviewer for his helpful comments and critical review. R. Silva-Ortigoza, M. Marcelino-Aranda, and M. Marciano-Melchor acknowledge financial support from Secretaría de Investigación y Posgrado del Instituto Politécnico Nacional (SIP-IPN), SNI-México, and the programs EDI and COFAA of IPN. The work of C. Márquez-Sánchez was supported by a CONACYT scholarship. G. Silva-Ortigoza and R. Bautista Quintero thank SNI-México for financial support. Also M. Marcelino-Aranda and G. Silva-Ortigoza acknowledge financial support from IPN through the *Proyectos de Investigación en Apoyo a la Consolidación de Profesores SNI del IPN.*

References

[1] R. Silva-Ortigoza, M. Marcelino-Aranda, G. Silva-Ortigoza et al., "Wheeled mobile robots: a review," *IEEE Latin America Transactions*, vol. 10, no. 6, pp. 2209–2217, 2012.

[2] R. M. Murray and S. S. Sastry, "Nonholonomic motion planning. Steering using sinusoids," *IEEE Transactions on Automatic Control*, vol. 38, no. 5, pp. 700–716, 1993.

[3] R. W. Brockett, "Asymptotic stability and feedback stabilization," in *Differential Geometric Control Theory*, R. W. Brockett, R. S. Millmann, and H. J. Sussmann, Eds., pp. 181–191, Birkhauser, Boston, Mass, USA, 1983.

[4] C. Canudas de Wit and O. J. Sordalen, "Exponential stabilization of mobile robots with nonholonomic constraints," *IEEE Transactions on Automatic Control*, vol. 37, no. 11, pp. 1791–1797, 1992.

[5] C. Samson, "Velocity and torque feedback control of a nonholonomic cart," in *Advanced Robot Control*, C. Canudas de Wit, Ed., vol. 162 of *Lecture Notes in Control and Information Sciences*, pp. 125–151, Springer, New York, NY, USA, 1991.

[6] C. Samson, "Control of chained systems application to path following and time-varying point-stabilization of mobile robots," *IEEE Transactions on Automatic Control*, vol. 40, no. 1, pp. 64–77, 1995.

[7] J.-M. Coron, "Global asymptotic stabilization for controllable systems without drift," *Mathematics of Control, Signals, and Systems*, vol. 5, no. 3, pp. 295–312, 1992.

[8] J. M. Coron and B. d'Andrea-Novel, "Smooth stabilizing time-varying control laws for a class of nonlinear systems. Application to mobile robots," in *Proceedings of the IFAC Symposium on Nonlinear Control Systems Design*, pp. 413–418, June 1992.

[9] J.-B. Pomet, "Explicit design of time-varying stabilizing control laws for a class of controllable systems without drift," *Systems and Control Letters*, vol. 18, no. 2, pp. 147–158, 1992.

[10] A. W. Divelbiss and J. T. Wen, "Trajectory tracking control of a car-trailer system," *IEEE Transactions on Control Systems Technology*, vol. 5, no. 3, pp. 269–278, 1997.

[11] E. Aranda-Bricaire, T. Salgado-Jiménez, and M. Velasco-Villa, *Control No Lineal Discontinuo De Un Robot Móvil*, Computación y Sistemas (Special edition), 2002.

[12] H. Sira-Ramírez and S. K. Agrawal, *Differentially Flat Systems*, Marcel Dekker, New York, NY, USA, 2004.

[13] M. Fliess, J. Levine, P. Martin, and P. Rouchon, "Flatness and defect of non-linear systems: introductory theory and examples," *International Journal of Control*, vol. 61, no. 6, pp. 1327–1361, 1995.

[14] J. A. Chacal B. and H. Sira-Ramirez, "On the sliding mode control of wheeled mobile robots," in *Proceedings of the IEEE International Conference on Systems, Man and Cybernetics*, pp. 1938–1943, October 1994.

[15] J.-M. Yang, I.-H. Choi, and J.-H. Kim, "Sliding mode control of a nonholonomic wheeled mobile robot for trajectory tracking," in *Proceedings of the IEEE International Conference on Robotics and Automation*, pp. 2983–2988, May 1998.

[16] R. Silva-Ortigoza, G. Silva-Ortigoza, V. M. Hernández-Guzmán, V. R. Barrientos-Sotelo, J. M. Albarrán-Jiménez, and V. M. Silva-García, "Trajectory tracking in a mobile robot without using velocity measurements for control of wheels," *IEEE Latin America Transactions*, vol. 6, no. 7, pp. 598–607, 2008.

[17] J. L. Avendaño-Juárez, V. M. Hernández-Guzmán, and R. Silva-Ortigoza, "Velocity and current inner loops in a wheeled mobile robot," *Advanced Robotics*, vol. 24, no. 8-9, pp. 1385–1404, 2010.

[18] M. Nystrom and M. Norrlof, "Path generation for industrial robots," Tech. Rep. LiTH-ISY-R-2529, Department of Electrical Engineering; Linkoping University, 2003.

[19] L. Biagiotti and C. Melchiorri, *Trajectory Planning for Automatic Machines and Robots*, Springer, 2008.

[20] B. Tondu and S. A. Bazaz, "Three-cubic method: an optimal online robot joint trajectory generator under velocity, acceleration, and wandering constraints," *International Journal of Robotics Research*, vol. 18, no. 9, pp. 893–901, 1999.

[21] R. L. Burden, J. D. Faires, and Numerical Analysis Brooks-Cole, Boston, Mass, USA, 9th edition, 2010.

[22] K. E. Atkinson, *An Introduction to Numerical Analysis*, John Wiley, New York, NY, USA, 2nd edition, 1989.

[23] S. C. Chapra and R. P. Canale, *Numerical Methods for Engineers*, McGraw-Hill, New York, NY, USA, 2nd edition, 1989.

[24] T. H. Lee, H. K. Lam, F. H. F. Leung, and P. K. S. Tam, "A fast path planning-and-tracking control for wheeled mobile robots," in *Proceedings of the IEEE International Conference on Robotics and Automation (ICRA '01)*, pp. 1736–1741, May 2001.

[25] T. H. Lee, H. K. Lam, F. H. F. Leung, and P. K. S. Tam, "A practical fuzzy logic controller for the path tracking of wheeled mobile robots," *IEEE Control Systems Magazine*, vol. 23, no. 2, pp. 60–65, 2003.

[26] M. Lepetic, G. Klnacar, I. Skrjanc, D. Matko, and B. Potocnik, "Path planning and path tracking for nonholonomic robots," in *Mobile Robots: New Research*, J. X. Liu, Ed., pp. 345–368, Nova Science, 2005.

[27] "MicroMo Electronics, Faulhaber Group," http://www.micromo.com.

6

Nonlinear Forced Response of Electromechanical Integrated Toroidal Drive to Coupled Excitation

Lizhong Xu and Fen Wang

Mechanical Engineering Institute, Yanshan University, Qinhuangdao 066004, China

Correspondence should be addressed to Lizhong Xu, xlz@ysu.edu.cn

Academic Editors: K. Prokeš and B. Yasilata

The electric excitation and the parameter excitation from mesh stiffness fluctuation are analyzed. The forced response equations of the drive system to the coupled excitations are presented. For the exciting frequencies far from and near natural frequencies, the forced responses of the drive system to the coupled excitations are investigated. Results show that the nonlinear forced responses of the drive system to the coupled excitations change periodically and unsteadily; the time period of the nonlinear forced responses depends on the frequencies of the electric excitation, the mesh parameter excitation, and the nonlinear natural frequencies of the drive system; in order to improve the dynamics performance of the drive system, the frequencies of the electric excitations should not be taken as integral multiple of the mesh parameter exciting frequency.

1. Introduction

Toroidal drive can transmit large torque in a very small size and is suitable for technical fields such as aviation and space flight [1–3]. As electrical and control techniques are utilized in mechanical engineering field widely, generalized composite drives become advancing edge of the mechanical science. So far, types of the generalized composite drives with integrated structure are still very limited.

The electromagnetic harmonic drive [4] and piezoelectric harmonic drive [5] are active drives in which the meshing forces between flexible gear and rigid one are controlled by electromagnetic force or piezoelectric one, and drive and power are integrated. Based on researching toroidal drive [6], the authors presented a kind of active generalized composite drive: electromechanical integrated toroidal drive. In the drive, the toroidal drive, power, and control are integrated [7].

The drive consists of four basic elements (Figure 1) (a) the central worm, (b) radially positioned planets, (c) a toroidal shaped stator, and (d) a rotor, which forms the central output shaft upon which the planets are mounted. The central worm is fixed and coils are mounted in helical grooves of its surface. The planets have permanent magnets instead of teeth. The N and S polar permanent magnets are mounted alternately on a planet. And the stator has helical permanent magnets instead of helical teeth. In the same manner as planet, the N and S polar helical permanent magnets are mounted alternately on the stator.

If a specific parameter relation is realized, N pole of one element will correspond to S pole of the other one all along. The attractive forces between N and S pole of the different elements are driving forces and the meshes without contact are realized. When the alternating voltage source is connected to the coils of the worm, a toroidal circular field is formed. It drives several planets to rotate about their own axial. And by means of magnetic forces between teeth of the planet and stator, the rotor is driven to rotate about its own axial. Thus, a power of low speed and large torque is output.

Compared with toroidal drive, the new drive is easy to produce, without wear, and does not need lubrication. It can be substituted for a servo system to simplify the structure of the existing electromechanical systems. Beside the above-mentioned fields that require compactness, the drive can be used in fields such as robots, which require accurate control.

The electromechanical integrated toroidal drive consists of a mechanical system, an electrical system and a coupled part. The mechanical vibration may occur in the mechanical

FIGURE 1: The electromechanical integrated toroidal drive.

system, and the electrical current oscillation may occur in the electrical system. With the coupled part, the mechanical vibration and the electrical current oscillation will influence on each other. Hence, the drive system is an electromechanical coupled dynamics system. The electromechanical coupled dynamics was first proposed for the motor [8]. Then, the electromechanical coupled dynamics of the electromechanical system consisting of the motor and mechanical system driven by the motor was developed. An electromechanical coupled dynamics model of the electromechanical system consisting of the several motors and mechanical system driven by these motors was proposed. Using the model, the natural frequency of the electromechanical system is analyzed [9]. The authors investigated nonlinear forced response of electromechanical integrated toroidal drive to voltage excitations [10]. However, not only the electric excitation occurs, but also the parameter excitations from mesh stiffness fluctuations occur. The nonlinear forced responses of the drive system to the coupled excitations consisting of the electric excitation and the parameter excitations have not been investigated yet. These nonlinear forced responses have important influence on the operating performance of the drive system. To design, evaluate, and control dynamics behavior of the drive system effectively, the nonlinear forced responses of the drive system to the coupled excitations should be developed.

In this paper, the electric excitation and the parameter excitation from mesh stiffness fluctuation are analyzed. The forced response equations of the drive system to the coupled excitations are presented. For the exciting frequencies far from and near natural frequencies, the forced responses of the drive system to the coupled excitations are investigated. The work can be used to predict the noise and dynamic load and are useful in maximizing the power density of the drive and reducing noise radiation.

2. Electric Excitation of the Drive System

The magnetic energy storage of the electric system for the drive system is

$$W = \frac{1}{2}\sum_{i=1}^{n}\lambda_{is}I_{is}, \tag{1}$$

where λ_{is} is magnetic linkage of the worm coils, $\lambda_{is} = \sum_{j=1}^{n} L_{ij}I_{is}$, L_{ij} is inductances of the ith phase worm coils, I_{is} is current of the ith phase worm coils, and n is phase number of the worm coils.

From (1), the electromagnetic torque on the planet is given as

$$T_p = -\frac{\partial W}{\partial \theta} = -\frac{1}{2}\sum_{j=1}^{n}\sum_{i=1}^{n}\frac{dL_{ij}}{d\theta}I_{is}I_{js}, \tag{2}$$

where θ is the relative rotating angle between planet and worm.

The torque T_p consists of the static torque T_{p0} and the dynamic torque δT_p. The current I_{is} consists of the static component I_i and the dynamic one δi_i. The angle θ consists of the static angle θ_0 and the dynamic angle $\delta\theta$. The torque T_p is written in series form as

$$\begin{aligned} T_p &= T_{p0} + \delta T_p \\ &= \frac{1}{2}\sum_{j=1}^{n}\sum_{i=1}^{n}\left[\left(\frac{\delta L_{ij}}{\delta\theta}\right)_{\theta=\theta_0} I_iI_j + \left(\frac{\delta^2 L_{ij}}{\delta\theta^2}\right)_{\theta=\theta_0} I_iI_j\delta\theta \right.\\ &\quad + \left(\frac{\delta^3 L_{ij}}{\delta\theta^3}\right)_{\theta=\theta_0} I_iI_j(\delta\theta)^2 + \left(\frac{\delta L_{ij}}{\delta\theta}\right)_{\theta=\theta_0} I_i\delta i_j \\ &\quad + \left(\frac{\delta L_{ij}}{\delta\theta}\right)_{\theta=\theta_0} I_j\delta i_i + \left(\frac{\delta L_{ij}}{\delta\theta}\right)_{\theta=\theta_0} I_i\left(\delta i_j\right)^2 \\ &\quad \left. + \left(\frac{\delta L_{ij}}{\delta\theta}\right)_{\theta=\theta_0} I_j(\delta i_i)^2 + \cdots\right]. \end{aligned} \tag{3}$$

If $\delta i_j = \delta i_i = 0$, from (3), one can give

$$T_{p0} = \frac{1}{2}\sum_{j=1}^{n}\sum_{i=1}^{n}\left(\frac{\delta L_{ij}}{\delta\theta}\right)_{\theta=\theta_0} I_iI_j, \tag{4}$$

$$\begin{aligned} \delta T_p &= \frac{1}{2}\sum_{j=1}^{n}\sum_{i=1}^{n}\left[\left(\frac{\delta^2 L_{ij}}{\delta\theta^2}\right)_{\theta=\theta_0} I_iI_j\delta\theta \right.\\ &\quad \left. + \left(\frac{\delta^3 L_{ij}}{\delta\theta^3}\right)_{\theta=\theta_0} I_iI_j(\delta\theta)^2 + \cdots\right]. \end{aligned} \tag{5}$$

Let ΔF_{wpi} denote the dynamic magnetic meshing force between a planet tooth and worm, $\Delta F_{wpi} = \delta T_p/R$. From (5), neglecting the high-order terms, one knows.

$$\Delta F_{wpi} = -\frac{R\delta\theta}{2R^2}\sum_{i=1}^{n}\sum_{j=1}^{n}\left[\left(\frac{\delta^2 L_{ij}}{\delta\theta^2}\right)_{\theta=\theta_0} I_iI_j + \left(\frac{\delta^3 L_{ij}}{\delta\theta^3}\right)_{\theta=\theta_0} I_iI_j\delta\theta\right]. \tag{6}$$

Here, $\zeta_{wpi} = R\delta\theta$; it is the dynamic relative displacement between planet and worm. Thus, (6) can be changed into the following form:

$$\Delta F_{wpi} = \left(k_{wpi} + \Delta k_{wpi}\right)\zeta_{wpi}. \quad (7)$$

Here, $k_{wpi} = (1/2R^2)\sum_{i=1}^{n}\sum_{i=1}^{n}(\delta^2 L_{ij}/\delta\theta^2)_{\theta=\theta_0} I_{is}I_{js}$, it is considered as linear electromagnetic mesh stiffness between a tooth of the planet and worm; $\Delta k_{wpi} = -(1/2R^2)\sum_{i=1}^{n}\sum_{j=1}^{n}(\delta^3 L_{ij}/\delta\theta^3)_{\theta=\theta_0} I_i I_j \delta\theta$, it is nonlinear component of the electromagnetic mesh stiffness.

Equation (3) shows that current fluctuation can produce electromagnetic torque fluctuation. The torque fluctuation from current fluctuation can be considered as equivalent exciting torque ΔT_e. Let the current periodically change as below:

$$\delta i_i = \delta i_j = \Delta I \cos(\omega_e t), \quad (8)$$

where ΔI is the magnitude of the fluctuation current, ω_e is the frequency of the fluctuation current, and t is time.

Substituting (8) into (3), neglecting high-order terms, yields

$$\Delta T_e = \sum_{j=1}^{n}\sum_{i=1}^{n}\left(\frac{\delta L_{ij}}{\delta\theta}\right)_{\theta=\theta_0} I_i \Delta I \cos\omega_e t. \quad (9)$$

Let ΔF_e denote the equivalent exciting force between a planet tooth and worm; thus

$$\Delta F_e = \frac{1}{R}\sum_{j=1}^{n}\sum_{i=1}^{n}\left(\frac{\delta L_{ij}}{\delta\theta}\right)_{\theta=\theta_0} I_i \Delta I \cos\omega_e t. \quad (10)$$

3. Parametric Excitation from Stiffness Fluctuation

In operation of the drive system, the number of meshing tooth pairs between the worm and the planets is variable. It can cause fluctuation of the mesh stiffness between them. It is dependent on the number z_1 of planet teeth and the conditional face angles of the worm ϕ_v. At $\phi_v = 100°$ and $z_1 = 8$, the changes of mesh stiffness along with the mesh tooth pair number are shown in Figure 2. In Figure 2, a typical stiffness variation through a mesh cycle of the drive system is given. The mesh stiffness at more contact regions is higher than that at less contact regions. If the drive rotates at appreciable speed, this time-varying stiffness will be a major excitation source of the drive system.

The mesh stiffness between a planet and worm is considered to consist of its mean value $\bar{k}$ and time-varying one $\Delta k(t)$. The average mesh stiffness between them through one periodic time can be given by

$$\bar{k}_{wpi} = \frac{4}{\pi}\left\{\int_0^{\pi/36} k_3 d\theta + \int_{\pi/36}^{2\pi/9} k_2 d\theta + \int_{2\pi/9}^{\pi/4} k_3 d\theta\right\}, \quad (11)$$

where $k_2 = 2(k_{wpi} + \Delta k_{wpi})$ and $k_3 = 3(k_{wpi} + \Delta k_{wpi})$; they are two and three teeth mesh stiffness, respectively.

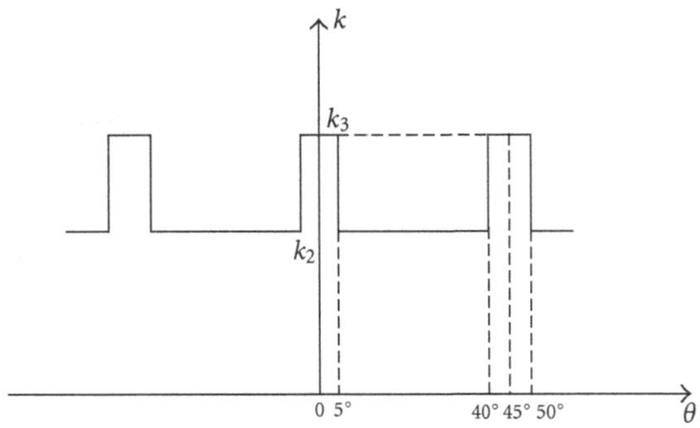

FIGURE 2: Changes of the mesh stiffness.

The periodical time-varying portion of the mesh stiffness can be defined in the Fourier series form as

$$\Delta k(t) = \sum_{n=1}^{\infty}\Delta k_n \cos n\omega_p t, \quad (12)$$

where $\Delta k_n = (2/l)\int_0^l k(t)\omega_p \cos n\omega_p t dt$, l is the period of the stiffness fluctuation. In Figure 2, $l = \pi/4$.

For worm and planet, the periodical time-varying portion of the mesh stiffness is

$$\begin{aligned}\Delta k_{nwpi} &= \frac{2}{l}\left\{\int_0^{\pi/36}\left(k_3 - \bar{k}_{wpi}\right)\omega_p \cos n\omega_p t dt\right. \\ &\quad + \int_{\pi/36}^{2\pi/9}\left(k_2 - \bar{k}_{wpi}\right)\omega_p \cos n\omega_p t dt \\ &\quad \left. + \int_{2\pi/9}^{\pi/4}\left(k_3 - \bar{k}_{wpi}\right)\omega_p \cos n\omega_p t dt\right\} \\ &= \frac{8}{\pi}\left[\left(k_3 - \bar{k}_{wpi}\right)\left(\sin\frac{\pi}{36} + \sin\frac{\pi}{4} - \sin\frac{2\pi}{9}\right)\right. \\ &\quad \left. + \left(k_2 - \bar{k}_{wpi}\right)\left(\sin\frac{2\pi}{9} - \sin\frac{\pi}{36}\right)\right].\end{aligned} \quad (13)$$

Substituting (13) into (12) yields

$$\Delta k_{wpi}(t) = \sum_{n=1}^{\infty}\Delta k_{nwpi}\cos n\omega t. \quad (14)$$

4. Forced Response Equation to Coupled Excitation

The dynamic model for the electromechanical integrated toroidal drive (see Figure 3) allows rotor and each planet to rotate about their own rotating axes and allows each planet to translate in x_i and z_i directions. The rotations are replaced by the corresponding translational mesh displacements as $u_j = r_j\theta_j$, $j = 1, \ldots, m, r$ (here, m is planet number, θ_j the rotation of planet or rotor, r_j is the rolling circle radius for planet and the radius of the circle passing through planet centers for the rotor). A displacement vector $\mathbf{q}_j$ and a mass matrix $\mathbf{m}_j$ are defined for each planet j as $\mathbf{q}_j = [u_j \ x_j \ z_j]^T$ and $\mathbf{m}_j = \mathrm{Diag}[J_j/r_j^2 \ m_j \ m_j]$. Here, J_j and m_j are polar mass moment of inertia and mass for planet j, respectively. $M_r(M_r = J_r/r_r^2)$ is equivalent mass of rotor corresponding to its displacement u_r. Thus, the motion equations of the drive system are

$$\begin{aligned}
&\frac{J_i}{r_i^2}\ddot{u}_i + \left(k_{wpi} + \Delta k_{wpi}\right) p_{wpi} \sin\gamma_{wpi} \\
&\quad + \Delta F_e \sin\gamma_{wpi} - k_{spi} p_{spi} \cos\gamma_{spi} = 0, \\
&m_i \ddot{x}_i + k_{cpxi} x_i = 0, \\
&m_i \ddot{z}_i - \left(k_{wpi} + \Delta k_{wpi}\right) p_{wpi} \cos\gamma_{wpi} \\
&\quad - \Delta F_e \cos\gamma_{wpi} - k_{spi} p_{spi} \sin\gamma_{spi} + k_{cpzi} p_{cpzi} = 0, \\
&M_r \ddot{u}_r - \sum_{i=1}^{m} k_{cpzi} p_{cpzi} = -\frac{T_r}{r_r} \quad (i = 1 \text{ to } m),
\end{aligned} \tag{15}$$

where p_{wpi}, p_{spi} and p_{cpzi} are relative displacements between planet-i and worm, stator, or rotor, respectively, $p_{wpi} = u_i \sin\gamma_{wpi} - z_i \cos\gamma_{wpi}$, $p_{spi} = -u_i \cos\gamma_{spi} - z_i \sin\gamma_{spi}$ and $p_{cpzi} = z_i - u_r$. γ_{wpi} and γ_{spi} are lead angles at contact points between planet-i and worm or stator, respectively, $\tan\gamma_{wpi} = 1/[i_{wp}(a/R - 1)]$, and $\tan\gamma_{spi} = 1/[i_{sp}(a/R + 1)]$. Here, a is center distance between worm and planet R is reference circle radius of planet i_{wp} and i_{sp} are speed ratios between planet and worm or stator, respectively. k_{wpi} and k_{spi} are mesh stiffness between planet-i and worm or stator, respectively, k_{czi} and k_{cxi} are planet support stiffness in z_i and x_i directions, respectively, and T_r is torque transmitted by rotor.

Equation (15) can be written in matrix form as

$$\mathbf{M}\ddot{\mathbf{X}} + \mathbf{K}\mathbf{X} = \mathbf{F} + \varepsilon\Delta\mathbf{F} + \varepsilon\Delta\mathbf{F}_e, \tag{16}$$

where $\mathbf{X}$ and $\mathbf{F}$ are displacement and static load vectors; respectively,

$$\mathbf{X} = \{\mathbf{q}_1 \;\cdots\;\cdots\; \mathbf{q}_m \;\; u_r\}^T,$$

$$\mathbf{F} = \{\mathbf{0} \;\cdots\;\cdots\; \mathbf{0} \;\; -T_r/r_r\}^T;$$

$\mathbf{M}$ and $\mathbf{K}$ are mass and stiffness matrix, respectively,

$$\mathbf{M} = \text{diag}[\mathbf{m}_1 \;\cdots\;\cdots\; \mathbf{m}_m \;\; M_r],$$

$$\mathbf{K} = \begin{bmatrix}
\mathbf{k}_{wp1} + \mathbf{k}_{sp1} + \mathbf{k}_{cp1} & \cdots & \mathbf{0} & \cdots & \mathbf{0} & \mathbf{k}_{c1} \\
 & \ddots & \vdots & \vdots & \vdots & \vdots \\
 & & \mathbf{k}_{wpi} + \mathbf{k}_{spi} + \mathbf{k}_{cpi} & \cdots & \vdots & \mathbf{k}_{ci} \\
 & & & \ddots & \vdots & \vdots \\
 & & & & \mathbf{k}_{wpm} + \mathbf{k}_{spm} + \mathbf{k}_{cpm} & \mathbf{k}_{cm} \\
\text{symmetric} & & & & & \sum_{i=1}^{m} k_{czi}
\end{bmatrix},$$

$$\mathbf{k}_{wpi} = k_{wpi}\begin{bmatrix} \sin^2\gamma_{wpi} & 0 & -\cos\gamma_{wpi}\sin\gamma_{wpi} \\ 0 & 0 & 0 \\ -\cos\gamma_{wpi}\sin\gamma_{wpi} & 0 & \cos^2\gamma_{wpi} \end{bmatrix},$$

$$\mathbf{k}_{spi} = k_{spi}\begin{bmatrix} \cos^2\gamma_{spi} & 0 & \sin\gamma_{spi}\cos\gamma_{spi} \\ 0 & 0 & 0 \\ \sin\gamma_{spi}\cos\gamma_{spi} & 0 & \sin^2\gamma_{spi} \end{bmatrix},$$

$$\mathbf{k}_{cpi} = \begin{bmatrix} 0 & 0 & 0 \\ 0 & k_{cxi} & 0 \\ 0 & 0 & k_{czi} \end{bmatrix}, \quad \mathbf{k}_{ci} = \left\{0 \;\; 0 \;\; -k_{czi}\right\}^T. \tag{17}$$

$\Delta\mathbf{F}$ is nonlinear component of the force vector, $\Delta\mathbf{F}_e$ is equivalent force vector caused by current fluctuation:

Consider

$$\Delta\mathbf{F} = \left(Bu_i^2 + Cu_i^3\right)\left[-\sin\gamma_{wpi} \;\; 0 \;\; \cos\gamma_{wpi} \;\; -\sin\gamma_{wpi} \;\; 0 \;\; \cos\gamma_{wpi} \;\; \cdots \;\; 0\right]^T, \quad u_i = R\delta\theta, \quad \varepsilon = \frac{r}{R}, \tag{18}$$

it is a disturbing parameter, $B = z_1 \sum_{i=1}^{m} \sum_{j=1}^{m} A \tan(z_1\theta - \phi_v/np) I_i I_j / 2Rr$, $C = -z_1^2 \sum_{i=1}^{m} \sum_{j=1}^{m} A I_i I_j / 2Rr$.

$$\Delta F_e = D\cos(\omega_e t)\left[\sin\gamma_{wpi}\ \ 0\ \ \cos\gamma_{wpi}\ \ \sin\gamma_{wpi}\ \ 0\ \ \cos\gamma_{wpi}\ \ \sin\gamma_{wpi}\ \ 0\ \ \cos\gamma_{wpi}\ \ \cdots\ \ 0\right]^T,$$
$$D = \frac{1}{2R}\sum_{i=1}^{n}\sum_{j=1}^{n}\left(\frac{\partial L_{ij}}{\partial\theta}\right)_{\theta=\theta_0} I_i \Delta I. \tag{19}$$

The total displacement of the each component consists of the static displacement and the dynamic one. Hence, the displacement vector is

$$\mathbf{X} = \overline{\mathbf{X}} + \Delta\mathbf{X}, \tag{20}$$

where $\overline{\mathbf{X}}$ and $\Delta\mathbf{X}$ are static and dynamic displacement vectors; respectively,

$$\Delta\mathbf{X} = \left\{\Delta\mathbf{q}_1\ \cdots\ \cdots\ \Delta\mathbf{q}_m\ \ \Delta u_r\right\}^T,$$
$$\Delta\mathbf{q}_j = \left[\Delta u_j\ \ \Delta x_j\ \ \Delta z_j\right]^T. \tag{21}$$

Considering the fluctuation of the mesh stiffness caused by the changes of the mesh tooth pair number, the stiffness matrix can be expressed as

$$\mathbf{K} = \overline{\mathbf{K}} + \Delta\mathbf{K}(t), \tag{22}$$

where $\overline{\mathbf{K}}$ is the mean stiffness matrix, and $\Delta\mathbf{K}(t)$ is time-varying stiffness matrix.

Substituting (20)–(22) into (16) yields

$$\overline{\mathbf{K}}\overline{\mathbf{X}} = \mathbf{F}, \tag{23}$$

$$\mathbf{M}\Delta\ddot{\mathbf{X}} + \overline{\mathbf{K}}\Delta\mathbf{X} = \varepsilon\Delta\mathbf{F} + \varepsilon\Delta\mathbf{F}_e + \varepsilon\Delta\mathbf{F}_p, \tag{24}$$

where $\Delta\mathbf{F}_p = -E\Delta\mathbf{K}(t)\overline{\mathbf{X}}$; it is considered as the equivalent exciting force vector caused by mesh stiffness excitation; $E = R/r$.

Equation (24) is just the forced response equation of the drive system to coupled excitation.

5. Solution of the Nonlinear Forced Response Equation

5.1. Far from Natural Frequencies. First, we consider to resolve the nonlinear forced response equation of the drive system to electric excitation (let $\Delta\mathbf{F}_p = 0$).

For simplicity purposes, (24) should be transmitted into equations independent on each other. Then, (24) is changed into the following form:

$$\Delta\ddot{\mathbf{X}}_N + \overline{\mathbf{K}}_N\Delta\mathbf{X}_N = \varepsilon\Delta\mathbf{F}_N + \varepsilon\Delta\mathbf{F}_{eN}, \tag{25}$$

where $\overline{\mathbf{K}}_N$ is the diagonal mean stiffness matrix, respectively. $\Delta\mathbf{X}_N$ is transmitted dynamic displacement vector. $\Delta\mathbf{F}_{eN}$ and $\Delta\mathbf{F}_N$ are transmitted exciting force vectors of the forces $\Delta\mathbf{F}_e$ and $\Delta\mathbf{F}$, respectively.

Matrix $\mathbf{K}_N$ and vector $\Delta\mathbf{X}_N$ are given by $\mathbf{K}_N = \mathbf{A}_N^T\mathbf{K}\mathbf{A}_N$ and $\Delta\mathbf{X}_N = \mathbf{A}_N^T\Delta\mathbf{X}$. The transmitted nonlinear force vector $\Delta\mathbf{F}_N$ and transmitted equivalent exciting force vector $\Delta\mathbf{F}_{eN}$ can be given as below:

$$\Delta\mathbf{F}_N = \mathbf{A}_N^T\Delta\mathbf{F} = \left(Bu_i^2 + Cu_i^3\right)\Delta\mathbf{P}_N,$$
$$\Delta\mathbf{F}_{eN} = \varepsilon D\cos(\omega_e t)\Delta\mathbf{P}_{eN}, \tag{26}$$

where $\mathbf{A}_N$ is the mode matrix of (24):

$$\mathbf{A}_N = \begin{bmatrix} A_1^{(1)} & A_2^{(1)} & \cdots & A_m^{(1)} \\ A_1^{(2)} & A_2^{(2)} & \cdots & A_m^{(2)} \\ \cdots & \cdots & \cdots & \cdots \\ A_1^{(m)} & A_2^{(m)} & \cdots & A_m^{(m)} \end{bmatrix}. \tag{27}$$

At $m = 3$, $\Delta\mathbf{P}_N = [P_{N1}\ P_{N2}\ \cdots\ P_{Ni}\ \cdots]^T$, $\mathbf{P}_{Ni} = -3A_{N1}^i\sin\gamma_{wpi} + 3A_{N3}^i\cos\gamma_{wpi}$ $(i = 1, 4, \ldots, 3m-2)$, $P_{Ni} = 0$ $(i = 2, 3, \ldots, 3m+1, i \neq 1, 4, \ldots, 3m-2)$; $\Delta\mathbf{P}_{eN} = [P_{eN1}\ 0\ 0\ P_{eN2}\ 0\ 0\ P_{eN3}\ 0\ 0\ 0]^T$, $P_{Ne} = 3A_{N1}^i\sin\gamma_{wpi} + 3A_{N3}^i\cos\gamma_{wpi}$ $(i = 1, 4, \ldots, 3m-2)$, $P_{Nei} = 0$ $(i = 2, 3, \ldots, 3m+1, i \neq 1, 4, \ldots, 3m-2)$.

Let

$$\Delta\mathbf{X}_N = \mathbf{X}_0 + \varepsilon\mathbf{X}_1 + \varepsilon^2\mathbf{X}_2 + \cdots, \tag{28}$$

$$\omega_i^2 = \omega_{0i}^2\left(1 + \varepsilon\sigma_1^i + \varepsilon^2\sigma_2^i + \cdots\right)\quad (i = 1, 2, 3), \tag{29}$$

where ω_i is natural frequency of ith order mode for the nonlinear drive system, and ω_{0i} is natural frequency of ith order rotational mode for the linear drive system.

Substituting (28) and (29) into (25), let sum of the coefficients with the same-order power of the parameter ε equal zero, following equations can be given

$$\ddot{x}_{N0} + \omega_i^2 x_{N0} = 0,$$
$$\ddot{x}_{N1} + \omega_i^2 x_{N1} = -\sigma_1\ddot{x}_{N0} + P_N B u_{0i}^2 + P_N C u_{0i}^3 + D\cos(\omega_e t)P_{Ne}$$
$$\ddot{x}_{N2} + \omega_i^2 x_{N2} = -\sigma_1\ddot{x}_{N1} - \sigma_2\ddot{x}_{N0} + P_N B\left(2u_{0i}u_{1i} + \sigma_1 u_{0i}^2\right)$$
$$+ P_N C\left(3u_{0i}^2 u_{1i} + \sigma_1 u_{0i}^3\right) + D\sigma_1\cos(\omega_e t)P_{Ne}$$
$$\cdots. \tag{30}$$

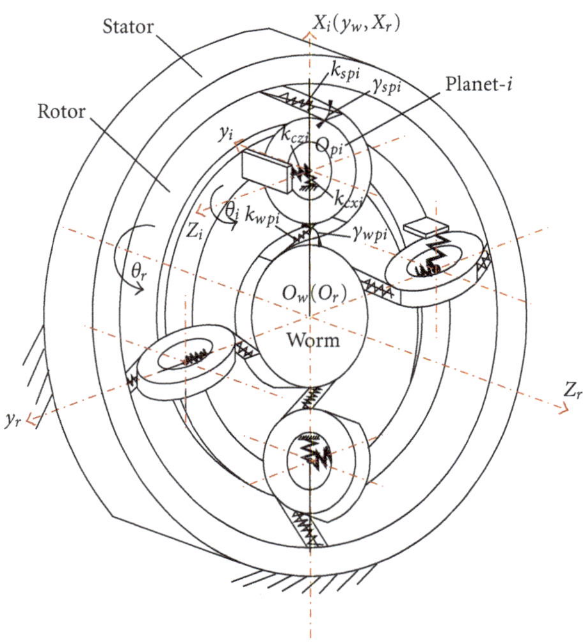

FIGURE 3: Dynamic model for a four-planet electromechanical integrated toroidal drive.

Here, initial conditions are $x^i_{N0}(0) = A^i_{N0}$, $\dot{x}^i_{N0}(0) = 0$.

The solution of zero-order equation under the above initial conditions is

$$x^i_0 = A^i_0 \cos \omega_i t \quad (i = 1, 2, 3). \tag{31}$$

Substituting (31) into the second equation of (30) yields

$$\begin{aligned}
\ddot{x}^1_{N1} + \omega_1^2 x^1_{N1} &= -\sigma_1^1 \ddot{x}^1_{N0} + P_{N1} B u^2_{0i} + P_{N1} C u^3_{0i} + D\cos(\omega_e t) P_{Ne}, \\
\ddot{x}^2_{N1} + \omega_2^2 x^2_{N1} &= -\sigma_1^2 \ddot{x}^2_{N0} + P_{N2} B u^2_{0i} + P_{N2} C u^3_{0i} + D\cos(\omega_e t) P_{Ne}, \\
\ddot{x}^3_{N1} + \omega_3^2 x^3_{N1} &= -\sigma_1^3 \ddot{x}^3_{N0} + P_{N3} B u^2_{0i} + P_{N3} C u^3_{0i} + D\cos(\omega_e t) P_{Ne}.
\end{aligned} \tag{32}$$

The rotational displacement u_i is

$$u_i = A^1_{N1} x^1_{Ni} + A^2_{N1} x^2_{Ni} + A^3_{N1} x^3_{Ni}. \tag{33}$$

Substituting (33) into (30) yields

$$\begin{aligned}
&\ddot{x}^1_{N1} + \omega_1^2 x^1_{N1} \\
&\quad = -\sigma_1^1 \ddot{x}^1_{N0} + P_{N1} B \left(A^1_{N1} x^1_{N0} + A^2_{N1} x^2_{N0} + A^3_{N1} x^3_{N0} \right)^2 \\
&\quad\quad + P_{N1} C \left(A^1_{N1} x^1_{N0} + A^2_{N1} x^2_{N0} + A^3_{N1} x^3_{N0} \right)^3 \\
&\quad\quad + D \cos(\omega_e t) P_{Ne1} \\
&\ddot{x}^2_{N1} + \omega_2^2 x^2_{N1} \\
&\quad = -\sigma_1^2 \ddot{x}^2_{N0} + P_{N2} B \left(A^1_{N1} x^1_{N0} + A^2_{N1} x^2_{N0} + A^3_{N1} x^3_{N0} \right)^2 \\
&\quad\quad + P_{N2} C \left(A^1_{N1} x^1_{N0} + A^2_{N1} x^2_{N0} + A^3_{N1} x^3_{N0} \right)^3 \\
&\quad\quad + D \cos(\omega_e t) P_{Ne2}, \\
&\ddot{x}^3_{N1} + \omega_3^2 x^3_{N1} \\
&\quad = -\sigma_1^3 \ddot{x}^3_{N0} + P_{N3} B \left(A^1_{N1} x^1_{N0} + A^2_{N1} x^2_{N0} + A^3_{N1} x^3_{N0} \right)^2 \\
&\quad\quad + P_{N3} C \left(A^1_{N1} x^1_{N0} + A^2_{N1} x^2_{N0} + A^3_{N1} x^3_{N0} \right)^3 \\
&\quad\quad + D \cos(\omega_e t) P_{Ne3}.
\end{aligned} \tag{34}$$

In order to remove secular item, let

$$\begin{aligned}
\sigma_1^1 &= -\frac{P_{N1} C}{\omega_1^2} \\
&\quad \times \left[\frac{3}{4} \left(A^1_{N1} A^1_{N0} \right)^3 + \frac{3}{2} \left(A^2_{N1} A^2_{N0} \right)^2 A^1_{N1} A^1_{N0} \right. \\
&\quad\quad \left. + \frac{3}{2} \left(A^3_{N1} A^3_{N0} \right)^2 A^1_{N1} A^1_{N0} \right], \\
\sigma_1^2 &= -\frac{P_{N2} C}{\omega_2^2} \\
&\quad \times \left[\frac{3}{4} \left(A^2_{N1} A^2_{N0} \right)^3 + \frac{3}{2} \left(A^1_{N1} A^1_{N0} \right)^2 A^2_{N1} A^2_{N0} \right. \\
&\quad\quad \left. + \frac{3}{2} \left(A^3_{N1} A^3_{N0} \right)^2 A^2_{N1} A^2_{N0} \right], \\
\sigma_1^3 &= -\frac{P_{N3} C}{\omega_3^2} \\
&\quad \times \left[\frac{3}{4} \left(A^3_{N1} A^3_{N0} \right)^3 + \frac{3}{2} \left(A^2_{N1} A^2_{N0} \right)^2 A^3_{N1} A^3_{N0} \right. \\
&\quad\quad \left. + \frac{3}{2} \left(A^1_{N1} A^1_{N0} \right)^2 A^3_{N1} A^3_{N0} \right].
\end{aligned} \tag{35}$$

Substituting (35) into (34), the solutions of the first-order equations can be obtained. As the equations of the solutions are relatively complicated, it is not given here.

In a same manner, the solution of nth order equation can be obtained as well. Substituting these solutions into (28) and (29), the solutions of the regular nonlinear forced response equations and natural frequencies of the drive system can be given. Then, the real solutions of the nonlinear forced responses can be calculated as below:

$$\Delta \mathbf{X} = \mathbf{A}_N \Delta \mathbf{X}_N. \tag{36}$$

5.2. Near Natural Frequencies. When the exciting frequency is near natural frequency, the nonlinear forced response equation can be resolved as below. Considering damping of the drive system, (25) can be changed into

$$\Delta \ddot{\mathbf{X}}_N + \mathbf{C}_N \Delta \dot{\mathbf{X}}_N + \mathbf{K}_N \Delta \mathbf{X}_N = \varepsilon \Delta \mathbf{F}_N + \varepsilon \Delta \mathbf{F}_{eN}, \tag{37}$$

where $\mathbf{C}_N$ is regular damping matrix, $\mathbf{C}_N = \mathbf{A}_N^T\mathbf{C}\mathbf{A}_N = \mathrm{diag}[C_{N1}\ C_{N2}\ C_{N3}\ C_{N4}]$ ($\mathbf{C}$ is damping matrix). $C_{Ni} = 2\zeta\omega_i$, ζ is relative damping coefficient.

When exciting frequency is near natural frequency, the exciting frequency can be written as

$$\omega_{ei}^2 = \omega_{0i}^2\left(1+\varepsilon\sigma_1^i\right) \quad (i=1,2,3). \tag{38}$$

Substituting (38) and (28) into (37), let sum of the coefficients with the same-order power of the parameter ε equal zero, the following equations can be given:

$$\begin{aligned}
&\ddot{\mathbf{X}}_0 + \mathbf{K}_N\mathbf{X}_0 = 0,\\
&\ddot{\mathbf{X}}_1 + \mathbf{K}_N\mathbf{X}_1 = -\sigma_1^i\ddot{\mathbf{X}}_0 - 2\zeta_1\dot{\mathbf{X}}_0\\
&\qquad + \Delta\mathbf{P}_N\left(B_iu_i^2 + C_iu_i^3\right) + \Delta\mathbf{P}_{eN}D_i\cos(\omega_{ei}t+\theta)\\
&\qquad \ldots,
\end{aligned} \tag{39}$$

where $C_{Ni} = 2\zeta\omega_i$,

$$\begin{aligned}
\zeta &= \varepsilon\zeta_1,\\
\omega_{ei}^2 &= \omega_i^2(1+\varepsilon\sigma_1),\\
\Delta x_{Ni} &= x_{N0i} + \varepsilon x_{N1i},\\
B &= B_i\omega_i^2,\\
C &= C_i\omega_i^2,\\
D &= D_i\omega_i^2.
\end{aligned} \tag{40}$$

The solution of zero-order equation under the above initial conditions is

$$x_{N0}^i = A_{N0}^i\cos\omega_{ei}t \quad (i=1,2,3). \tag{41}$$

Substituting (41) into the second equation of (43) yields

$$\begin{aligned}
\ddot{x}_{N1}^1 + \omega_{e1}^2x_{N1}^1 &= -\sigma_1\ddot{x}_{N0}^1 - 2\zeta_1\dot{x}_{N0}^1 + P_{N1}B_1u_{01}^2 + P_{N1}C_1u_{01}^3\\
&\quad + D_1P_{Ne1}\cos(\omega_{e1}t+\theta),\\
\ddot{x}_{N1}^2 + \omega_{e2}^2x_{N1}^2 &= -\sigma_1\ddot{x}_{N0}^1 - 2\zeta_1\dot{x}_{N0}^2 + P_{N2}B_2u_{01}^2 + P_{N2}C_2u_{01}^3\\
&\quad + D_2P_{Ne2}\cos(\omega_{e2}t+\theta),\\
\ddot{x}_{N1}^3 + \omega_{e3}^2x_{N1}^3 &= -\sigma_1\ddot{x}_{N0}^1 - 2\zeta_1\dot{x}_{N0}^3 + P_{N3}B_3u_{01}^2 + P_{N3}C_3u_{01}^3\\
&\quad + D_3P_{Ne3}\cos(\omega_{e3}t+\theta).
\end{aligned} \tag{42}$$

Substituting rotational displacement $u_i = A_{N1}^1x_{Ni}^1 + A_{N1}^2x_{Ni}^2 + A_{N1}^3x_{Ni}^3$ into (42) yields

$$\begin{aligned}
&\ddot{x}_{N1}^1 + \omega_{e1}^2x_{N1}^1\\
&\quad = -\sigma_1\ddot{x}_{N0}^1 - 2\zeta_1\dot{x}_{N0}^1 + P_{N1}B_1\left(A_{N1}^1x_{N0}^1 + A_{N1}^2x_{N0}^2 + A_{N1}^3x_{N0}^3\right)^2\\
&\qquad + P_{N1}C_1\left(A_{N1}^1x_{N0}^1 + A_{N1}^2x_{N0}^2 + A_{N1}^3x_{N0}^3\right)^3\\
&\qquad + D_1P_{Ne1}(\cos\omega_{e1}t\cos\theta + \sin\omega_{e1}t\sin\theta),\\
&\ddot{x}_{N1}^2 + \omega_{e2}^2x_{N1}^2\\
&\quad = -\sigma_1\ddot{x}_{N0}^2 - 2\zeta_1\dot{x}_{N0}^2 + P_{N2}B_2\left(A_{N1}^1x_{N0}^1 + A_{N1}^2x_{N0}^2 + A_{N1}^3x_{N0}^3\right)^2\\
&\qquad + P_{N2}C_2\left(A_{N1}^1x_{N0}^1 + A_{N1}^2x_{N0}^2 + A_{N1}^3x_{N0}^3\right)^3\\
&\qquad + D_2P_{Ne2}(\cos\omega_{e2}t\cos\theta + \sin\omega_{e2}t\sin\theta),\\
&\ddot{x}_{N1}^3 + \omega_{e3}^2x_{N1}^3\\
&\quad = -\sigma_1\ddot{x}_{N0}^3 - 2\zeta_1\dot{x}_{N0}^3 + P_{N3}B_3\left(A_{N1}^1x_{N0}^1 + A_{N1}^2x_{N0}^2 + A_{N1}^3x_{N0}^3\right)^2\\
&\qquad + P_{N3}C_3\left(A_{N1}^1x_{N0}^1 + A_{N1}^2x_{N0}^2 + A_{N1}^3x_{N0}^3\right)^3\\
&\qquad + D_3P_{Ne3}(\cos\omega_{e3}t\cos\theta + \sin\omega_{e3}t\sin\theta).
\end{aligned} \tag{43}$$

In order to remove secular item, let

$$\begin{aligned}
\sigma_1^1A_{N0}^1 + P_{N1}C_1A_{N0}^1P_1' + D_1P_{Ne1}\cos\theta &= 0,\\
2\zeta_1A_{N0}^1 + D_1P_{Ne1}\sin\theta &= 0,
\end{aligned} \tag{44}$$

where $P_1' = (1/A_{N0}^1)[(3/4)(A_{N1}^1A_{N0}^1)^3 + (3/2)(A_{N1}^2A_{N0}^2)^2A_{N1}^1A_{N0}^1 + (3/2)(A_{N1}^3A_{N0}^3)^2A_{N1}^1A_{N0}^1]$.

From (44), it is known that

$$(\sigma_1^1 + P_{N1}C_1P_1')^2 + (2\zeta_1)^2 = \left(\frac{D_1P_{Ne1}}{A_{N0}^1}\right)^2. \tag{45}$$

Thus

$$\begin{aligned}
s_i^2 &= 1 - \varepsilon P_{Ni}C_iP_i' - 2\zeta^2\\
&\quad \pm\sqrt{\left(\frac{D_i'}{A_{N0}^i}\right)^2 - 4\zeta^2(1-\varepsilon P_{Ni}C_iP_i' - \zeta^2)},
\end{aligned} \tag{46}$$

where $s_i = \omega_{ei}/\omega_i$ and $D_i' = \varepsilon D_iP_{Nei}$.

From (46), the changes of the nonlinear vibrating magnitudes along with exciting frequencies can be given.

6. Results and Discussions

When exciting frequency is far from natural frequency, from the above equations, the nonlinear forced vibrations for the drive system are analyzed. The parameters of the numerical example are shown in Table 1. Figure 4 shows changes of the forced vibrations of the transmitted variables along with nonlinear parameter ε. From Figure 4, the following are known.

TABLE 1: Parameters of the example system.

ε	ω_e (rad/s)	a/R	i_{wpi}	NI_s (A)	r (mm)	I (A)	C	D	R (mm)	z_1	L (H)
0.25	1000	2	8	100	25	180	8707.2	−0.00066	100	8	1×10^{-3}

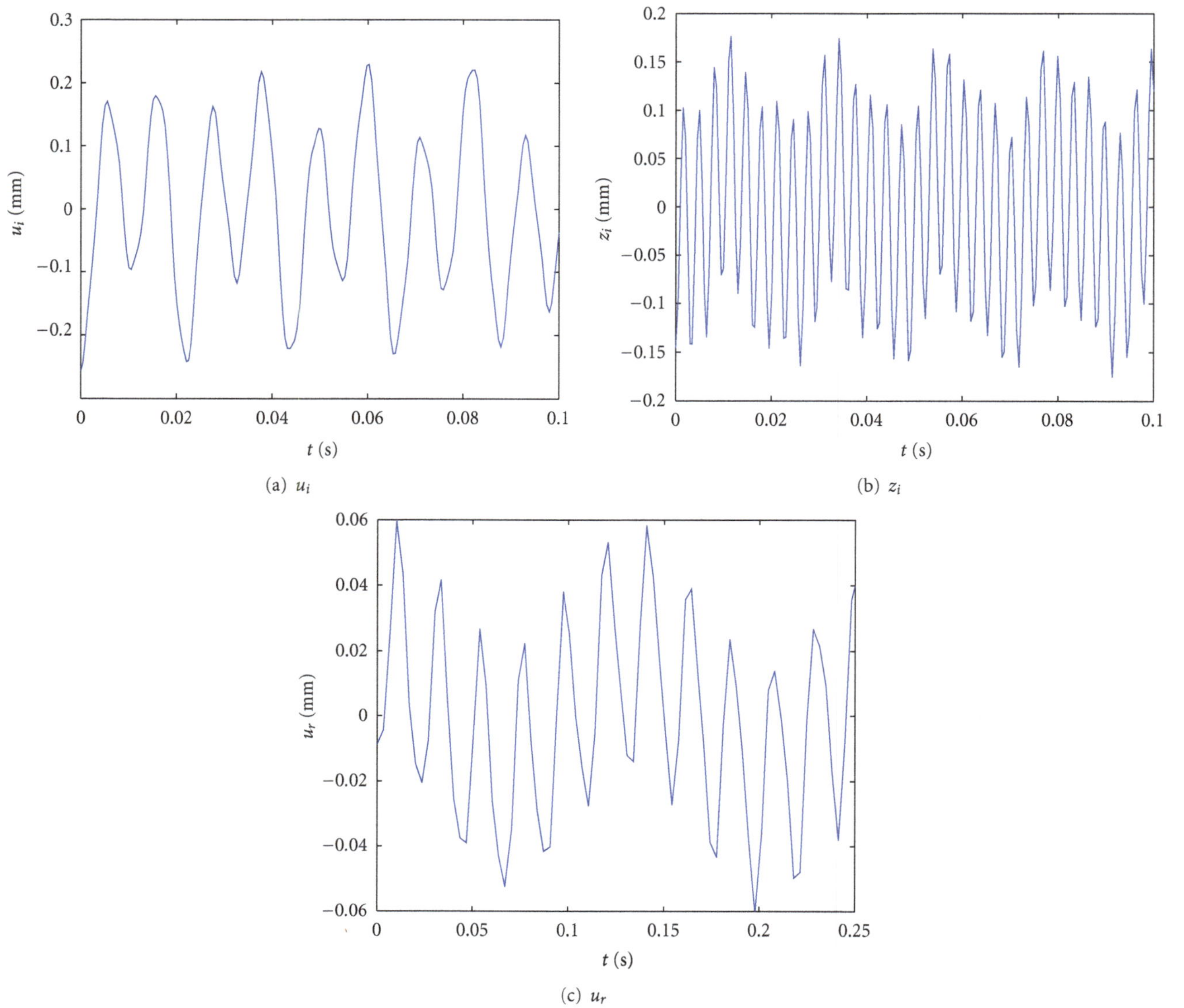

FIGURE 4: Nonlinear forced vibration to coupled excitations far from natural frequencies.

(1) Under the coupled excitations, the nonlinear forced responses of the drive system change periodically and unsteadily. The time period of the nonlinear forced responses depends on the frequencies of the electric excitation the mesh parameter excitation, and the nonlinear natural frequencies of the drive system.

(2) The vibrating amplitudes of the nonlinear forced responses of the drive system to the coupled excitations are larger than those of the nonlinear forced responses to the single excitation. The vibrating amplitudes of the nonlinear forced responses of the planet are larger than that of the rotor. It is because the exciting frequencies are near to the vibrating frequency of the planet modes.

(3) The vibrating amplitudes of the tangent vibration for the planet are larger than that of the axial vibration for the planet, and the frequency of the tangent vibration for the planet is smaller than that of the axial vibration for the planet.

(4) The unstable periodic vibrations are harmful for the drive operation. In order to increase the dynamics performance of the drive system, the electric excitations should be avoided or the frequencies of the

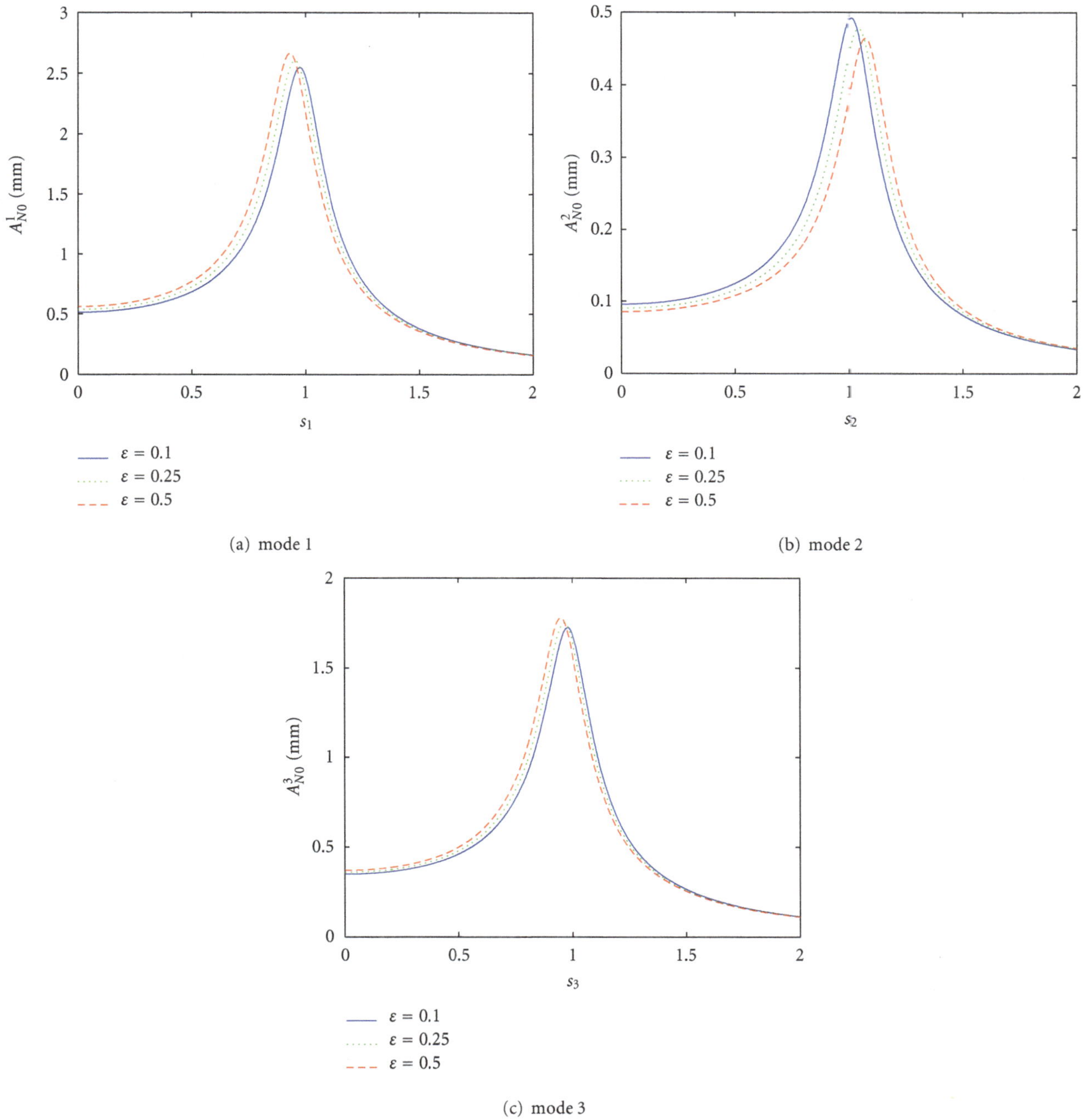

FIGURE 5: Nonlinear forced vibration to coupled excitations near natural frequencies.

electric excitations should not be taken as integral multiple of the mesh parameter exciting frequency.

When exciting frequency is near natural frequency, changes of the nonlinear vibrating magnitudes along with exciting frequencies and the drive parameters are given in Figure 5. From Figure 5, the following are known.

(1) As exciting frequency increase, the vibrating magnitudes of all the modes increases, and at point $s_i = \omega_{ei}/\omega_i \approx 1$, the vibrating magnitudes get to the maximum, and then they decrease with increasing exciting frequency. For different modes, the previously mentioned curves bend toward the direction of the exciting frequency increase. The results are typical nonlinear character of the drive system.

(2) For the first and third modes, as nonlinear parameter ε increases, their vibrating magnitudes increase. For the second mode, as nonlinear parameter ε increases, its vibrating magnitude decreases. The vibrating magnitudes are large for the first and third modes and small for the second mode. The vibrating magnitude for the first mode is larger than that for the third mode.

(3) For a given exciting frequency, the vibrating amplitudes of the nonlinear forced responses of the drive system to the coupled excitations for modes 1 and 3 are larger than those of the nonlinear forced responses to the single excitation; for mode 2, the vibrating amplitude of the nonlinear forced responses of the drive system to the coupled excitations is smaller than that of the nonlinear forced responses to the single excitation.

(4) Under the coupled excitations, the nonlinearity has obvious effects on the relationship between the frequency and the amplitudes; it should be considered.

7. Conclusions

In this paper, the electric excitation and the parameter excitation from mesh stiffness fluctuation are analyzed. The forced response equations of the drive system to the coupled excitations are presented. For the exciting frequencies far from and near to natural frequencies, the forced responses of the drive system to the coupled excitations are investigated. Results show the following.

(1) Under the coupled excitations, the nonlinear forced responses of the drive system change periodically and unsteadily.

(2) The time period of the nonlinear forced responses depends on the frequencies of the electric excitation the mesh parameter excitation, and the nonlinear natural frequencies of the drive system.

(3) In order to improve the dynamics performance of the drive system, the frequencies of the electric excitations should not be taken as integral multiple of the mesh parameter exciting frequency.

Acknowledgment

This project is supported by the National Natural Science Foundation of China (no. 51075350).

References

[1] M. R. Kuehnle, "Toroidgetriebe," Urkunde uber die Erteilung des deutschen Patents, 1301682, 1966.

[2] M. R. Kuehnle, H. Peeken, C. Troeder, and S. Cerniak, "The Toroidal drive," *Mechanical Engineering*, vol. 103, no. 2, pp. 32–39, 1981.

[3] M. R. Kuehnle, "Toroidal transmission and method and apparatus for making and assembling same," United States Patent, 5863273, 1999.

[4] Z. Delin and L. Huamin, "Side surface harmonic stepper motor," *Journal of Mechanical Engineering*, vol. 29, no. 5, pp. 96–98, 1993 (Japanese).

[5] O. Barth, "Harmonic piezodrive—miniaturized servo motor," *Mechatronics*, vol. 10, no. 4, pp. 545–554, 2000.

[6] L. Xu, Z. Huang, and Y. Yang, "Contact stress for toroidal drive," *Journal of Mechanical Design, Transactions of the ASME*, vol. 125, no. 1, pp. 165–168, 2003.

[7] L. Xu and J. Huang, "Torques for electromechanical integrating toroidal drive," *Proceedings of the Institution of Mechanical Engineers, Part C*, vol. 219, no. 8, Article ID C15704, pp. 801–811, 2005.

[8] A. Shaltout and M. Al-Omoush, "Reclosing torques of large induction motors with stator trapped flux," *IEEE Transactions on Energy Conversion*, vol. 11, no. 1, pp. 84–90, 1996.

[9] H. Y. Kanaan, K. Al-Haddad, and G. Roy, "Analysis of the electromechanical vibrations in induction motor drives due to the imperfections of the mechanical transmission system," *Mathematics and Computers in Simulation*, vol. 63, no. 3–5, pp. 421–433, 2003.

[10] L. Xu and X. Hao, "Nonlinear forced vibration for electromechanical integrated toroidal drive," *International Journal of Applied Electromagnetics and Mechanics*, vol. 28, no. 3, pp. 351–369, 2008.

7

Optimal Control for a Parallel Hybrid Hydraulic Excavator Using Particle Swarm Optimization

Dong-yun Wang[1] **and Chen Guan**[2]

[1] *School of Engineering, Zhejiang Normal University, Zhejiang 321005, China*
[2] *School of Mechanical Engineering, Zhejiang University, Zhejiang 321005, China*

Correspondence should be addressed to Dong-yun Wang; zsdwdy@zjnu.cn

Academic Editors: Y. Lu and S. Torii

Optimal control using particle swarm optimization (PSO) is put forward in a parallel hybrid hydraulic excavator (PHHE). A power-train mathematical model of PHHE is illustrated along with the analysis of components' parameters. Then, the optimal control problem is addressed, and PSO algorithm is introduced to deal with this nonlinear optimal problem which contains lots of inequality/equality constraints. Then, the comparisons between the optimal control and rule-based one are made, and the results show that hybrids with the optimal control would increase fuel economy. Although PSO algorithm is off-line optimization, still it would bring performance benchmark for PHHE and also help have a deep insight into hybrid excavators.

1. Introduction

With the purposes of reducing fuel consumption and emissions, colleges, institutes, and companies are trying to develop a new type of excavator for the next generation. In 2003, Hitachi first produced one hybrid wheel loader which was the first hybrid construction machinery around the world. Until now, a large portion of construction machinery producers including Kobelco, Komatsu, and Caterpillar are involved in the field of hybrid excavators.

From the system configuration point of view, there are three types of hybrid excavators depending upon different power-train configurations. One is series hybrid excavator, another is the parallel one, and the remaining is the split one. To get detailed information regarding the properties of all these different configurations, readers would refer to the literature [1, 2]. In summary, compared with other two, the parallel configuration is considered as the one with low cost and better performance. One parallel hybrid hydraulic excavator has an additional motor that is connected to the engine via the same shaft; in addition, one storage unit (usually it is battery or capacitor) is introduced into power-train as to receive extra energy from engine to charge itself when the motor is in generation mode or deplete its energy to the pump when the motor is in motor mode. The main concepts of hybrids are to keep the engine's working points located in high fuel economy range that would badly decrease the fuel consumption regardless of working condition. Wang et al. did simulation research and evaluated energy saving effect in parallel hybrid excavators in 2005 and also presented a way to recover potential energy from excavators [2]. The parallel model, system dynamic analysis, rule-based control strategy, fuzzy logic control strategy, and component sizing are put forward by Wang et al. [3–7]. Despite of all these efforts in the field, few papers focus on the nonlinear constraints optimal control problem for PHHE.

As described in [1], the electric motor provides a fast response to the torque and speed demand with high efficiency in a large angular velocity scale. So in this paper, PSO algorithm is mainly used to get the optimal working points of the diesel engine. So the obtained group of engine's working points is globally optimized at each sampling time. This program automatically moves on to the next time interval once it finishes the previous sampling time. Thus, the global optimization for the whole cycle could be achieved. Here, PSO algorithm was first invented by Kennedy and Eberhart in 1995, that is, a stochastic global optimization approach

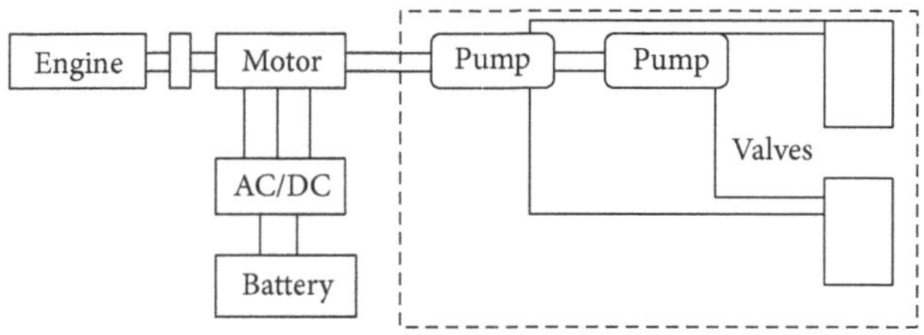

FIGURE 1: A schematic of a parallel hybrid hydraulic excavator.

capable of obtaining the global optimum [8]. Compared with other kinds of derivative-free algorithms such as DIRECT, simulated annealing (SA), and genetic algorithm (GA), PSO's strength lies in its simplicity, being easily coded and requiring few algorithm parameters to define convergence behavior. Also, one could use gradient-based algorithms such as sequential quadratic programming (SQP) to find the local minima, but it would not search the entire space, so the global optimum could not be obtained by this method.

The structure of the paper is as follows. Section 2 describes the model of the parallel hybrid hydraulic excavator and components' parameters in detail. The optimal control problem is formulated in Section 3, followed by inequality/equality constraints. PSO algorithm is then coded to deal with this specific application. The obtained set of optimum points will be implemented in the aforementioned model, and the analysis of the optimal control will be presented based on comparisons. Finally, Section 4 provides an extension on further research and a summary of this paper.

2. The Plant Model

A schematic of a parallel hybrid hydraulic excavator is shown in Figure 1. The motor connects to the pump and engine together through the same shaft, and all the required power at the pumps is satisfied by the engine and motor. So with this configuration, it has three different types of modes listed as follows. (1) Motor assist hybrid mode: the engine and the motor would drive the final load together, and the motor is working under motor mode. (2) Motor generation hybrid mode: only engine delivers power to the load, and part of the power from the engine will be transmitted to the motor then will be charged into the battery. (3) Pure engine mode: only the engine works, and the motor will be rotated along with the shaft with no torque generated. The equations of the powertrain are presented as follows:

$$
\begin{gathered}
J_{\text{eng}} \frac{dw_{\text{eng}}}{dt} = \tau_{\text{eng\_}T} - \tau_{\text{eng}}, \\
J_{\text{mc}} \frac{dw_{\text{mc}}}{dt} = \tau_{\text{mc\_}T} - \tau_{\text{mc}}, \\
\tau_{\text{eng}} + \tau_{\text{mc}} = \tau_p, \\
w_{\text{eng}} = w_{\text{mc}} = w_p, \\
P_{p\_\text{in}} = \tau_p w_p = \frac{p_x Q_x + p_2 Q_2}{\eta_p},
\end{gathered}
\tag{1}
$$

where J_{eng}, τ_{eng}, w_{eng}, and $\tau_{\text{eng\_}T}$ are inertia, speed, torque, and total torque of the engine, respectively, and J_{mc}, w_{mc}, τ_{mc}, and $\tau_{\text{mc\_}T}$ are inertia, speed, torque, and total torque of the motor. P_{p_in}, τ_p, w_p, p_x, Q_x, and η_p are input power, torque, speed, pressure, flow, and efficiency of the pumps. Since in current configuration of the excavator, there are 2 pumps connected in parallel by which bucket cylinder, arm cylinder, swing hydraulic motor, and boom cylinder are driven.

In addition, the dynamic equations of the battery are analyzed as follows:

$$
\begin{gathered}
i = \frac{U_{\text{oc}} - \sqrt{U_{\text{oc}} - 2P_{\text{bat}} R_{\text{bat}}}}{2R_{\text{bat}}}, \\
\text{SOC} = -\frac{U_{\text{oc}} - \sqrt{U_{\text{oc}} - 2P_{\text{bat}} R_{\text{bat}}}}{2R_{\text{bat}} C_{\text{bat}}}, \\
P_{\text{bat}} = \frac{\tau_{\text{mc}} w_{\text{mc}}}{\eta_{\text{mc}}^k},
\end{gathered}
\tag{2}
$$

where U_{oc}, P_{bat}, R_{bat}, C_{bat}, i, and SOC are open-circuit voltage, required power, inner resistor, maximum capacity, current, and state of charge of the battery, respectively. k indicates the motor's current mode and 1 means that the motor is under motor mode, while 0 indicates that the motor is under generation mode.

In summary, there are several static look-up tables with which the fuel rate of the engine, open-circuit voltage of the battery, resistor of the battery, and motor's efficiencies are obtained by former experimental data:

$$
\begin{gathered}
\dot{m}_{\text{fuel}} = \dot{m}_{\text{fuel}} \left(\text{th}_{\text{eng}}, \tau_{\text{eng}} \right), \\
U_{\text{oc}} = U_{\text{oc}} \left(\text{SOC}, \text{temp} \right), \\
R_{\text{bat}} = R_{\text{bat}} \left(k, \text{SOC}, \text{temp} \right), \\
\eta_{\text{mc}} = \eta_{\text{mc}} \left(k, \tau_{\text{mc}}, w_{\text{mc}} \right),
\end{gathered}
\tag{3}
$$

where $\dot{m}_{\text{fuel}}$ and th_{eng} are engine's fuel rate and throttle angle, and temp is the temperature value of the battery pack.

In Figure 2, the required power of the pumps is collected from experiments while the 5-ton excavator is under heavy digging condition (the rock excavation, e.g.). This power chart is considered as the required load in our model; then engine, motor, and battery would be controlled in order to meet the requirements. In Table 1, the components' parameters are listed there.

3. Optimal Control

For every second, $P_{p\_\text{in}}(t)$ is supposed to be known, and with the shaft speed $w_p(t)$ as feedback, the required torque $\tau_p(t)$

TABLE 1: Components' parameters of a PHHE.

Components' parameters	Specific information
Engine	
Type	ZN485Q, turbocharged diesel ICE
Rating power	25 kW at 2200 r/min
Maximum torque	124.8 N·m at 1860 r/min
Motor	
Type	Custom-built PMSM
Rating power	15.7 kW
Maximum torque	150 N·m
Rating voltage	250 V
Voltage range	216~360 V
Speed range	0~4000 r/min
Battery	
Type	NIMH, QNFG 8
Each cell's voltage	1.2 V
Number of cells	200
Energy capacity	8 Ah
Maximum discharge power	19.2 kW
Hydraulic pumps	
Type	2 variable pumps and 1 gear pump
Maximum flow	(2 ∗ 44 + 32) L/min
Working pressure	21 MPa

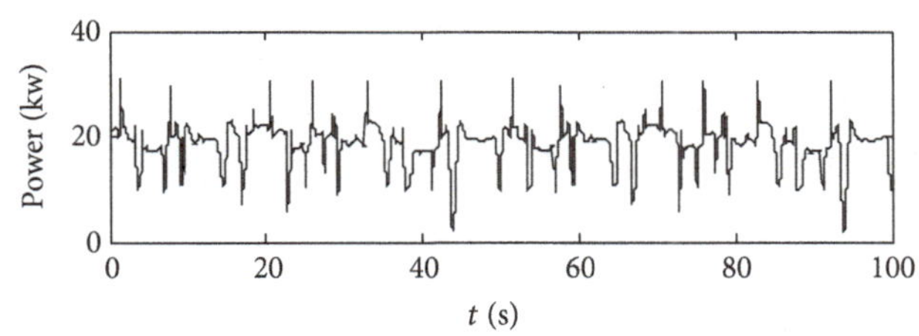

FIGURE 2: The required power at the pumps under a heavy digging cycle.

could be calculated. Thus, in all, these variables are considered as inputs to the power-train system. Then, the optimal control problem could be addressed in a mathematical way like the following:

$$\min \quad J = \sum_{t=0}^{t=\tau} \left(\int_{\tau}^{\tau+1} \dot{m}_{\text{fuel}} \left(\text{th}_{\text{eng}}(t), \tau_{\text{eng}}(t) \right) dt + f\left(\text{SOC}(t) \int_{\tau}^{\tau+1} \dot{m}_{\text{bat}}(t)\, dt \right) \right), \quad (4)$$

$$\dot{m}_{\text{bat}} = \frac{\tau_{\text{mc}} w_{\text{mc}} / \eta_{\text{mc}}^{k}}{\delta} \cdot \vartheta.$$

TABLE 2: Constraints parameter in optimal control formulation.

Constraints parameter	Value
$\text{th}_{\text{eng\_min}} \sim \text{th}_{\text{eng\_max}}$	0.5~1
$\tau_{\text{eng\_min}} \sim \tau_{\text{eng\_max}}$	0~124.8 N·m
$w_{\text{mc\_min}} \sim w_{\text{mc\_max}}$	−4000~4000 r/min
$\tau_{\text{mc\_min}} \sim \tau_{\text{mc\_max}}$	Subject to battery's SOC and motor's current speed
$\text{SOC}_{\text{min}} \sim \text{SOC}_{\text{max}}$	0.2~0.92
ϑ	240 g/kWh
δ	2.78×10^{-7}

Subject to:

$$\begin{aligned}
&\text{th}_{\text{eng_min}} \le \text{th}_{\text{eng}}(t) \le \text{th}_{\text{eng_max}}, \\
&\tau_{\text{eng_min}} \le \tau_{\text{eng}}(t) \le \tau_{\text{eng_max}} \left(w_{\text{eng}}(t) \right), \\
&w_{\text{mc_min}} \le w_{\text{mc}}(t) \le w_{\text{mc_max}}, \\
&\tau_{\text{mc_min}} \left(w_{\text{mc}}(t), \text{SOC}(t) \right) \\
&\qquad \le \tau_{\text{mc}}(t) \le \tau_{\text{mc}_{\text{max}}} \left(w_{\text{mc}}(t), \text{SOC}(t) \right), \\
&\text{SOC}_{\text{min}} \le \text{SOC}(t) \le \text{SOC}_{\text{max}},
\end{aligned} \quad (5)$$

where $\dot{m}_{\text{bat}}$ is equivalent fuel rate of the battery, δ is a gain to convert unit Ws to kWh, and ϑ is the average fuel rate of the engine. $\text{th}_{\text{eng\_min}}$ and $\text{th}_{\text{eng\_max}}$ are minimal and maximal throttle angle of the engine, $\tau_{\text{eng\_min}}$ and $\tau_{\text{eng\_max}}$ are minimal and maximal engine's torque, $w_{\text{mc\_min}}$ and $w_{\text{mc\_max}}$ are speed thresholds for the motor, and $\tau_{\text{mc\_min}}$ and $\tau_{\text{mc\_max}}$ are torque boundaries of the motor which are function of w_{mc} and SOC. SOC_{min} and SOC_{max} are limitations for battery's SOC.

The previous formulation is based on the concept of equivalent fuel consumption, and the objective is to find the global optimum by searching the entire possible space. The engine's fuel rate is indexed by the throttle angle and torque of the engine. For any given required power at each one-second period, engine's different working points will bring about different amounts of fuel consumption. So here we define the throttle angle and torque as the input variables. For a single time interval, the optimum will be found with lowest equivalent fuel consumption. Then moving to the next time interval, so a set of optimal points would be achieved, and for every potential point, it is necessary to check its feasibility so as to meet all the constraints as shown in Table 2.

In objective function, weight is introduced, that is, function of battery's SOC, to maintain SOC in narrow range (see Figure 3). Once SOC is 0.6, the weight is equal to 1, and when SOC increases, the weight would be slightly changed to 0.5. Furthermore, if SOC changes in the opposite direction, our program will finally increase the weight to 2 that would bring about impacts of using battery less and operating engine in high output power level.

As described in Figure 4, the plant block will give the required torque demand to the PSO block (the required

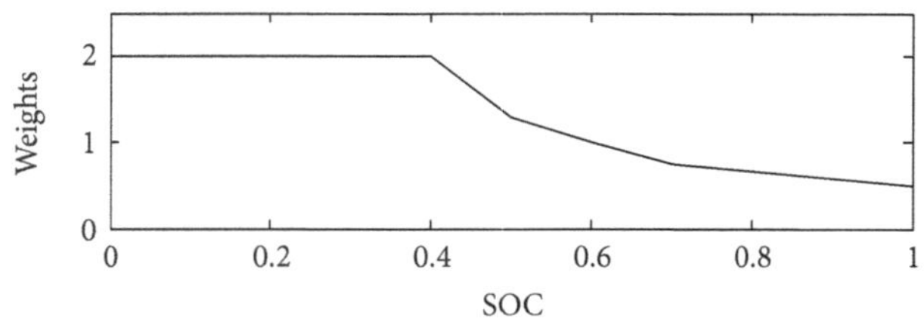

FIGURE 3: SOC weighting factor $f(\mathrm{SOC})$ for optimal control.

power demand at the pumps); then PSO will initialize a group of particles with random positions and velocities that are located in the predefined ranges. Here, our space is defined by engine's throttle angle and torque, and the outputs of PSO block are commands for the engine and motor; then these commands will be delivered to the checking feasibility block, and if only one command in the set does not satisfy the constraints, a relative high value will be assigned to the corresponding fitness function. In other words, these unfeasible points will not been selected to generate next population. After the first generation is totally completed, PSO uses the following two equations to reproduce next points:

$$v_d^i(k+1) = w(k)\, v_d^i(k) + c_1 \cdot \text{rand} \cdot \left(p\text{Best}^i(k) - x_d^i(k)\right) + c_2 \cdot \text{rand} \cdot \left(g\text{Best}(k) - x_d^i(k)\right)$$
$$x_d^i(k+1) = x_d^i(k) + v_d^i(k+1), \tag{6}$$

where i, d, and k represent the number of particles, dimensions of each particle, and iterations, respectively. v and x are the velocity and position of each particle, and $p\text{Best}^i(k)$ indicates after k iterations, the position of one point with best fitness value located in number i particles. So this is a local minimum which is obtained only by one group of particles. $g\text{Best}(k)$ means, after k iterations, the position of one point with best fitness value for all the particles, so this is a global minimum. Obviously, after a relatively large number of iterations $p\text{Best}^i(k)$ and $g\text{Best}(k)$ will approach to the same position.

As listed in Table 3, all the parameters related to the PSO have been defined. Here, the iteration k is a stopping criterion. We need to find out how many generations are needed for each optimization process to finally reach the optimum point. One case of optimization process is then launched in order to observe how the global best fitness value evolves when the number of iteration is increasing. In Figure 5, when iteration arrives at 80, it seems that the improvement will not be easily achieved. So we would state that the optimization process will approach its optimum after 80 iterations, and it is reasonable to set 80 iterations as the stopping criterion in PSO.

After each time interval, the optimum set of engine's throttle angle and torque could be recorded as well as the SOC value at the end of this time interval. So for the next time interval, the previous SOC value will be used as the

TABLE 3: Parameters in PSO.

PSO parameters	Value
$w(k)$	0.9~0.4, with a reduction of 1/120 for each k
c_1	1.4
c_2	1.4
d	2, throttle angle: 0.5~1; engine torque: 0~124.8 N·m
i	20
k	80

initial SOC value. After 100 seconds, we could get a series of optimum points which are optimized for the given working condition in the specific time periods (see Figures 6 and 7). In Figure 6, the optimum throttle angle is plotted, since we get the optimum results in a period of one second, and also this is command signal, so one throttle angle value will last at 1 second and will be changed to another value steeply. In Figure 7, the optimum engine torque is located approximately from 85 to 115 N·m, and the engine will deliver the average required torque to the final load. Then, these two groups of optimum points will be used as control commands in model-based study. Some extensions would then be obtained.

In Figures 8–10, these are results got from using the aforementioned optimum points of the engine. Figure 8 depicts the motor's performance in terms of its real-time torque (black curve), maximum propelling torque (red curve), maximum regenerating torque (yellow curve), and battery's constrains on the motor (pink curve and green curve). So we could get a conclusion that the motor is operated properly in reasonable range that meets all the constraints. In Figure 9, the engine's torque is plotted; since in engine's model, engine's dynamic characteristics (e.g., engine's response time) are taken into consideration, so the engine's real-time torque does have some difference from the engine's torque commands. Also we could calculate the real-time engine's speed which is equal to the motor's speed and pumps' speed using the engine's speed tuning characteristic curve.

In Figures 11 and 12, the comparisons are studied. For these two control strategies, under the same circumstance, that is, the same initial SOC value, same system configuration, same working cycle, and so on, we would compare their results, especially in terms of fuel economy and system efficiency. In Figure 11, compared with rule-based one, a slight fuel economy could be achieved, but not too much, because at that point, the engine is operated constantly with 80% throttle angle which is suitable for that working cycle once using rule-based controller, but the reality is that the working condition for excavators is very bad, and the rule-based one would not fit for all the conditions. In addition, rule-based control will have the drawbacks as follows: (1) it selects SOC as its signal to decide engine's output power level, and compared with load alteration, SOC value is supposed to have some delays; (2) during the transitions of different working cycles, it cannot have high-level time response. In Figure 12, SOC value is increased more if using optimal control, which means that the

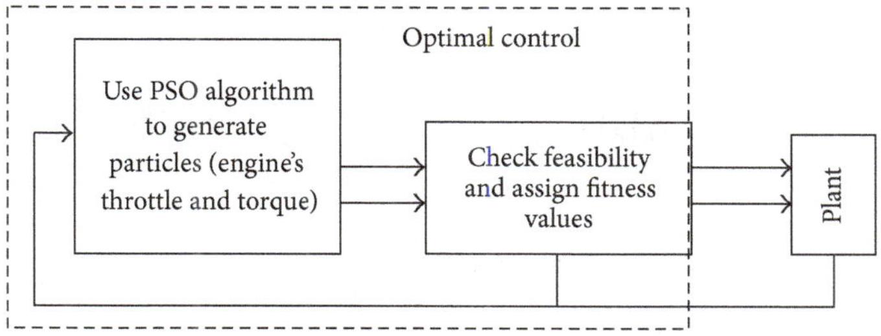

FIGURE 4: Structure of the optimal control.

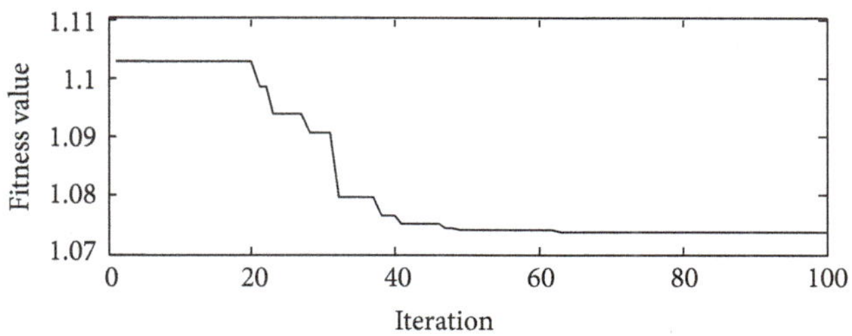

FIGURE 5: Best values for each iteration.

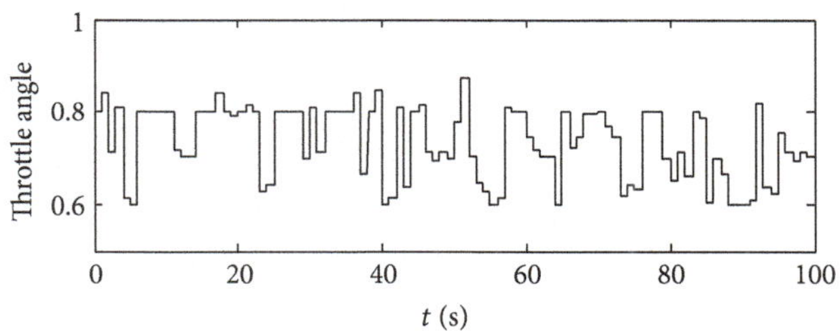

FIGURE 6: A group of optimized throttle angle demands for the engine.

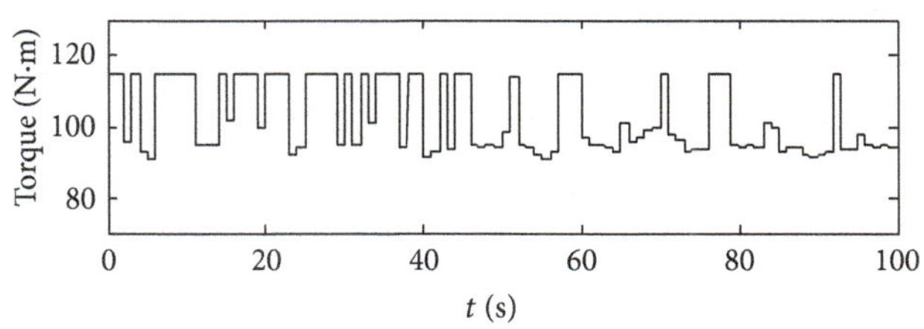

FIGURE 7: A group of optimized torque demands for the engine.

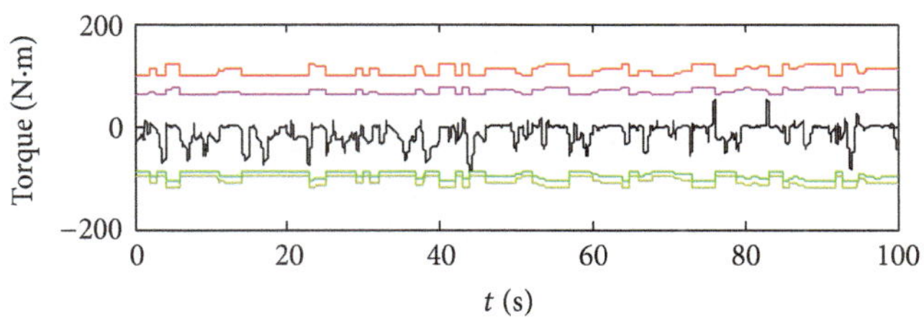

FIGURE 8: The motor's performance using optimum engine's points as commands.

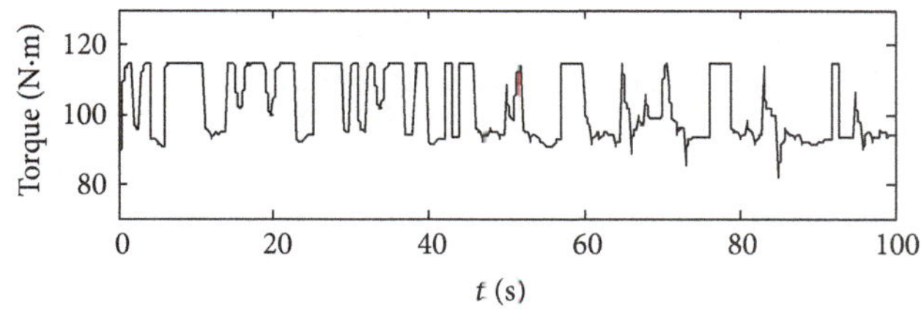

FIGURE 9: The engine's real-time torque along 100-second period.

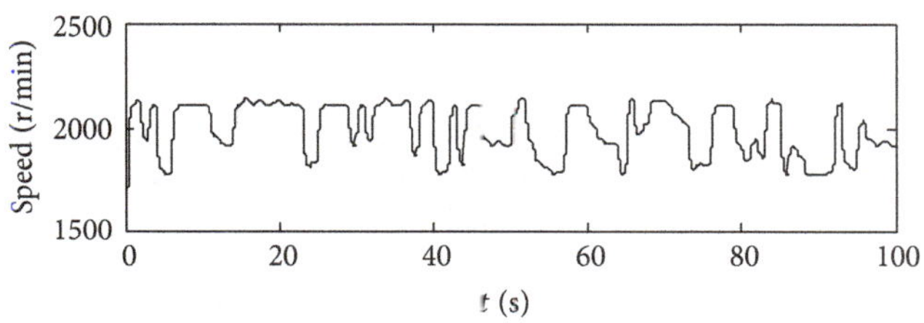

FIGURE 10: The engine's real-time speed along 100-second period.

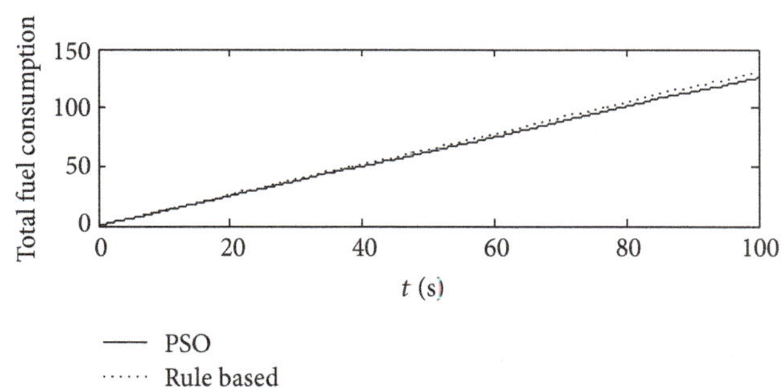

FIGURE 11: The comparisons of total fuel consumption between PSO optimal control and rule-based one.

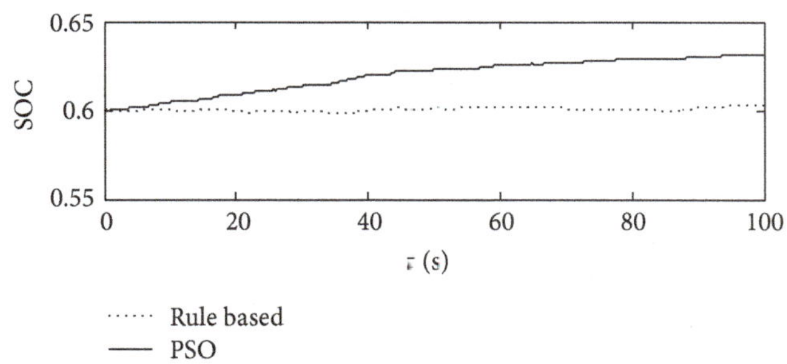

FIGURE 12: The comparisons of SOC alteration between PSO optimal control and rule-based one.

battery is receiving much more energy from the engine, but referring to Figure 5, the throttle angle of the optimal control is obviously lighter than that of the rule-based one, and it indicates that the optimal control decreases the throttle angle for the engine but increases the efficiency of the system.

4. Conclusion

In this paper, the optimal control problem in parallel hybrid hydraulic excavators is first brought forward on the basis of analysis of components' dynamic in power-train and parameters. One may reach a conclusion that higher fuel economy would be achieved easily with hybrid configuration, but cannot imagine how much the system's efficiency and fuel economy would be increased with hybrids or exactly potential room for using this configuration. The optimal control results would give a benchmark in this field, so other control strategies could compare their effects with the previously stated standard one, and it has been clearly demonstrated that parallel hybrid system with optimal control is able to increase the fuel economy compared with rule-based control strategy in all kinds of working cycles. This comparison would give us a right direction in helping redesign components' size and revise control strategies.

The current work is done only under heavy digging working condition, so the results of rule-based control strategy are not that bad, but extending to the blended working conditions, the rule-based one would never change its throttle angle quickly enough, so we will see a large difference between these two strategies. For further research work, there at least exist two aspects which could be studied deeply to update our results. One is to focus on the blended working conditions, trying to find rules on how to adjust the engine's throttle angle to get high fuel economy. Finally, to develop real-time application for PHHE that would become available. The other one is to use the optimal control results as guidance to revise rule-based control strategy accordingly.

Acknowledgment

The authors received the financial support of the National Natural Science Foundation (51205368) and the Natural Science Foundation of Zhejiang Province (Grant no. LQ12E05003).

References

[1] D. Wang, C. Guan, S. Pan, M. Zhang, and X. Lin, "Performance analysis of hydraulic excavator powertrain hybridization," *Automation in Construction*, vol. 18, no. 3, pp. 249–257, 2009.

[2] D. Wang, X. Lin, and Y. Zhang, "Fuzzy logic control for a parallel hybrid hydraulic excavator using genetic algorithm," *Automation in Construction*, vol. 20, no. 5, pp. 581–587, 2011.

[3] X. Lin, S. Pan, and D. Wang, "Dynamic simulation and optimal control strategy for a parallel hybrid hydraulic excavator," *Journal of Zhejiang University Science A*, vol. 9, no. 5, pp. 624–632, 2008.

[4] Q. Xiao, Q. Wang, Y. Zhang, and Q. Fu, "Study on modeling and control strategy of hybrid system in hydraulic excavator," *Journal of Zhejiang University*, vol. 41, no. 3, pp. 480–528, 2007.

[5] Q. Xiao and Q. Wang, "Parameter matching method for hybrid power system of hydraulic excavator," *China Journal of Highway and Transport*, vol. 21, no. 1, pp. 121–126, 2008.

[6] T. Kwon, S. Lee, S. Sul et al., "Power control algorithm for hybrid excavator with supercapacitor," *IEEE Transactions on Industry Applications*, vol. 46, no. 4, pp. 1447–1455, 2010.

[7] D. Wang, X. Lin, and Y. Zhang, "Fuzzy logic control for a parallel hybrid hydraulic excavator using genetic algorithm," *Automation in Construction*, vol. 20, no. 5, pp. 581–587, 2011.

[8] J. Kennedy and R. Eberhart, "Particle swarm optimization," in *Proceedings of the IEEE International Conference on Neural Networks*, pp. 1942–1948, IEEE Service Center, Piscataway, NJ, USA, December 1995.

8

Design and Performance Evaluation of a Rotary Magnetorheological Damper for Unmanned Vehicle Suspension Systems

Jae-Hoon Lee,[1] Changwan Han,[1] Dongsu Ahn,[2] Jin Kyoo Lee,[2] Sang-Hu Park,[1] and Seonghun Park[1]

[1] *School of Mechanical Engineering, Pusan National University, Jangjeon-dong, Geumjeong-gu, Busan 609-732, Republic of Korea*
[2] *Suspension Systems Engineering Team, Doosan Corporation Mottrol, 456-3 Nae-dong, Seongsan-gu, Changwon, Gyeongnam 642-050, Republic of Korea*

Correspondence should be addressed to Seonghun Park; paks@pusan.ac.kr

Academic Editors: U. Lee and A. K. Saha

We designed and validated a rotary magnetorheological (MR) damper with a specified damping torque capacity, an unsaturated magnetic flux density (MFD), and a high magnetic field intensity (MFI) for unmanned vehicle suspension systems. In this study, for the rotary type MR damper to have these satisfactory performances, the roles of the sealing location and the cover case curvature of the MR damper were investigated by using the detailed 3D finite element model to reflect asymmetrical shapes and sealing components. The current study also optimized the damper cover case curvature based on the MFD, the MFI, and the weight of the MR damper components. The damping torques, which were computed using the characteristic equation of the MR fluid and the MFI of the MR damper, were 239.2, 436.95, and 576.78 N·m at currents of 0.5, 1, and 1.5 A, respectively, at a disk rotating speed of 10 RPM. These predicted damping torques satisfied the specified damping torque of 475 N·m at 1.5 A and showed errors of less than 5% when compared to experimental measurements from the MR damper manufactured by the proposed design. The current study could play an important role in improving the performance of rotary type MR dampers.

1. Introduction

Unmanned vehicles for combat can run and petrol automatically to defend against enemy attack and intrusion. Unmanned vehicles can drive at a speed of 50 km/h, although on the off-road, and have approximately a weight of 1 ton with various machine guns and sensors to attack and detect enemies as well as to navigate a path [1, 2]. Therefore, it is crucial that the vehicle body and equipped sensors of unmanned vehicles are under stable operating conditions by using robust suspension systems.

A damper is a major component of a vehicle suspension system that provides comfort and steering controllability of the vehicle [3]. Although electrically controlled suspension systems have been used to improve the dynamic performance of vehicles, such systems are still limited due to discontinuous damping forces, structural complexity, and high cost. To address these limitations, semiactive suspension systems with electronic-rheological (ER) and magneto-rheological (MR) fluids have been widely studied [4–7]. An MR fluid has variable yield shear stresses that depend on magnetic field intensity [8]. This characteristic is used to develop the design and control algorithms for various application systems [9–12]. Moreover, weight reduction and improvements in the performance of mechanical systems that use MR fluids have been studied recently [13–17].

The facts that a rotary MR damper (a rotary damper with an MR fluid) has a simple structure with a fast response speed driven by electronic control circuits [13, 18, 19] and that its low friction and wear increase the running life and stability of other mechanical systems as well as its own can satisfy the damper requirements for unmanned vehicle suspension systems [20–22]. Many previous studies have evaluated the MR damper performance by calculating the magnetic field intensity (MFI) through finite element (FE) analysis [13, 16, 18, 21, 23–30]. However, these studies have focused on reciprocating type MR dampers, not rotary types. Even previous studies regarding rotary type MR dampers have only focused on designing the overall shapes and rectangular cross-section dimensions of MR dampers, using simple 2D or 3D geometric models with a symmetric shape between the upper and lower sides for a rotating axis [10, 13, 21, 29–31].

Unlike related studies in the literature, the present study attempted to investigate the effect of sealing locations and cover case curvatures on the performance of a rotary type MR damper, thus optimizing these design parameters. Moreover, the asymmetric nature of the MR damper used in the study requires that a detailed 3D FE model needs to be generated and that the damping torque needs to be calculated individually from the upper and lower surfaces between the disk and MR fluid. Therefore, the first objective of this study is to investigate the sealing location and the optimized cover case curvature of a rotary type MR damper which could maximize the magnetic field intensity (MFI) as well as to minimize the weight of the MR damper with the magnetic flux density (MFD) lower than the saturation limit. The second objective is to determine if the damping torque capacity of the proposed MR damper design would be suitable for unmanned vehicle suspension systems and further to validate it through comparisons of the analytical predictions and experimental measurements of the damping torque capacity for the proposed optimal design.

2. Design Procedure

The main issues in the design of a rotary MR damper are to reduce the size and weight of its components to allow installation within a limited space and to determine the optimal dimensions of the components such that the MR fluid has the maximum magnetic field strength. Reducing the size and weight of a rotary MR damper decreases the required driving energy of an MR damper-installed system with a corresponding increase in the efficiency of the system. In addition, optimization of the component dimensions contributes to maximizing the damping torque capacity of an MR damper. The proposed rotary MR damper is required to produce a damping torque of more than 475 N·m using 400 coil turns and an input current of 1.5 A at a rotating speed of 10 RPM. Our detailed design procedure is shown in Figure 1.

Using ANSYS commercial finite element (FE) analysis software (V11, ANSYS Inc., USA), we constructed a three dimensional (3D) magnetostatic finite element model based on an initial design of the rotary MR damper [13, 18, 27]. The MFD and MFI values of the damper components were evaluated and compared to the saturation ranges of operation for the component materials. Changes in the design and MFD evaluations using magnetostatic FE analysis were repeated until the MFD values of all the components were shown to be unsaturated.

The damping torque of the proposed rotary MR damper was computed using the characteristic equation of the MR fluid and the MFI of the rotary MR damper. When the MR damper could not achieve the required damping torque capacity, further changes in the design were made. For the final design of the rotary MR damper, the damping torque was experimentally measured to validate the design procedure and the accuracy of our analytical methods.

3. Characteristic Equation of the MR Damper

Damping torque is generated by changes in friction between the rotating disk and MR fluid due to changes in the magnetically dependent viscosity of the MR fluid [8]. The magnetic field of the MR fluid is controlled electronically by the magnitude of the current (which depends on the number of solenoid coil turns) and structurally by the shape and dimensions of the ferromagnetic material around the coil.

To achieve the desired damping torque capacity, the rotary MR damper is composed of single or multiple rotating disks connected to a rotating shaft, coils that generate a magnetic field, a sealed case to prevent the MR fluid from leaking outside by enclosing the coils, and various sealing rings (Figure 2). The case is divided into two parts, a flange and a cover, and can be detached and reattached. Both the case and the rotating disk have design limitations with respect to the external shape, depending on the details of their installation in a system. The design limitations are identified by the red line shown in Figure 2.

The total damping torque (T_{total}) of the rotary MR damper is given by

$$T_{\text{total}} = T_{\text{vis}} + T_{\text{yd}}, \tag{1}$$

where T_{vis} is the torque due to the viscosity of the MR fluid itself without applied current and T_{yd} is the torque due to increased viscosity with the increasing magnetic field strength of the MR fluid by the applied current.

MRF-1400CG (LORD Corporation, USA) was used as the MR fluid with viscosity (η) of 0.250 Pa·s [32]. T_{vis} is proportional to the rotational speed (ω) and the cross-sectional area of the MR fluid and is inversely proportional to the gap (h) between the rotational disk and the cover. T_{vis} can be approximated by [26]

$$T_{\text{vis}} = \pi\eta\frac{\omega}{2h}\left\{(r_i + t_a)^4 - (r_i)^4\right\}, \tag{2}$$

where r_i is the radius between the axis of rotation and the MR fluid and t_a is the radial distance of the MR fluid from the center axis exposed to the magnetic field.

T_{yd} is calculated as follows [9]:

$$T_{\text{yd}} = \frac{2\pi}{3}\tau(H)\left\{(r_i + t_a)^3 - (r_i)^3\right\}, \tag{3}$$

where $\tau(H)$ is the yield shear stress of the MR fluid and is a function of the MFI. $\tau(H)$ is given by the following experimentally derived equation from Carlson [9] and is proportional to the MFI of the gap between the rotational disk and the cover (H_{MR}):

$$\tau(H) = C \times 271{,}700 \times \Phi^{1.5239} \times \tanh\left(6.33 \times 10^{-6} H_{\text{MR}}\right), \tag{4}$$

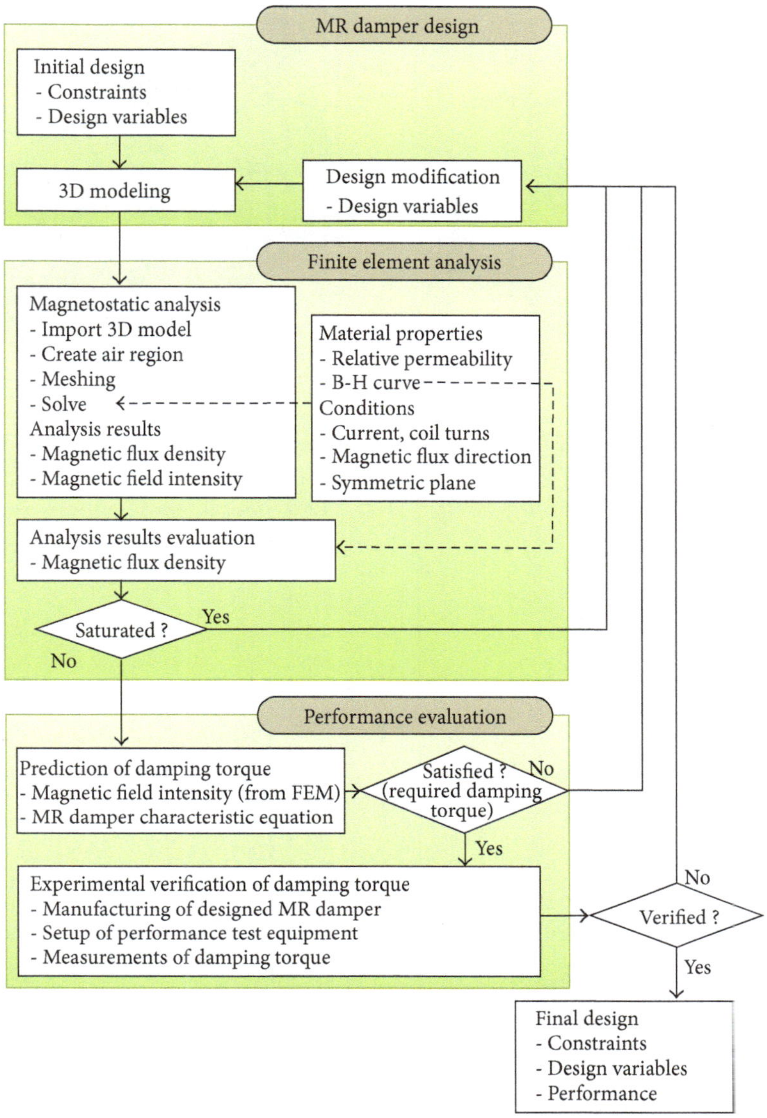

FIGURE 1: Design procedure of the rotary MR damper.

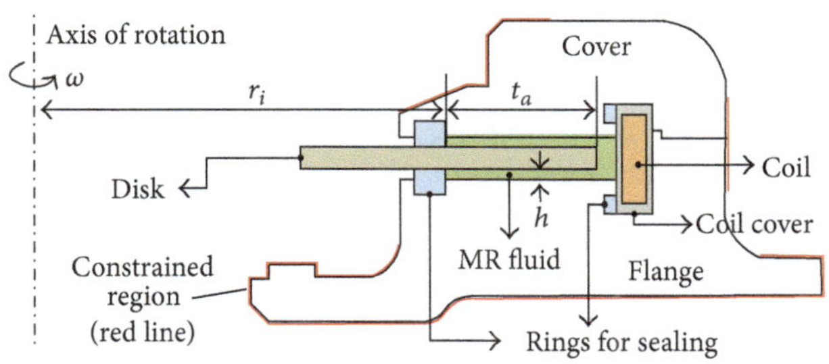

FIGURE 2: Cross-section of the rotary MR damper.

where Φ is the iron particle fraction (0.4) (which is the volume ratio of iron particles to the MR fluid) and C is a coefficient dependent on the carrier fluid of the MR fluid ($C = 1$ in this study because the carrier fluid of MRF-140CG is hydrocarbon).

To design an efficient rotary MR damper with a high damping torque capacity, high MFI and unsaturated MFD values of the damper components and fluid are required.

4. Damping Torque Evaluation by Finite Element Analysis

4.1. Magnetostatic Finite Element Analysis. A 3D FE model of the rotary MR damper was constructed using a one-sixth segment of the structure by considering axial symmetry, as shown in Figure 3(a). Most previous studies used 2D FE models to reduce the time and cost of generating and modifying geometric models as well as performing FE analysis. In this study, however, a 3D model with details of sealing components was generated due to the asymmetric nature of the damper for a vertical rotating axis, because those 2D models of previous studies can be only used when changes in the geometry are ignored in the circumferential direction

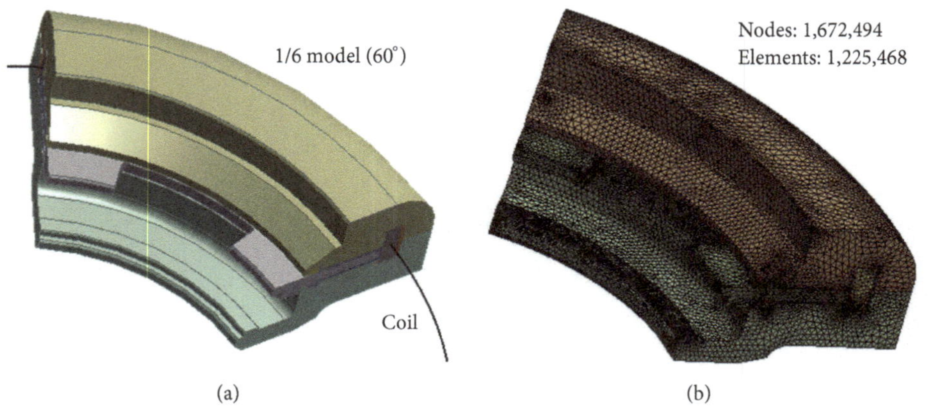

FIGURE 3: 3D model of the rotary MR damper for FE analysis. (a) One-sixth MR damper model. (b) Finite element (FE) model.

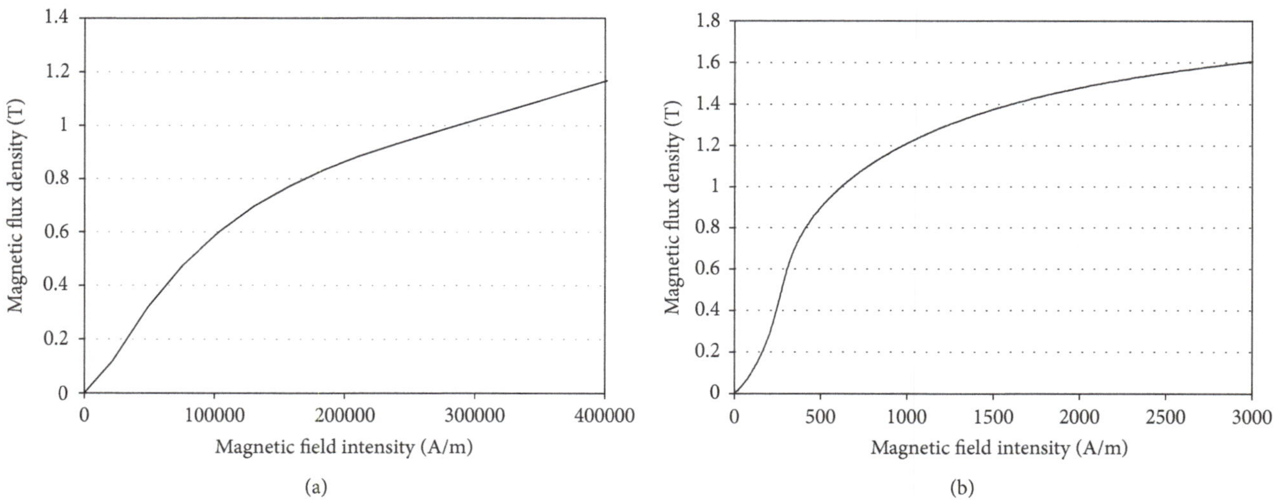

FIGURE 4: B-H curves of MR fluid and ferromagnetic materials. (a) MR Fluid (MRF-140CG). (b) Ferromagnetic material (SS-400).

[13, 33]. The coil turns used to generate electric current were modeled as line elements, and other components were modeled as solid elements. In addition, an external air region surrounding the entire MR damper was generated. The final FE model was constructed using tetrahedral meshes with approximately 120 million second-order elements and 160 million nodes, as shown in Figure 3(b).

The rotary MR damper components can be categorized into two material groups according to their magnetization properties. The rotational disk, flange, and cover are usually made of structural steel that can be easily magnetized by electric current and can be classified as ferromagnetic. Other components of the MR damper are not magnetizable and can be classified as paramagnetic. The relative permeability of paramagnetic materials, such as rubber for the sealing rings and copper for the coils, is close to 1, while that of the ferromagnetic materials is not a constant value and varies nonlinearly with the MFI. The nonlinear relative permeability values of the MR fluid (MRF-140CG) and cover case (SS-400) materials can be calculated from the slopes of the B-H curves (the MFD versus MFI curves) published in the literature (Figure 4) [32]. The relative permeability of ferromagnetic materials is much higher than paramagnetic materials, which contributes to a reduction in the volume of the rotary MR damper while allowing increased magnetic field strength. The relative permeability values of other paramagnetic components, including the surrounding air, are defined as 1.

In the magnetostatic FE analysis, 400 coil turns and an electric current of 1.5 A were modeled in the coil region (Figure 5). The magnetic field that formed in concentric circles around the coil where electric currents were applied was assumed to be parallel to the two symmetric planes (the edges cut diametrically to create a one-sixth model), since the coil direction is perpendicular to the symmetric planes. The direction of the magnetic field was also assumed to be parallel to the external surface of the surrounding air because most of the magnetic field is produced in the ferromagnetic material, not in the air (which is modeled as paramagnetic material). In addition, the air is located far from the coil. The boundary condition that the magnetic field is parallel to the FE model's surface is necessary for the magnetostatic FE analysis to be performed because it prevents the magnetic field from leaking outside the model's surface.

A magnetostatic FE analysis of the initial rotary MR damper design (Model A), as described above, resulted in a maximum MFD value of 1.66 T near the coil case of the

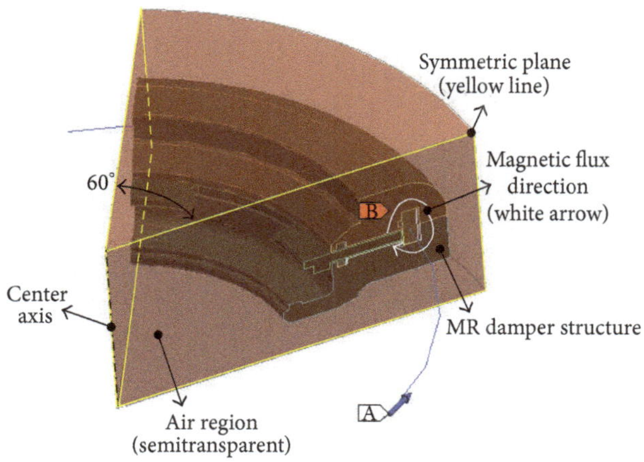

A: current: 1.5 A, 400 turns
B: magnetic flux direction: parallel to both the exterior surface of the air and the symmetric plane

FIGURE 5: Load and boundary conditions.

upper cover case. The sealing area of the flange also showed a high MFD value of 1.62 T (Figure 6(a)). At an applied current of 1.5 A, these MFD values are very close to the saturation limit of SS-400 material (1.6 T), which can cause unstable magnetic behavior of the material. Therefore, the MR damper design needs to be changed to avoid the saturation limit by increasing the cross-sectional area of the ferromagnetic material and thereby deintensifying the MFD. The maximum MFI value was 132,805 mA/mm and occurred in the MR fluid (Figure 6(b)). In this study, the MFI can be used to compute the yield stress of the MR fluid, thus producing the MR damper damping torque capacity. To evaluate damping torque capacity, however, both MFI values of the upper and lower contact areas between the MR fluid and disk should be used, because the MFI value of the upper side is different from that of the lower side due to the asymmetrical shape of the MR damper. In the initial design, the saturated MFD value at the sealing area was overcome by locating the disk and MR fluid positions in the middle of the coil (Figure 7) and by increasing the gap between the MR fluid and the sealing point (panel A in Figure 6(a)). As a result of these changes in Model A, Model B was created.

Although Model B overcame the sealing area saturation problem of the MFD, the MFD value near the outer surface of the MR damper cover case (~1.7 T) was still saturated, as shown in Figure 7. Therefore, the cross-sectional area of this region further changed within the design limitations of the inner surface curvature of the cover case and the heights of the cover and flange; the design limitations were caused by the dimensions of a rotary MR damper installation space. To change the cross-sectional area, the curvature (R) of the upper cover case was selected to be a design parameter, and the corresponding 3D model was constructed (Figure 8). The generated 3D model was cut as 1/72 of the whole model from Model A and simplified by excluding some geometries on disks, sealing parts, and flanges to reduce the calculation time of FE analysis. The parameter R value was changed between 5 mm and 25 mm. When the curvature R is approximately 5 mm, the MFD value of the cover case structure was lowest by overcoming the saturation limit, 1.6 T (Figure 9(a)). Moreover, the average MFI on the MR fluid, which was extracted from the surface contacted with disks, was highest around $R = 16$ mm (Figure 9(b)). However, the MFI of all components increased with increasing curvature values, while the mass of the cover case decreased with increasing curvatures as expected (Figure 9). Based on these results, a value of $R = 15$ mm is determined for the optimal design of the damper cover case curvature, which could give the best performance for the MFD on the cover case and the MFI on the MR fluid as well as a tradeoff between the MFI of all components and the mass of the cover case. Therefore, the outer surface radius of the cover with the saturated MFD was reduced from 25 mm (Model A and Model B) to 15 mm (Model C) in order to increase the cross-section of the saturated region and thereby decrease the MFD.

The magnetostatic FE analysis of Model C, obtained using the same electric current and boundary conditions as Model A and B, exhibited a maximum MFD value of 1.57 T in the middle of the cover and flange (Figure 10(a)). This value was lower than the saturation limit of the SS-400 material (~1.6 T). The maximum MFI of Model C, which is higher than that of Model B, was 157,109 mA/mm. The higher MFI value resulted in a higher damping torque capacity, as shown by (4). The first objective of this study is that the maximum MFD should be lower than the saturation limits of the component materials, and this objective was achieved using Model C. Therefore, we analyzed Model C to determine if the damping torque capacity would be suitable for unmanned vehicle suspension systems, which is the second objective of our study.

4.2. Damping Torque Evaluation. The damping torque capacity of the rotary MR damper was determined from the MFI magnitude on the contact plane between the MR fluid and the disk (this also directly affects the shear stress on the disk's surface). To calculate the damping torque of the MR damper, the MFI values of the upper and lower contact lines between the MR fluid and the disk were extracted from Models B and C (Figure 11) and plotted as a function of the radial distance from the center axis (Figure 12). By changing the outer radius of the cover from $R = 25$ mm (Model B) to $R = 15$ mm (Model C), the average MFI value increased by 22,800 mA/mm. Thus, the damping torque of Model C was expected to be higher than that of Model B.

The MFI values were different between the upper and lower lines although at the same curvature size of the damper cover. Because previous investigators have generally designed MR dampers with symmetrical shapes between the upper and lower sides, both extracted MFI values at the upper and lower lines are equal to each other, and thus either one of the two MFI values calculated at the upper and lower lines can be used. In this study, however, the asymmetry in the MR damper against the horizontal plane led to two different MFI values between the upper and lower lines, and the use

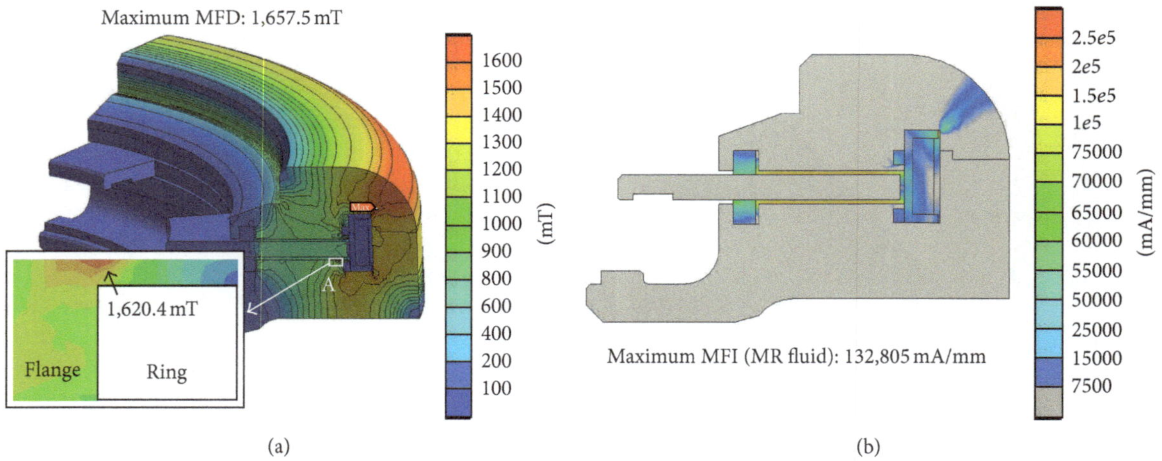

FIGURE 6: FEA results of the rotary MR damper (Model A). (a) Magnetic flux density (MFD). (b) Magnetic field intensity (MFI).

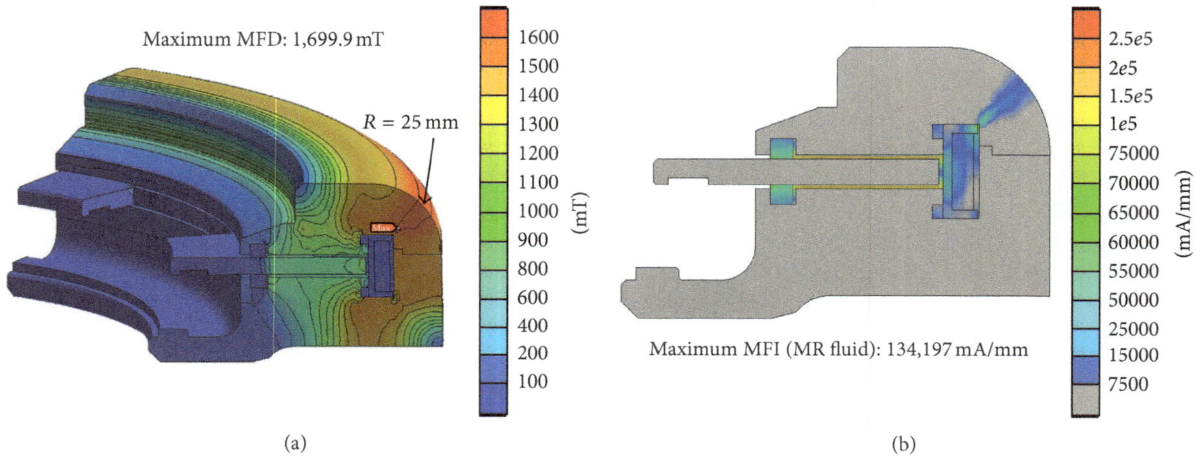

FIGURE 7: FEA results of the rotary MR damper (Model B, $R = 25$ mm). (a) Magnetic flux density (MFD). (b) Magnetic field intensity (MFI).

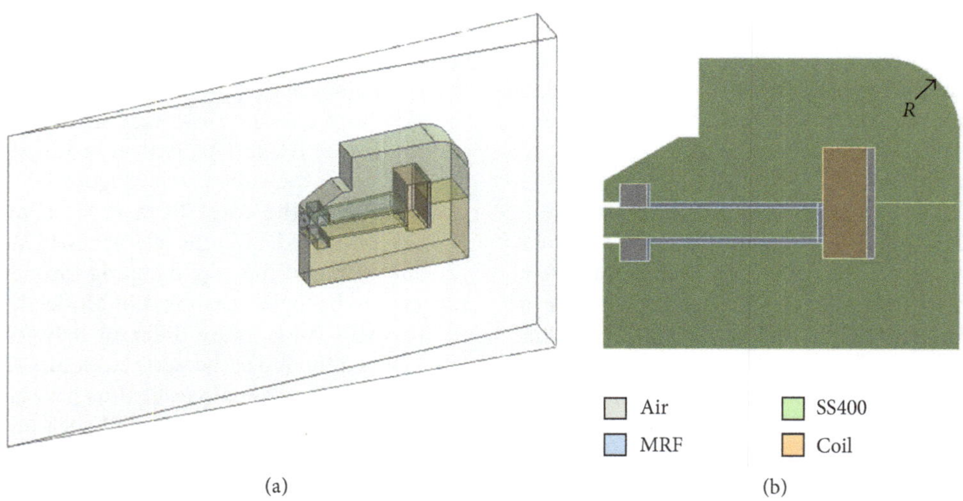

FIGURE 8: 3D model of the rotary MR damper for FE analysis with the curvature, $R = 5\sim25$ mm. (a) Simplified MR damper model (1/72 symmetric model). (b) Cross-section and component materials.

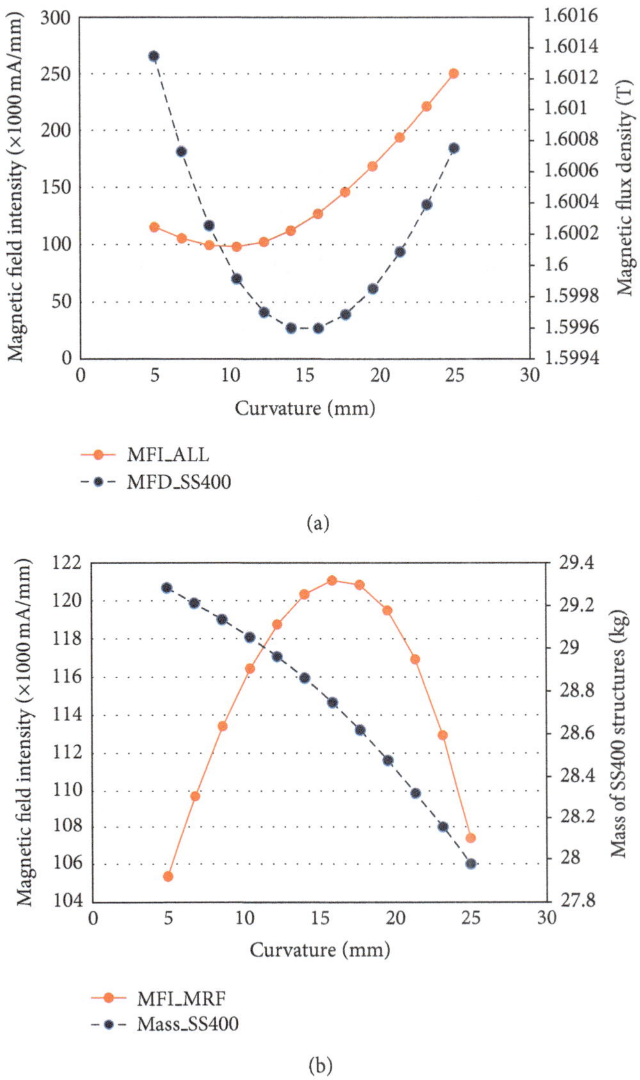

FIGURE 9: FEA results of the rotary MR damper with the curvature, $R = 5{\sim}25$ mm. (a) Maximum magnetic field intensity (MFI) of all components and magnetic flux density (MFD) of the SS400 cover case structure. (b) Average magnetic field intensity (MFI) of the MR fluid (MRF) and mass of the SS400 cover case structure.

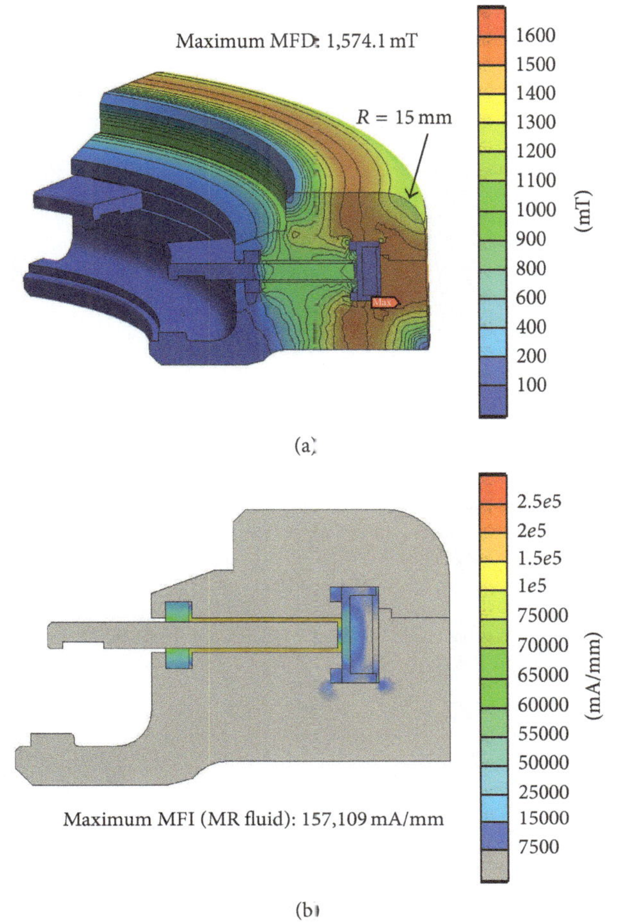

FIGURE 10: FEA results of the rotary MR damper (Model C, $R = 15$ mm). (a) Magnetic flux density (MFD). (b) Magnetic field intensity (MFI).

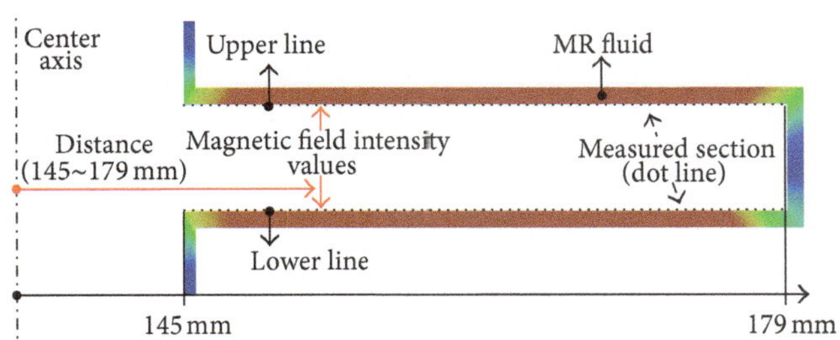

FIGURE 11: Magnetic field intensity (MFI) extracted from the upper and lower contact lines between the MR fluid and the disk.

of both different MFI values enhanced the accuracy of the damping torque evaluation in the present study. There was a tendency for the MFI of the upper to be higher than that of the lower with decreasing damper curvature size as well as both MFI values of the upper and lower to increase with increasing distances from the center axis.

The damping torques of the rotary MR damper calculated using both MFI values of the upper and lower contact region (Figure 12) and the characteristic equation of the MR fluid (1) were 239.2, 436.95, and 576.78 N·m at currents of 0.5, 1, and 1.5 A, respectively, for a disk rotating speed of 10 RPM. The damping torque satisfied the required value of damping torque of 475 N·m at 1.5 A at a rotating speed of 10 RPM, and its magnitude increased with increasing current, although the increment was gradually reduced with increasing current.

5. Experimental Validation of the Proposed Rotary MR Damper

The damping torque was experimentally measured by changing the current using the custom-built testing apparatus shown in Figure 13. This apparatus consisted of a servomotor (HC-RFS153, Mitsubishi, Japan) and a decelerator (PGX142, Dae-sung PTS, Korea) connected in series with a torque sensor (T22, Hottinger Baldwin Messtechnik GmbH, Germany). To measure the damping torque of the rotary MR

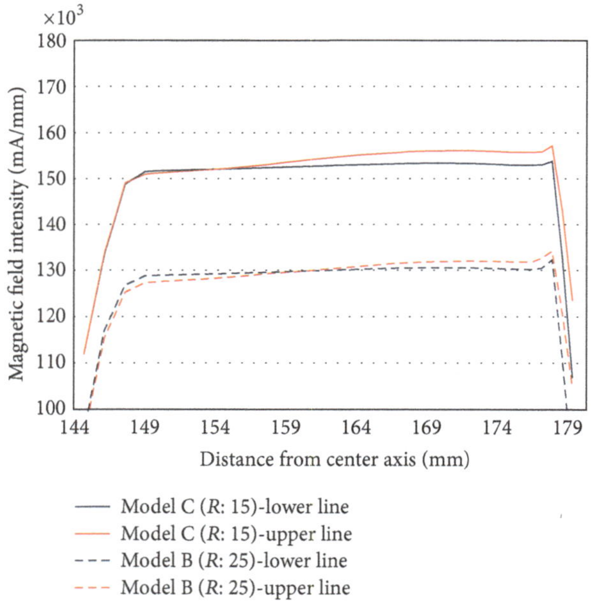

Figure 12: Magnetic field intensity (MFI) as a function of the radial distance from the center axis.

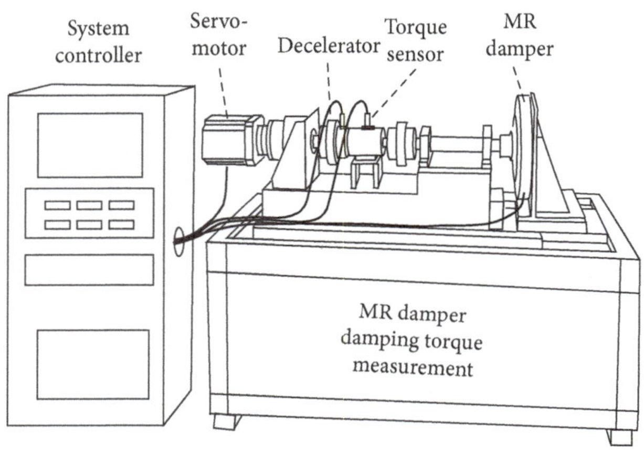

Figure 13: Apparatus for performing damping torque measurement tests.

damper, the servo-motor was controlled at a constant speed of 10 RPM using a motor driver (MR-J2S-200A, Mitsubishi, Japan) and a Programmable Logic Controller (PLC) (FX3u-16 M, Mitsubishi, Japan). A torque sensor installed between the servo-motor and the rotary MR damper was used to measure the resisting torque of the servo-motor to overcome a reduction in motor speed, due to an increase in the viscosity of the MR damper with current supplied by an MR damper controller (TMS320LF2406 DSP, Texas Instruments, USA). The experimentally measured damping torques were 251, 428, and 555 N·m at currents of 0.5, 1, and 1.5 A, respectively, and the values were comparable with the analytically predicted values with errors of 4.70, −2.09, and −3.92% at 0.5, 1, and 1.5 A, respectively (Figure 14).

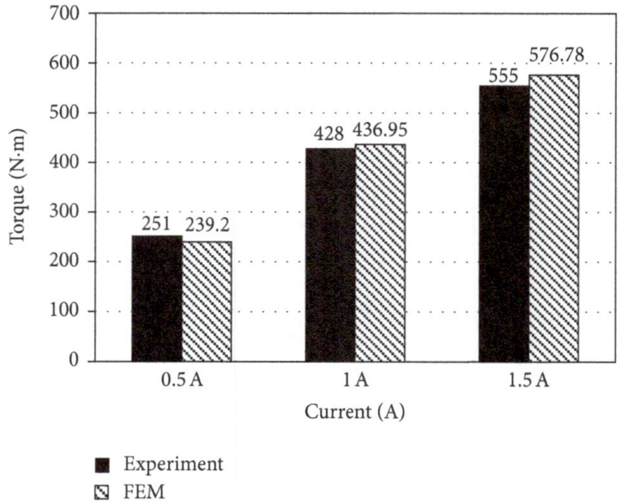

Figure 14: Comparison of the analytical and experimental damping torque values.

6. Conclusions

Using magnetostatic 3D FE analysis and the MR fluid characteristic equation, we successfully designed a rotary MR damper with a specified damping torque capacity and unsaturated magnetic flux density (MFD) for unmanned vehicle suspension systems. The present study revealed the effect of the sealing location and the cover case curvature on the performance of the rotary type MR damper and optimized the curvature of the MR damper cover case. The small cross-sectional area of the sealing region in the flange, due to the narrow gap between the rotation disk and the sealing ring, resulted in the saturated MFD. This saturated sealing region MFD was reduced below the saturation limit of the material by locating the disk and MR fluid positions in the middle of the coil and by increasing the gap between the MR fluid and the sealing point. Moreover, the MFD was very sensitive to the curvature of the MR damper cover case, and the saturated MFD in the radial region between the coil case and the MR damper cover could be addressed by adjusting the curvature of the damper cover case. This curvature was further optimized by considering the MFD, the magnetic field intensity (MFI), and the weight of the MR damper components.

The damping torque of the rotary MR damper was predicted using the MFI of the MR damper and the characteristic equation of the MR fluid, with errors of less than 5% when compared to experimental measurements from the rotary MR damper, which was manufactured by our proposed design, using a custom-made testing apparatus. Here, the accuracy of the analytically predicted damping torque was improved due to the use of both MFI values of the asymmetric upper and lower contact areas between the MR fluid and disk as well as the use of the much detailed 3D MR damper model on FE analysis. The results of the present study could play an important role in the design of rotary

MR dampers by reducing product development time and cost while maintaining a high level of performance.

Acknowledgments

This research was supported by the New & Renewable Energy of the Korea Institute of Energy Technology Evaluation and Planning (KETEP) grant funded by the Korea government Ministry of Knowledge Economy (no. 20113020020010) and by Leading Foreign Research Institute Recruitment Program (no. K20701002274-12E0100-05710) and Basic Science Research Program (2010-0021759) through the National Research Foundation of Korea (NRF) Grant funded by the Ministry of Education, Science and Technology (MEST).

References

[1] J. H. Choi, Y. W. Park, J. Kim, T. S. Choe, and J. B. Song, "Federated-filter-based unmanned ground vehicle localization using 3D range registration with digital elevation model in outdoor environments," *Journal of Field Robotics*, vol. 29, no. 2, pp. 298–314, 2012.

[2] S. H. Joo and J. Lee, "A high-speed autonomous navigation based on real time traversability for 6 × 6 skid vehicle," *Journal of Institute of Control, Robotics and Systems*, vol. 18, no. 3, pp. 251–257, 2012.

[3] I. P. Kang and W. K. Baek, "Roll characteristics evaluation due to the steering of a suv with mr dampers," *Journal of the Korean Society for Power System Engineering*, vol. 13, pp. 26–32, 2009.

[4] S. B. Choi, H. S. Lee, and Y. P. Park, "H∞ control performance of a full-vehicle suspension featuring magnetorheological dampers," *Vehicle System Dynamics*, vol. 38, no. 5, pp. 341–360, 2002.

[5] K. G. Sung and S. B. Choi, "Design and control of a mr shock absorber for electronic control suspension," *Journal of the Korean Society for Precision Engineering*, vol. 28, pp. 31–39, 2011.

[6] S. So and K. Cho, "A study on the variable damping characteristics of the continuous controlled semi-active suspension system and the effect analysis of the vehicle motion performance," *Transaction of the Korean Society of Automotive Engineers*, vol. 7, pp. 190–198, 1999.

[7] A. Turnip, S. Park, and K. S. Hong, "Sensitivity control of a MR-damper semi-active suspension," *International Journal of Precision Engineering and Manufacturing*, vol. 11, no. 2, pp. 209–218, 2010.

[8] M. R. Jolly, J. W. Bender, and J. D. Carlson, "Properties and applications of commercial magnetorheological fluids," *Journal of Intelligent Material Systems and Structures*, vol. 10, no. 1, pp. 5–13, 1999.

[9] D. Carlson, "Magnetorheological fluids," in *Smart Materials*, CRC Press, New York, NY, USA, 2008.

[10] J. D. Carlson, D. M. Catanzarite, and K. A. S. Clair, "Commercial magneto-rheological fluid devices," *International Journal of Modern Physics B*, vol. 10, no. 23-24, pp. 2857–2865, 1996.

[11] J. D. Carlson and M. R. Jolly, "MR fluid, foam and elastomer devices," *Mechatronics*, vol. 10, no. 4, pp. 555–569, 2000.

[12] G. Yang, B. F. Spencer, J. D. Carlson, and M. K. Sain, "Large-scale MR fluid dampers: modeling and dynamic performance considerations," *Engineering Structures*, vol. 24, no. 3, pp. 309–323, 2002.

[13] K. Karakoc, E. J. Park, and A. Suleman, "Design considerations for an automotive magnetorheological brake," *Mechatronics*, vol. 18, no. 8, pp. 434–447, 2008.

[14] Q. H. Nguyen and S. B. Choi, "Optimal design of a vehicle magnetorheological damper considering the damping force and dynamic range," *Smart Materials and Structures*, vol. 18, no. 1, Article ID 015013, 2009.

[15] Q. H. Nguyen, J. C. Jeon, and S. B. Choi, "Optimal design of an hybrid magnetorheological brake for middle-sized motorcycles," *Applied Mechanics and Materials*, vol. 52-54, pp. 371–377, 2011.

[16] H. Gavin, J. Hoagg, and M. Dobossy, "Optimal design of MR dampers," in *Proceedings of the U.S. Japan Workshop on Smart Structures for Improved Seismic Performance in Urban Region*, pp. 225–236, Seattle, Wash, USA, 2001.

[17] W. K. Baek, J. S. Lee, and J. H. Sohn, "Dynamic characteristics modeling for a mr damper using artifical neural network.," *Transaction of the Korean Society of Automotive Engineers*, vol. 12, pp. 170–176, 2004.

[18] N. Takesue, J. Furusho, and Y. Kiyota, "Fast response MR-fluid actuator," *JSME International Journal C*, vol. 47, no. 3, pp. 783–791, 2004.

[19] K. Karakoc, E. J. Park, and A. Suleman, "Improved braking torque generation capacity of an eddy current brake with time varying magnetic fields: a numerical study," *Finite Elements in Analysis and Design*, vol. 59, pp. 66–75, 2012.

[20] O. Ashour, C. A. Rogers, and W. Kordonsky, "Magnetorheological fluids: materials, characterization, and devices," *Journal of Intelligent Material Systems and Structures*, vol. 7, no. 2, pp. 123–130, 1996.

[21] U. Lee, D. Kim, N. Hur, and D. Jeon, "Design analysis and experimental evaluation of an MR fluid clutch," *Journal of Intelligent Material Systems and Structures*, vol. 10, no. 9, pp. 701–707, 2000.

[22] J. An and D. S. Kwon, "Modeling of a magnetorheological actuator including magnetic hysteresis," *Journal of Intelligent Material Systems and Structures*, vol. 14, no. 9, pp. 541–550, 2003.

[23] K. H. Gudmundsson, F. Jonsdottir, and F. Thorsteinsson, "A geometrical optimization of a magneto-rheological rotary brake in a prosthetic knee," *Smart Materials and Structures*, vol. 19, no. 3, Article ID 035023, 2010.

[24] S. Kciuk, R. Turczyn, and M. Kciuk, "Experimental and numerical studies of MR damper with prototype magnetorheological fluid," *Journal of Achievements in Materials and Manufacturing Engineering*, vol. 39, no. 1, article 8, 2010.

[25] W. H. Li, H. Du, and N. Q. Guo, "Finite element analysis and simulation evaluation of a magnetorheological valve," *International Journal of Advanced Manufacturing Technology*, vol. 21, no. 6, pp. 438–445, 2003.

[26] W. H. Li and H. Du, "Design and experimental evaluation of a magnetorheological brake," *International Journal of Advanced Manufacturing Technology*, vol. 21, no. 7, pp. 508–515, 2003.

[27] W. H. El-Aouar, *Finite element analysis based modeling of magneto rheological dampers [M.S. thesis]*, Department of Mechanical Engineering, Virginia Polytechnic Institute and State University, Blacksburg, Va, USA, 2002.

[28] A. Giorgetti, N. Baldanzini, M. Biasiotto, and P. Citti, "Design and testing of a MRF rotational damper for vehicle applications," *Smart Materials and Structures*, vol. 19, no. 6, Article ID 065006, 2010.

[29] E. J. Park, D. Stoikov, L. F. da Luz, and A. Suleman, "A performance evaluation of an automotive magnetorheological brake design with a sliding mode controller," *Mechatronics*, vol. 16, no. 7, pp. 405–416, 2006.

[30] E. J. Park, L. F. da Luz, and A. Suleman, "Multidisciplinary design optimization of an automotive magnetorheological brake design," *Computers and Structures*, vol. 86, no. 3–5, pp. 207–216, 2008.

[31] B. Liu, W. H. Li, P. B. Kosasih, and X. Z. Zhang, "Development of an MR-brake-based haptic device," *Smart Materials and Structures*, vol. 15, no. 6, pp. 1960–1966, 2006.

[32] Lord_Corporation, "Magneto rheological fluid mrf-140cg—lord technical data," Lord Corporation, 2008.

[33] Q. H. Nguyen, Y. M. Han, S. B. Choi, and N. M. Wereley, "Geometry optimization of MR valves constrained in a specific volume using the finite element method," *Smart Materials and Structures*, vol. 16, no. 6, pp. 2242–2252, 2007.

9

Multiobjective Optimization Design of a Fractional Order PID Controller for a Gun Control System

Qiang Gao,[1] Jilin Chen,[1] Li Wang,[1] Shiqing Xu,[2] and Yuanlong Hou[1]

[1] *School of Mechanical Engineering, Nanjing University of Science and Technology, Nanjing 210014, China*
[2] *Research Institute, North Heavy Industries Group Corp. LTD, Inner Mongolia, Baotou 014033, China*

Correspondence should be addressed to Qiang Gao; qgao5914@163.com

Academic Editors: J. P. S. Catalão and P. Lonetti

Motion control of gun barrels is an ongoing topic for the development of gun control equipments possessing excellent performances. In this paper, a typical fractional order PID control strategy is employed for the gun control system. To obtain optimal parameters of the controller, a multiobjective optimization scheme is developed from the loop-shaping perspective. To solve the specified nonlinear optimization problem, a novel Pareto optimal solution based multiobjective differential evolution algorithm is proposed. To enhance the convergent rate of the optimization process, an opposition based learning method is embedded in the chaotic population initialization process. To enhance the robustness of the algorithm for different problems, an adapting scheme of the mutation operation is further employed. With assistance of the evolutionary algorithm, the optimal solution for the specified problem is selected. The numerical simulation results show that the control system can rapidly follow the demand signal with high accuracy and high robustness, demonstrating the efficiency of the proposed controller parameter tuning method.

1. Introduction

Gun control equipments (GCEs) have been extensively believed to be one of the key components of fire control systems (FCSs); the motion robustness and the motion accuracy of the gun barrel are regarded as the two main challenges associated with the developments of GCEs possessing excellent performances [1, 2]. The motion control of gun barrels is an ongoing topic due to certain extremely complicated segments with strong nonlinearities and uncertainties [3–5], such as the time-varying parameters induced by the varying working conditions, the random external applied loads and the complex friction forces between the cannon and the trunnion, and so forth. To eliminate these nonlinearities that induced negative effects, a dominant method is the application of the well-known proportional-integral-derivative (PID) control strategy [3, 4, 6]. However, due to the inherent nonlinearities existing in gun control systems as mentioned above, it is hard for the linear PID control strategy to achieve excellent control behaviors, and consequently the unsuitable PID controller has significantly limited the dynamic performances of the GCEs. Facing this dilemma, a more robustness and efficient control method should be further explored for the gun control systems.

Fractional order PID (FOPID), which was first proposed by Podlubny, is the extended version of conventional integer order PID (IOPID) [7]. FOPID possesses unique characteristics of infinite dimensions, memory effects, low sensitiveness to external disturbances, and so forth, when comparing with IOPID. Moreover, abundant dynamics, high robustness, and fine tracking accuracy of control systems can be obtained when FOPID is applied [8–10]. One of the most tough problems for practical applications of FOPID is the determination of the controller parameters, which highly influences the stability and tracking performances of servosystems. However, there are no universal methods for optimally determining these parameters due to the complexity of fractional order operations [11, 12]. Up to date, various tuning methods have been proposed. For practical applications, these methods could mainly be classified into two sorts, namely, frequency domain based methods [12, 13] and time domain based evolutionary optimization based methods [14–16]. In frequency domain, the loop-shaping based method has recently been proposed and updated by Luo and Chen

[12], Li et al. [17], Luo and Chen [18], and Luo et al. [19]. In this method, the following three items are specified to achieve desired performances of the control system, namely, the gain crossover frequency, the phase margin, and the flat phase constraint at the specified crossover frequency to guarantee system robustness. However, this method could only be implemented on fractional order PI or PD control systems for the reason that the three constraint relations could only be used to determine three parameters. Therefore, this method would not be suitable for typical FOPID control systems which generally possess five unknown parameters to be tuned.

Recently, multiobjective optimization based tuning method has been proposed, and it will be a very promising method for designing optimal controllers, and various design objectives have been employed in both frequency domain and time domain [20, 21]. In this paper, a typical FOPID is employed for motion control of gun control system to enhance tracking accuracy and system robustness. To optimally determine the parameters of FOPID, an evolutionary optimization scheme (EOS) is proposed from loop-shaping perspective. The remainder of this paper could be summarized as follows. In Section 2, the physical model of the gun control system is developed; The basic principles of FOPID and the corresponding parameter determination principle are introduced in Section 3. The improved adaptive multi-objective differential evolution algorithm is then detailed in Section 4. Section 5 conducts the numerical simulation, and the results are carefully discussed. The main conclusions of this paper are drawn in Section 6.

2. Modelling the AC Servosystem for GCEs

The schematic of the AC servosystem utilized in certain sorts of GCEs is presented in Figure 1. Where β_d and β represent the desired angle position and the real angle position of the cannon, respectively. U is the control voltage; K_a is the amplify gain; K_d is the motor torque factor. T_d, T_L, and T_f are the motor torque, load torque disturbance, and friction torque disturbance, respectively. R and L represent the resistance and inductance of the motor armature, circuit, respectively. E_e is the counter-electromotive force (CEMF) of the motor armature and C_e denotes the CEMF coefficient. J is the total moment of inertia to the rotor; B is the viscous friction coefficient; ω_d is the angular velocity of the motor, i is the moderating ratio, and s denotes the Laplace operator.

Generally, the current time constant is much smaller than the mechanical time constant; the delay of the current response can be neglected and it yields

$$\frac{1}{Ls+R} = \frac{1}{R}\frac{1}{Ls/R+1} \approx \frac{1}{R}. \quad (1)$$

The motor torque T_d is given as follows:

$$T_d = -\frac{K_d C_e}{R}\omega_d + \frac{K_d K_a}{R}U. \quad (2)$$

According to the equilibrium equation of the torques, we can obtain

$$T_d - T_L - T_f = Ji\ddot{\beta} + Bi\dot{\beta}. \quad (3)$$

Substituting (2) into (3) yields

$$Ji\ddot{\beta} + Bi\dot{\beta} = -\frac{K_d C_e}{R}\omega_d + \frac{K_d K_a}{R}U - T_L - T_f. \quad (4)$$

When the motor torque and load torque disturbance are ignored, the govern principle of the AC servosystem can be obtained:

$$\ddot{\beta} + \left(\frac{B}{J} + \frac{K_d C_e}{JR}\right)\dot{\beta} = \frac{K_d K_a}{iJR}U. \quad (5)$$

The transfer function of the AC servosystem could be obtained by taking Laplace transformation of (4), which could be given as

$$P(s) = \frac{\beta(s)}{U(s)} = \frac{K_d K_a}{i}\frac{1}{s(JRs + BR + K_d C_e)}. \quad (6)$$

3. The Fractional Order PID Controller

3.1. A Preliminary to FOPID. According to the works of Podlubny, the $PI^{\mu}D^{\lambda}$ controller is introduced, where μ and λ denote the order of an integrator and a differentiator, respectively. The control law of such a controller can be written as [7, 9, 22]

$$\begin{gathered} u(t) = k_p e(t) + k_i D_t^{-\mu} e(t) + k_d D_t^{\lambda} e(t), \\ D_t^{\alpha} e(t) = {}_0 D_t^{\alpha} e(t), \end{gathered} \quad (7)$$

where k_p, k_i, and k_d are proportion, integrator, and differentiator gain, respectively. ${}_{t_0}D_t^{\alpha} f(t)$ is the noninteger order fundamental operator, and it is defined as [8, 23]

$${}_{t_0}D_t^{\alpha} f(t) = \begin{cases} \dfrac{d^{\alpha}}{dt^{\alpha}}, & \Re(\alpha) > 0 \\ 1, & \Re(\alpha) = 0 \\ \displaystyle\int_{t_0}^{t} (d\tau)^{\alpha}, & \Re(\alpha) > 0, \end{cases} \quad (8)$$

where t_0 and t are the limits of the operation, and α is the order.

Generally, there are two common definitions of the operator, known as Grunwald-Letnikov (G-L) definition and Riemann-Liouville (R-L) definition. The G-L definition is commonly utilized to directly conduct numerical computations, which is given as follows [9, 22]:

$${}_{t_0}D_t^{\alpha} f(t) = h^{-\alpha} \sum_{j=0}^{[(t-t_0/h)]} b_j f(t - jh), \quad (9)$$

where $b_0 = 1$, $b_j = [1 - (1+\alpha)/j]b_{j-1}$, h is calculation step. Substituting (9) into (7), the discrete FOPID control law can be obtained:

$$\begin{aligned} u(k) = {} & k_p e(k) + k_i h_s^{-\mu} \sum_{j=0}^{(t/h)} b_j e(k-1) \\ & + k_d h_s^{\lambda} \sum_{j=0}^{(t/h)} q_j e(k-1). \end{aligned} \quad (10)$$

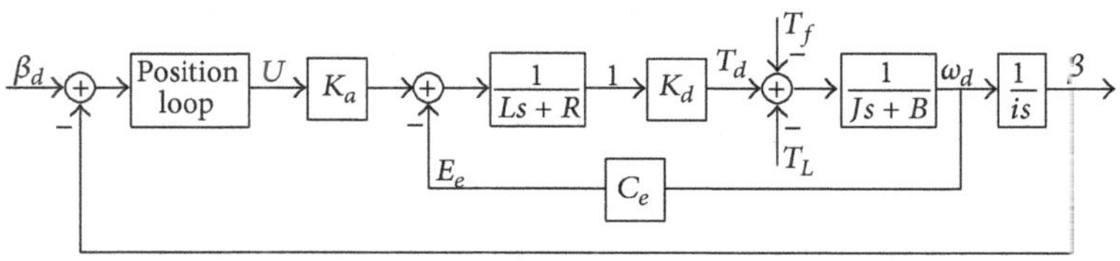

FIGURE 1: Schematic of the AC servosystem of GCEs.

As is discussed above, the FOPID has five parameters ($k_p, k_i, k_d, \mu, \lambda$) to be tuned. This adds more flexibility to controller design, and more dynamics behaviour can be obtained. However, this may simultaneously enhance the complexity in the selection of optimal control parameters.

By taking Laplace transformation of (7), the transfer function of FOPID could be obtained as

$$C(s) = k_p + \frac{k_i}{s^{\mu}} + k_d s^{\lambda},$$
$$C(j\omega) = k_p + \frac{k_i}{(j\omega)^{\mu}} + k_d (j\omega)^{\lambda}. \quad (11)$$

Since $j^{\alpha} = (e^{j(\pi/2)})^{\alpha} = \cos(\alpha\pi/2) + j\sin(\alpha\pi/2)$, the transfer function of FOPID could be rewritten as

$$C(j\omega) = \left[k_p + \frac{k_i}{\omega^{\mu}}\cos\left(\frac{\pi}{2}\mu\right) + k_d\omega^{\lambda}\cos\left(\frac{\pi}{2}\lambda\right)\right] + j\left[k_d\omega^{\lambda}\sin\left(\frac{\pi}{2}\lambda\right) - \frac{k_i}{\omega^{\mu}}\sin\left(\frac{\pi}{2}\mu\right)\right]. \quad (12)$$

The phase and gain of the FOPID could be further given as

$$|C(j\omega)| = \sqrt{\left[k_p + \frac{k_i}{\omega^{\mu}}\cos\left(\frac{\pi}{2}\mu\right) + k_d\omega^{\lambda}\cos\left(\frac{\pi}{2}\lambda\right)\right]^2 + \left[k_d\omega^{\lambda}\sin\left(\frac{\pi}{2}\lambda\right) - \frac{k_i}{\omega^{\mu}}\sin\left(\frac{\pi}{2}\mu\right)\right]^2},$$
$$\operatorname{Arg}[C(j\omega)] = \arctan\frac{k_d\omega^{\lambda}\sin((\pi/2)\lambda) - (k_i/\omega^{\mu})\sin((\pi/2)\mu)}{k_p + (k_i/\omega^{\mu})\cos((\pi/2)\mu) + k_d\omega^{\lambda}\cos((\pi/2)\lambda)}. \quad (13)$$

3.2. Frequency Domain Analysis of FOPID Gun Control System. As for the AC servosystem, the phase and gain of the plant in (6) can be given by

$$P(j\omega) = \frac{K_d K_a}{iJR}\frac{JR}{(j\omega)^2 JR + (BR + K_d C_e)(j\omega)},$$
$$|P(j\omega)| = \frac{K_d K_a}{i\omega}\frac{1}{\sqrt{\omega^2 (JR)^2 - (BR + K_d C_e)}},$$
$$\operatorname{Arg}[P(j\omega)] = -\frac{\pi}{2} + \arctan\frac{(BR + K_d C_e)}{JR\omega}. \quad (14)$$

The transfer function of the open-loop control system can be given as

$$G(s) = P(s)C(s). \quad (15)$$

The gain and phase of the open-loop system can be given as

$$|G(j\omega)| = |P(j\omega)C(j\omega)|,$$
$$\operatorname{Arg}[G(j\omega)] = \operatorname{Arg}[P(j\omega)] + \operatorname{Arg}[C(j\omega)]. \quad (16)$$

3.3. FOPID Design Specifications in Frequency Domain. Here, three specifications to be met by the FOPID controller are applied [12, 17–19]: phase margin specification, robustness to gain variations, and gain crossover frequency specification. To guarantee the robustness and stability of the control system, an extra-constrain, namely, output disturbance rejection capacity, is also employed [11]. The specifications and the constrain will be detailed below.

3.3.1. Phase Margin Specification. Consider

$$\operatorname{Arg}[P(j\omega_c)C(j\omega_c)] = -\pi + \phi_m, \quad (17)$$

where ω_c is the gain crossover frequency interested, and ϕ_m is the phase margin required.

3.3.2. Robustness to Gain Variations. Consider

$$\left.\frac{\mathrm{dArg}\left[P(j\omega)C(j\omega)\right]}{d\omega}\right|_{\omega=\omega_c} = 0. \quad (18)$$

With this condition, the phase Bode plot is flat at the gain crossover frequency. It means that the system is more robust to gain changes, and the overshoots of the response are almost the same.

3.3.3. Gain Crossover Frequency Specification. At the gain crossover frequency point, the amplitude of the open-loop transfer function should be zero,

$$\left|G(j\omega_c)\right|\Big|_{\mathrm{dB}} = \left|P(j\omega_c)C(j\omega_c)\right|\Big|_{\mathrm{dB}} = 0. \quad (19)$$

3.3.4. Output Disturbance Rejection Capacity. A constraint on the sensitivity function S can be defined

$$\left|S(j\omega)\right|\Big|_{\mathrm{dB}} = \left|\frac{1}{1+P(j\omega)C(j\omega)}\right|\Bigg|_{\mathrm{dB}} \le A\mathrm{d}B, \quad \forall\omega \ge \omega_s \ \mathrm{rad/s}, \quad (20)$$

with A the desired value of the sensitivity function for frequencies $\omega \ge \omega_s$ rad/s (desired frequency range).

3.4. Determination of Optimal Parameters. The three specifications may give constrains to parameter selection process; however, the five control parameters could not be optimally determined by (17)–(19). Thus, to solve this problem, a multiobjective optimization scheme is established in this paper to help the parameter determination process. In this scheme, the tuning process could be formulated as follows:

$$\begin{aligned} \min \quad & J_1 = \left|G(j\omega_c)\right|\Big|_{\mathrm{dB}} + \left.\frac{\mathrm{dArg}\left[G(j\omega)\right]}{d\omega}\right|_{\omega=\omega_c} + \mathrm{Arg}\left[G(j\omega_c)\right] + \pi - \phi_m, \\ & J_2 = \left|\max\left(\left|S(j\omega_i)\right|\right)\Big|_{\mathrm{dB}} - A\mathrm{d}B\right|, \quad i = 1,2,\ldots 5, \\ \text{s.t.} \quad & 0 < \mu, \quad \lambda < 1, \quad k_p, k_i, k_d > 0, \end{aligned} \quad (21)$$

where, ω_i denotes selected frequency in the working bandwidth.

As it is evident in (21), it is a nonlinear multiobjective optimization problem. The complexity of this set of nonlinear equations is very significant, especially when fractional orders of differential operations are introduced, and finding out the analytical solution is not trivial. Heuristic optimization seeks good feasible solutions to a set of optimization problems in circumstances where the complexities of the problem or the limited time available for solution do not allow exact solution. It would be suitable to solve the complicated parameter determination problem faced in this paper. Motivated by this, an improved chaotic-self-adaptive multiobjective differential evolutionary algorithm is proposed for this problem, which will be detailed in the next section.

4. Improved Multiobjective Differential Evolutionary Algorithm

Differential evolution (DE) is an efficient evolutionary optimization algorithm motivated by natural selection. Compared with other evolutionary algorithms, DE is a simple yet powerful optimizer with fewer parameters and has much stronger ability in global searching. DE generates new offsprings by forming a noisy replica (trial vector) of each parent individual (target vector) of the population. The population is successfully improved by three basic operators: mutation, crossover, and selection [24–26].

As is discussed above, we shall often face problems with several objectives, and certain of them even contradict each other in that there is no single solution which simultaneously optimizes all functions. Instead, one has a set of optimal solutions. To solve the specified problems, there generally exist two efficient approaches. One is the weighted-sum approach where the multiobjective is converted to the single-objective problem by weighting sum of the objectives. This approach highly depends on decision-maker's preferences [27, 28]. Another approach is based on a set of solutions called Pareto-optimal solutions. This approach would be much more comprehensive and objective. Thus, the Pareto domain concept is embedded in DE to formulate a multiobjective differential evolution algorithm. To enhance robustness and adaptability of the algorithm for different optimization situations, an adapting law of the corresponding control parameters is also proposed.

4.1. Definition of Pareto Dominance. Before discuss the employed MOCDE algorithm, the definition of Pareto dominance which is used to define Pareto-optimal points is firstly given as [29–31].

Definition 1. Let $x = (x_1, x_2, \ldots, x_k)$ and $y = (y_1, y_2, \ldots, y_k)$ be two vectors. Then, x dominates y (denoted by $x \prec y$) if and only if (1) $x_i \le y_i$ for all $i = 1, \ldots, m$; and (2) $x_i < y_i$ for at least one i.

Definition 2. We say that a vector of decision variables $x \in \chi \subset \mathbb{R}^n$ is nondominated with respect to χ, if there does not exist another $x' \in \chi$ such that $f(x') \prec f(x)$.

Definition 3. We say that a vector of decision variables $x^* \in \mathfrak{J} \subset \mathbb{R}^n$ ($\mathfrak{J}$ is the feasible region) is Pareto optimal if it is nondominated with respect to $\mathfrak{J}$.

We thus wish to determine the Pareto-optimal set from the set $\mathfrak{J}$ of all the decision variable vectors that satisfy Definitions 2 and 3.

4.2. Population Initialization. The initial population should better cover the entire search space as much as possible, so it is generated within the search space which is constrained by the prescribed lower and upper parameter bounds using the well known one-dimensional logistic chaotic map. The initial population is set as [24]

$$P^0 = \left(x_1^{0,C}, x_2^{0,C}, x_3^{0,C}, \ldots, x_N^{0,C}\right), \tag{22}$$

where the points $x_i^{0,C}$ are determined by

$$x_{i+1}^{0,C} = 4x_i^{0,C}\left(1 - x_i^{0,C}\right), \quad x_i^{0,C} \in (1, 0). \tag{23}$$

As is known, the initial distribution features of the population will have significant effects on convergent characteristics; namely, an initial population covers area with better solutions will possess faster convergence. Motivated by this, Rahnamayan et al. proposed an *opposition based population initialization (OPI)* method for DE to enhance the convergent speeds [32]. In this paper, we extended this initialization concept for a multiobjective optimization process. The opposition partner of x_i^0 can be first expressed as [32, 33]

$$x_i^{0,OP} = u_i + l_i - x_i^{0,C}, \tag{24}$$

where u_i and l_i denote the upper and lower boundary of the corresponding variable.

By means of Pareto dominance concept, the better individuals of the initial population can be given as

$$\begin{aligned} x_i^0 &= x_i^{0,OP}, \quad \text{if } x_i^{0,OP} \prec x_i^{0,C}, \\ x_i^0 &= x_i^{0,C}, \quad \text{otherwise.} \end{aligned} \tag{25}$$

4.3. Mutation Operation. For each target vector x_i^k at the kth generation, an associated mutant vector $\widehat{X}^k = (\widehat{x}_1^k, \widehat{x}_2^k, \widehat{x}_3^k, \ldots, \widehat{x}_N^k)$ should be generated via certain mutation operators. In the process of mutation operation, the diversity of the population and the convergence rate of DE should be ensured. In conventional DE, the mutation operation will rely on the optimal individual of the last offspring to enhance the convergence rate. However, as for multiobjective optimization, there is no optimal individual, instead there will be a set of better individuals. So, a new mutation operation is proposed for the problem, which will be given as

$$\begin{aligned} \widehat{x}_i^k = {} & x_i^k + F_i^k\left(x_d^k - x_i^k\right) \\ & + \alpha F_i^k\left(x_n^k - x_{r1}^k\right) + (1 - \alpha) F_i^k\left(x_{r2}^k - x_{r3}^k\right), \end{aligned} \tag{26}$$

where x_d^k and x_n^k denote the randomly selected dominant solution and the nondomain solution in the parent populations in the kth generation for the ith individual, respectively. x_{ri}^k denotes the randomly selected individual in the parent populations. α represents the weighting factor, F_i^k denotes a mutation scale factor for the ith individual at the kth generation. In conventional DE, F_i^k is a constant. To enhance the robustness and adaptability of the proposed algorithm, an adaptive scheme is employed for the determination of this factor. Thus, it can be expressed as [34]

$$F_i^k = \begin{cases} F^L + \text{rand}_1\left(F^U - F^L\right), & \text{if } \tau < \text{rand}_2 \\ F_i^{k-1}, & \text{otherwise,} \end{cases} \tag{27}$$

where F^U and F^L denote the preset upper and lower boundary of the factor.

4.4. Crossover Operation. After the mutation phase, the crossover operation is applied to each pair of the generated mutant vector $\widehat{X}^k$ and its corresponding target vector X^k to generate a trial vector [24]:

$$Y^k = \left(y_1^k, y_2^k, y_3^k, \ldots, y_N^k\right), \tag{28}$$

$$y_i^k = \begin{cases} y_i^k = x_i^k, & \text{rand}(\cdot) \le C_R \\ \widehat{x}_i^k, & \text{otherwise,} \end{cases} \tag{29}$$

where rand(·) is a randomly chosen real number in the range (0, 1) and CR is a user-specified crossover factor.

4.5. Selection Operation. Evaluate the candidate Y^k and its parent population X^k. There may exist three different conditions of the evaluation results; namely, the candidates dominate the parents and they are all the better ones. As for the selection operation, if the candidate dominates the parent, the candidate will replaces the parent. If the parent dominates the candidate, then the candidate will be discarded. Otherwise, the candidate is added in the population. If the population has more than PopSize individuals, truncate it according to the Filtrating Strategy proposed in [26].

5. Parameter Tuning Results and Discussion

5.1. Optimal Tuning of Controller Parameters. As for the description of the AC servosystem, the system parameters are chosen as follows: $J = 0.0352\,\text{kg}\cdot\text{m}^2$, $K_d = 0.195\,\text{N}\cdot\text{m/A}$, $C_e = 0.195\,\text{V/(rad·s}^{-1})$, $i = 3.5$, $R = 0.07\,\Omega$, $B = 0.000143\,\text{V/(rad}\cdot\text{s}^{-1})$. As for the optimization process, the population size is chosen as 30; the weighting factor α is chosen as 0.3;

the upper and lower boundary of the mutation factor F^U and F^L are set as 0.9 and 0.3, respectively; and the crossover factor is chosen as 0.3. As for the design specifications of the controller, the interested crossover frequency is set as 4 Hz with respect to practical motions of the gun control servosystem. The required phase margin is set as $\phi_m = \pi/4$; to describe the output disturbance rejection capacity of the control system, five specified frequencies within the range of 0.1 Hz to 4 Hz with constant interval are employed to formulate (21).

Figure 2 illustrates the Pareto-optimal solutions of this specified problem. It is evident that a compromise between the two performance requirements should be made, and the selected solution is highlighted in red color. As shown in Figure 2, the Pareto-optimal solutions have well-distributed spatial positions, demonstrating the efficiency of the employed *Filtrating Strategy* during the selection operation of the evolution process. The obtained optimal parameters of the controller corresponding to the optimal solution which is marked with red color in Figure 2 are $k_p = 0.2246$, $k_d = 0.4652$, $k_i = 0.5681$, $\mu = 0.8948$, $\lambda = 0.2162$.

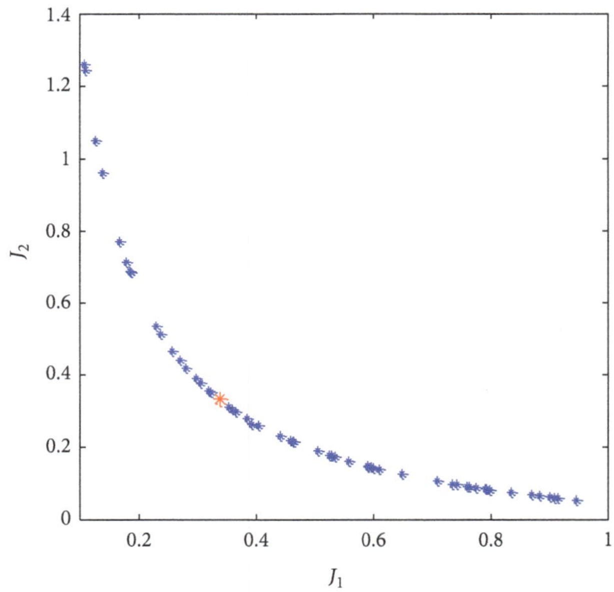

FIGURE 2: The Pareto-optimal solutions.

5.2. Performance Evaluation of the Optimal Control System. Figure 3 illustrates the Bode diagram of the open-loop control system. As is shown in Figure 3, the gain crossover frequency of the system is about 12.69 rad/s with a phase margin of 50.6°. A flat feature can be observed at the gain crossover frequency, and the change rate of the magnitude is about 0.13. It should be noticed that there exists a small shift of the desired performance due to the reason that a compromise has been made between the selected performances. In general, the obtained results in frequency domain validate efficiency of the proposed parameter determination process.

To give a more comprehensive evaluation of the performances of the control system, detailed investigations in the time domain have been conducted. Figure 4(a) illustrates the normal step response of the control system whereas the step value is set as 90 mil. To describe the robustness of the control system, a harmonic external disturbance with 7 mil amplitude and 0.5 Hz frequency is added to the control system; the corresponding positioning error is illustrated in Figure 4(b). From the response in Figure 4(a), the response time is about 0.23 s, and "zero" positioning error is obtained at about 0.88 s. A slight overshoot of about 7.78% is obtained. The results show that the control system can rapidly response to external demands, and fast petitioning criterion can be achieved. As is shown in Figure 4(b), the steady error of the system with external disturbance is about 1.1 mil, which is about 7.86% of the PV value of the disturbance. The results show that external disturbances can be well attenuated by the obtained optimal controller, demonstrating high robustness and strong disturbance rejection capacity of the control system.

To investigate the tracking performances of the control system, a harmonic signal with 90 mil amplitude and 1 Hz frequency is employed as the demand trajectory. To avoid repetition, Figure 5 just illustrates the tracking error of the control system. From the results shown in Figure 5, the tracking error is about 1.4 mil, which is about 0.78% of the full span of the demand signal. The tracking results show that the control system possesses high tracking accuracy, and it can be utilized for the high precision adjustment of the barrel of the gun.

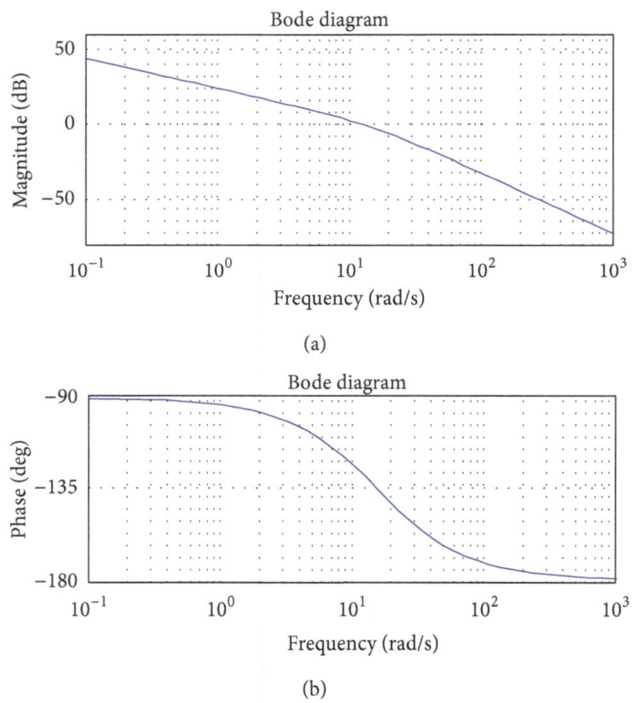

FIGURE 3: Bode diagram of $G(s)$.

6. Conclusion

In this paper, the fractional order PID control strategy is employed for a gun control system. To achieve optimal parameters of the controller, a multiobjective optimization scheme is developed from the loop-shaping perspective. To solve

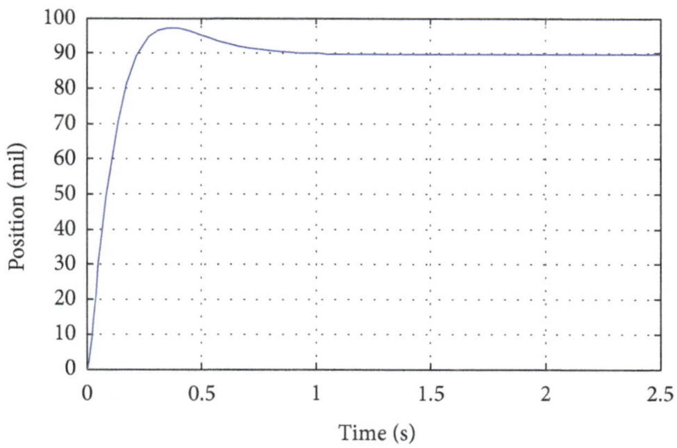

(a) Normal step response of the control system

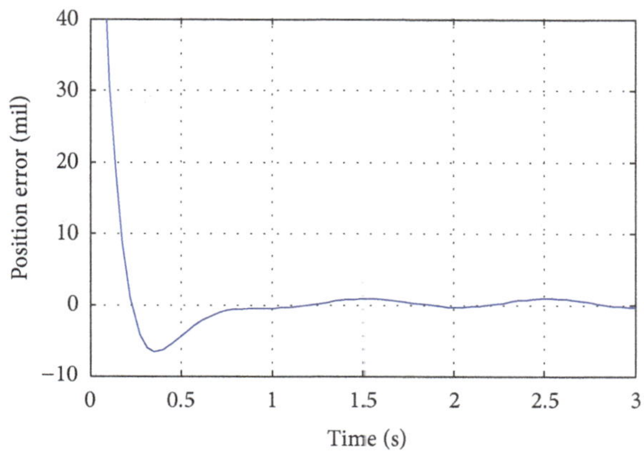

(b) Positioning error with external disturbance

FIGURE 4: Step response of the control system.

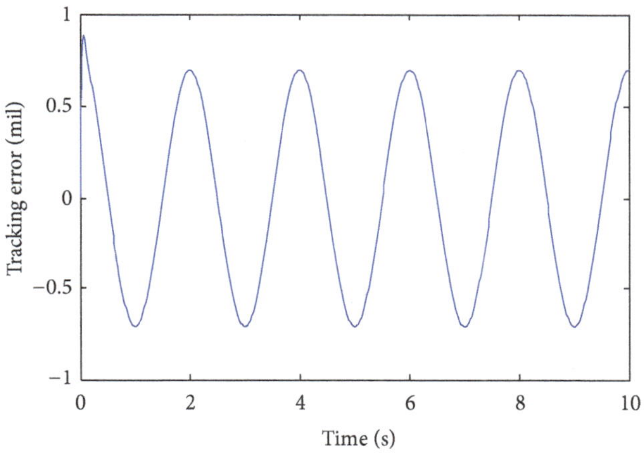

FIGURE 5: Tracking error of the control system.

the nonlinear multiobjective optimization problem, a novel Pareto optimal solution based multiobjective differential evolution algorithm is proposed. To enhance the convergent rate of the optimization process, an opposition based learning method is embedded in the chaotic population initialization process. To enhance the robustness of the algorithm for different problems, an adapting scheme of the mutation operation is further employed.

By means of numerical simulation, the Pareto fronts are obtained, and the corresponding optimal solution for the specified problem is selected. The results in frequency domain show that a compromise is well made between the selected objectives, demonstrating the efficiency of the proposed controller parameter tuning method. As for the investigations in time domain, the step response time of the control system is about 0.23 s and the tracking accuracy of the control system can reach up to 0.78% of the full span. Moreover, a strong attenuation of external disturbances can be obtained. All the results demonstrate that the control system can rapidly follow the demand signal with high accuracy and high robustness, and it will be very promising for engineering practices.

Acknowledgment

This work was supported by National Program on Key Basic Research Project of China (Contract no. 61311603).

References

[1] G. Kumar, P. Y. Tiwari, V. Marcopoli, and M. V. Kothare, "A study of a gun-turret assembly in an armored tank using model predictive control," in *Proceedings of the American Control Conference (ACC '09)*, pp. 4848–4853, June 2009.

[2] Q. Gao, Z. Sun, G. Yang, R. Hou, L. Wang, and Y. Hou, "A novel active disturbance rejection-based control strategy for a gun control system," *Journal of Mechanical Science and Technology*, vol. 26, pp. 4141–4148.

[3] X. Hao, Y. Wei, Z. Wendong, D. Xiaomin, and Y. Haiyang, "The application of dual-PID regulation based on sliding mode control in a tank artillery stabilizer," in *Proceedings of the 4th International Conference on Intelligent Computation Technology and Automation (ICICTA '11)*, pp. 731–733, March 2011.

[4] L. J. Shen, Y. Bao, and J. Cai, "Adaptive control of uncertain gun control system of tank," *Applied Mechanics and Materials*, vol. 88, pp. 88–92, 2011.

[5] L. Feng, X. J. Ma, Z. F. Yan, and H. Li, "Method of adaptive fuzzy sliding mode control of gun control system of tank," *Electric Machines and Control*, vol. 11, no. 1, pp. 65–69, 2007.

[6] C. Wang, Y. Xia, M. Fu, and Z. Zhu, "Application of active disturbance rejection control in tank gun control system," in *Proceedings of the 5th International Conference on Cybernetics and Intelligent Systems (CIS '11)*, pp. 13–18, 2011.

[7] I. Podlubny, "Fractional-order systems and PI/sup/spl lambda// D/sup/spl mu//-controllers," *IEEE Transactions on Automatic Control*, vol. 44, no. 1, pp. 208–214, 1999.

[8] Y. Q. Chen, I. Petráš, and D. Xue, "Fractional order control: a tutorial," in *Proceedings of the American Control Conference (ACC '09)*, pp. 1397–1411, June 2009.

[9] X. Zhou, Z. Zhu, S. Zhao, and D. Luo, "A novel hybrid control strategy for trajectory tracking of fast tool servo," in *Proceedings of the 2nd International Conference on Mechanical and Electronics Engineering (ICMEE '10)*, pp. V2325–V2328, August 2010.

[10] A. J. Calderón, B. M. Vinagre, and V. Feliu, "Fractional order control strategies for power electronic buck converters," *Signal Processing*, vol. 86, no. 10, pp. 2803–2819, 2006.

[11] C. A. Monje, B. M. Vinagre, V. Feliu, and Y. Chen, "Tuning and auto-tuning of fractional order controllers for industry applications," *Control Engineering Practice*, vol. 16, no. 7, pp. 798–812, 2008.

[12] Y. Luo and Y. Chen, "Fractional order [proportional derivative] controller for a class of fractional order systems," *Automatica*, vol. 45, no. 10, pp. 2446–2450, 2009.

[13] D.J. Wang and X.L. Gao, "H_{∞} design with fractional-order PD^{μ} controllers," *Automatica*, vol. 48, no. 5, pp. 974–977, 2012.

[14] Y. Tang, M. Cui, C. Hua, L. Li, and Y. Yang, "Optimum design of fractional order $PI^{\lambda}D^{\mu}$ controller for AVR system using chaotic ant swarm," *Expert Systems with Applications*, vol. 39, no. 8, pp. 6887–6896, 2012.

[15] A. Biswas, S. Das, A. Abraham, and S. Dasgupta, "Design of fractional-order $PI^{\lambda}D^{\mu}$ controllers with an improved differential evolution," *Engineering Applications of Artificial Intelligence*, vol. 22, no. 2, pp. 343–350, 2009.

[16] C. H. Lee and F. K. Chang, "Fractional-order PID controller optimization via improved electromagnetism-like algorithm," *Expert Systems with Applications*, vol. 37, no. 12, pp. 8871–8878, 2010.

[17] H. Li, Y. Luo, and Y. Chen, "A fractional order proportional and derivative (FOPD) motion controller: tuning rule and experiments," *IEEE Transactions on Control Systems Technology*, vol. 18, no. 2, pp. 516–520, 2010.

[18] Y. Luo and Y. Q. Chen, "Stabilizing and robust fractional order PI controller synthesis for first order plus time delay systems," *Automatica*, vol. 48, no. 9, pp. 2159–2167, 2012.

[19] Y. Luo, Y. Chen, and Y. Pi, "Experimental study of fractional order proportional derivative controller synthesis for fractional order systems," *Mechatronics*, vol. 21, no. 1, pp. 204–214, 2011.

[20] A. Hajiloo, N. Nariman-zadeh, and A. Moeini, "Pareto optimal robust design of fractional-order PID controllers for systems with probabilistic uncertainties," *Mechatronics*, vol. 22, no. 6, pp. 788–801, 2012.

[21] M. Zamani, M. Karimi-Ghartemani, N. Sadati, and M. Parniani, "Design of a fractional order PID controller for an AVR using particle swarm optimization," *Control Engineering Practice*, vol. 17, no. 12, pp. 1380–1387, 2009.

[22] Q. Gao, G.-L. Yang, L.-j. Ji, Z. Sun, and R.-M. Hou, *A Semi-Adaptive Fractional Order PID Control Strategy for a Certain Gun Control Equipment*, ASME Press, 2011.

[23] X. Zhou, Z. Zhu, S. Zhao, J. Lin, and J. Dou, "An improved adaptive feedforward cancellation for trajectory tracking of fast tool servo based on fractional calculus," *Procedia Engineering*, pp. 315–320, 2011.

[24] Z. Zhu, X. Zhou, Q. Liu, and S. Zhao, "Multi-objective optimum design of fast tool servo based on improved differential evolution algorithm," *Journal of Mechanical Science and Technology*, vol. 25, pp. 3141–3149, 2011.

[25] S. Das and P. N. Suganthan, "Differential evolution: a survey of the state-of-the-art," *IEEE Transactions on Evolutionary Computation*, vol. 15, no. 1, pp. 4–31, 2011.

[26] B. Alatas, E. Akin, and A. Karci, "MODENAR: multi-objective differential evolution algorithm for mining numeric association rules," *Applied Soft Computing Journal*, vol. 8, no. 1, pp. 646–656, 2008.

[27] H. Huang, P. Zeng, G. Cheng, and L. Ma, "Cross-scale modeling and optimization of a cantilever-type micro/nano valve based on nonlocal stress theory," *Nanoscience and Nanotechnology Letters*, vol. 4, pp. 993–998, 2012.

[28] T. Robic and B. Filipic, "DEMO: differential evolution for multiobjective optimization," in *Proceedings of the 3rd International Conference on Evolutionary MultiCriterion Optimization (EMO '05)*, pp. 520–533, Springer, 2005.

[29] E. Mezura-Montes, M. Reyes-Sierra, and C. A. Coello, "Multi-objective optimization using differential evolution: a survey of the state-of-the-art," *Studies in Computational Intelligence*, vol. 143, pp. 173–196, 2008.

[30] H. A. Abbass, R. Sarker, and C. Newton, "PDE: a Pareto-frontier differential evolution approach for multi-objective optimization problems," in *Proceedings of the Congress on Evolutionary Computation*, pp. 971–978, May 2001.

[31] M. Ali, P. Siarry, and M. Pant, "An efficient differential evolution based algorithm for solving multi-objective optimization problems," *European Journal of Operational Research*, vol. 217, pp. 404–416.

[32] S. Rahnamayan, H. R. Tizhoosh, and M. M. A. Salama, "A novel population initialization method for accelerating evolutionary algorithms," *Computers and Mathematics with Applications*, vol. 53, no. 10, pp. 1605–1614, 2007.

[33] R. S. Rahnamayan, H. R. Tizhoosh, and M. M. A. Salama, "Opposition-based differential evolution," *IEEE Transactions on Evolutionary Computation*, vol. 12, no. 1, pp. 64–79, 2008.

[34] J. Brest, S. Greiner, B. Bošković, M. Mernik, and V. Zumer, "Self-adapting control parameters in differential evolution: a comparative study on numerical benchmark problems," *IEEE Transactions on Evolutionary Computation*, vol. 10, no. 6, pp. 646–657, 2006.

10

Decoupling Suspension Controller Based on Magnetic Flux Feedback

Wenqing Zhang, Jie Li, Kun Zhang, and Peng Cui

College of Mechatronics Engineering and Automation, National University of Defense Technology, Changsha, Hunan 410073, China

Correspondence should be addressed to Wenqing Zhang; zwql97566@163.com

Academic Editors: Q. Dai, C. Hajiyev, and Y. Zhang

The suspension module control system model has been established based on MIMO (multiple input and multiple output) state feedback linearization. We have completed decoupling between double suspension points, and the new decoupling method has been applied to CMS04 magnetic suspension vehicle in national mid-low-speed maglev experiment field of Tangshan city in China. Double suspension system model is very accurate for investigating stability property of maglev control system. When magnetic flux signal is taken back to the suspension control system, the suspension module's antijamming capacity for resisting suspension load variety has been proved. Also, the external force interference has been enhanced. As a result, the robustness and stability properties of double-electromagnet suspension control system have been enhanced.

1. Introduction

Maglev vehicle has minor noise, no pollution safety, and comfort many other advantages, so the maglev traffic has a bright prospect in the future. Maglev vehicle concludes several complex subsystems [1] with coupling relations. The vehicle's suspension and guidance depend on magnetic force, and its traction relies on linear motion actuator. According to the production method of magnetic force, maglev system can be divided into two types: EDS and EMS. EMS can make vehicle suspend on the guideway keeping a certain gap, but suspension system in itself is not stable, so it needs active control. Maglev control technology is one of crucial techniques in suspension system. Maglev vehicle is a typical nonlinear system, and suspension module concludes two coupled suspension points which are connected by a rigid joint. The coupling problem is not obvious on straight guideway, but when maglev vehicle arrives at the vertical curve and the curved guideway, the coupling problem can make maglev system unstable. So how to work out this problem is an imperative thing for maglev engineers. Lin et al. have studied the design of a robust discrete-time sliding mode control (DT-SMC) [2] for a high-precision electrohydraulic actuator (EHA) recently. Ngoc [3] first proved an explicit criterion for positive linear time-varying differential systems with distributed delay. A new approach can keep stabilization of time-varying nonlinear systems with distributed input delay by feedback of plant's state in paper [4]. The stability bound [5] of the singular perturbation parameter can be obtained by solving a set of matrix inequalities. Control redundancies are proposed [6], for example, the controllable redundancy, observable redundancy, stabilization redundancy, and detectable redundancy. The fuzzy sliding mode control approach [7] can be applied to deal with the output tracking of the uncertain system. Nonlinear value control based on FDM (feedback domination method) is simple and has robustness [8] to some kinds of unknown disturbances. The nonlinear feedback principle is developed [9] using the direct-axis and the quadrature-axis stator voltage as manipulated variables. Roy and Asada have investigated nonlinear feedback control of a gravity assisted under actuated manipulator [10] with application to aircraft assembly. A simple methodology [11] to evaluate harmonic distortion in the frequency domain for circuits and systems made up of a nonlinear high-gain path with a nonlinear feedback network is presented. Think that suspension force is a single function about magnetic flux density, and the relation between suspension force and magnetic density's square is linear [12]. So we take magnetic flux density back to maglev system for promoting the robustness and stability. CMS04 maglev vehicle designed by NUDT (National University of Defense Technology) is showed in Figure 1.

2. Foundation of Flux Feedback Control Suspension Model

EMS middle- and low-speed maglev vehicle applies modular suspension bogies, and one module concludes two suspension points. Double suspension points' suspension task is accomplished by one suspension controller.

Assume the following:

(1) the leakage flux of magnetic winding is neglected;

(2) the magnetic resistances of the ferrite core and rail are neglected; namely, magnetic potential falls on air gap $z(t)$ evenly;

(3) the inclination angle of magnetic rigid body is minuscule, and active point of magnetic force is considered invariable;

(4) the action line of load forces is superposed with direction of suspension gap measured;

(5) distribution of mass of the bracket is even, and the masses of two magnets are equal, so the action point of gravity is superposed on the geometry center O.

Based on above assumptions, the force analysis and geometrical relationship of maglev system are showed in Figure 2.

Signs of Figure 2 are illuminated in Table 1.

2.1. Mathematical Model

2.1.1. Kinematics Equations. When electromagnet module is in condition of balance, we can obtain the equation $\sum F_i = 0$ and $\sum M_i = 0$. The kinematics equations of maglev system can be got as follows:

$$\frac{1}{2}M \cdot (\ddot{z}_1 + \ddot{z}_2) = Mg - F_1 - F_2 + f_1 + f_2,$$
$$\frac{I}{L} \cdot (\ddot{z}_1 - \ddot{z}_2) = f_1 \cdot \frac{1}{2}l - f_2 \cdot \frac{1}{2}l - \frac{1}{4}F_1 \cdot l + \frac{1}{4}F_2 \cdot l. \quad (1)$$

Extract acceleration signals from (1):

$$\ddot{z}_1 = \frac{-2L - 3l}{2ML} \cdot F_1 + \frac{-2L + 3l}{2ML} \cdot F_2 + \frac{L + 3l}{ML} \cdot f_1 + \frac{L - 3l}{ML} \cdot f_2 + g,$$
$$\ddot{z}_2 = \frac{-2L + 3l}{2ML} \cdot F_1 + \frac{-2L - 3l}{2ML} \cdot F_2 + \frac{L - 3l}{ML} \cdot f_1 + \frac{L + 3l}{ML} \cdot f_2 + g. \quad (2)$$

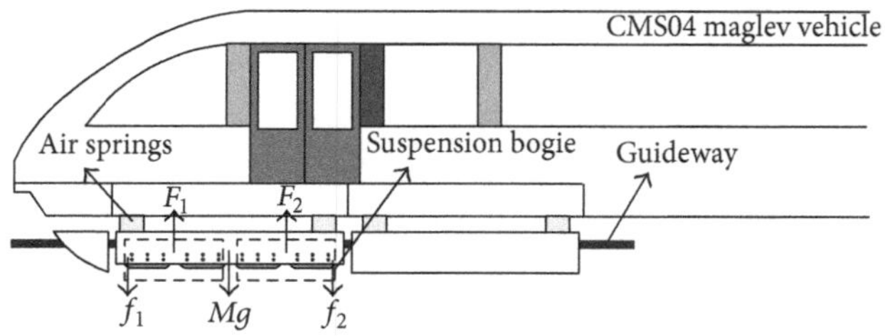

FIGURE 1: CMS04 mid-low-speed maglev vehicle.

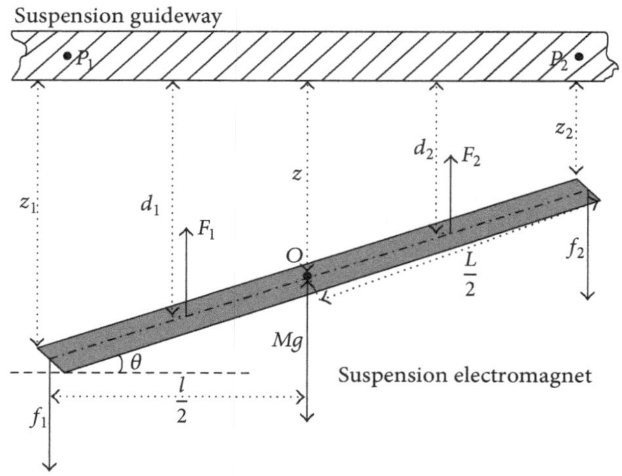

FIGURE 2: Suspension system's force analysis map.

Set $A = (-2L - 3l)/2ML$, $B = (-2L + 3l)/2ML$, $C = (L + 3l)/ML$, and $D = (L - 3l)/ML$.

Simplify (2):

$$\ddot{z}_1 = A \cdot F_1 + B \cdot F_2 + C \cdot f_1 + D \cdot f_2 + g,$$
$$\ddot{z}_2 = B \cdot F_1 + A \cdot F_2 + D \cdot f_1 + C \cdot f_2 + g. \quad (3)$$

2.1.2. Electromagnetics Equations. We can get geometrical relationship of parameters in Figure 2. The relation between measurement positions z_1, z_2 and real physical positions d_1, d_2 is

$$d_1 = \frac{1}{2}(z_1 + z) = \frac{3}{4}z_1 + \frac{1}{4}z_2,$$
$$d_2 = \frac{1}{2}(z_2 + z) = \frac{1}{4}z_1 + \frac{3}{4}z_2. \quad (4)$$

Magnetic force equations are as follows:

$$F_1 = \frac{B_1^2 S}{\mu_0},$$
$$F_2 = \frac{B_2^2 S}{\mu_0}. \quad (5)$$

TABLE 1: Symbols in double suspension points system.

Signs	Signification
N	Number of magnetic coils
R	Magnetic resistance
μ_0	Magnetic permeability of atmosphere $4\pi \times 10^{-7}$ H/m
g	Acceleration of gravitation 9.8 m/s^2
θ	Separation angle between electromagnet axis line and horizon
F_1	Electromagnetic force of suspension point P_1
N_1	Disturbing force of suspension point P_1
L	Total length of electromagnet
$i_1(t)$	Current of magnetic coil in suspension P_1
$u_1(t)$	Voltage of magnetic coil in suspension point P_1
$z_1(t)$	Measured position value in suspension point P_1
$d_1(t)$	Average gap value between guideway and suspension point P_1
S	Magnetic pole area
B	Suspension gap flux density
M	Total mass of suspension system
I	Rotation inertia of electromagnet for O point
d	Distance from guideway to O point
F_2	Electromagnetic force of suspension point P_2
N_2	Disturbing force of suspension point P_2
l	Disturbing force of arm relative to O point
$i_2(t)$	Current of magnetic coil in suspension point P_2
$u_2(t)$	Voltage of magnetic coil in suspension point P_2
$z_2(t)$	Measured position value in suspension point P_2
$d_2(t)$	Average gap value between guideway and suspension point P_2

2.1.3. Electricity Equations. Voltage balance equations of electromagnet winding are denoted in (6):

$$u_1 = \frac{2Rd_1}{\mu_0 N} B_1 + NS\dot{B}_1,$$
$$u_2 = \frac{2Rd_2}{\mu_0 N} B_2 + NS\dot{B}_2. \tag{6}$$

Substituting (4) into (6), we get

$$u_1 = \frac{R(3z_1 + z_2)}{2\mu_0 N} \cdot B_1 + NS\dot{B}_1,$$
$$u_2 = \frac{R(z_1 + 3z_2)}{2\mu_0 N} \cdot B_2 + NS\dot{B}_2. \tag{7}$$

From (7), we obtain

$$\dot{B}_1 = \varphi_1 = -\frac{R(3z_1 + z_2)}{2\mu_0 SN^2} \cdot B_1 + \frac{1}{NS} \cdot u_1,$$
$$\dot{B}_2 = \varphi_2 = -\frac{R(z_1 + 3z_2)}{2\mu_0 SN^2} \cdot B_2 + \frac{1}{NS} \cdot u_2. \tag{8}$$

On all accounts, the dynamic law of maglev system can be determined by the following equations:

$$\ddot{z}_1 = A \cdot \frac{B_1^2 S}{\mu_0} + B \cdot \frac{B_2^2 S}{\mu_0} + C \cdot f_1 + D \cdot f_2 + g,$$
$$\ddot{z}_2 = B \cdot \frac{B_1^2 S}{\mu_0} + A \cdot \frac{B_2^2 S}{\mu_0} + D \cdot f_1 + C \cdot f_2 + g,$$
$$\dot{B}_1 = \varphi_{i1} = -\frac{R(3z_1 - z_2)}{2\mu_0 SN^2} \cdot B_1 + \frac{1}{NS} \cdot u_1,$$
$$\dot{B}_2 = \varphi_{i2} = -\frac{R(z_1 + 3z_2)}{2\mu_0 SN^2} \cdot B_2 + \frac{1}{NS} \cdot u_2. \tag{9}$$

2.2. State-Space Equations. Choose $x = [x_1, x_2, x_3, x_4, x_5, x_6]^T = [z_1, \dot{z}_1, z_2, \dot{z}_2, B_1, B_2]^T$ as state variable vectors, and the state-space model of maglev system is given as follows:

$$\dot{x}_1 = x_2,$$
$$\dot{x}_2 = A \cdot \frac{x_5^2 S}{\mu_0} + B \cdot \frac{x_6^2 S}{\mu_0} + C \cdot f_1 + D \cdot f_2 + g,$$
$$\dot{x}_3 = x_4,$$
$$\dot{x}_4 = B \cdot \frac{x_5^2 S}{\mu_0} + A \cdot \frac{x_6^2 S}{\mu_0} + D \cdot f_1 + C \cdot f_2 + g,$$
$$\dot{x}_5 = -\frac{R \cdot (3x_1 + x_3)}{2\mu_0 SN^2} x_5 + \frac{1}{NS} u_1,$$
$$\dot{x}_6 = -\frac{R \cdot (x_1 + 3x_3)}{2\mu_0 SN^2} x_6 + \frac{1}{NS} u_2. \tag{10}$$

Simplify (10):

$$\dot{x} = f(x) + g(x) \cdot u,$$
$$y = h(x), \tag{11}$$

where

$$f(x) = \begin{bmatrix} x_2 \\ A \cdot \frac{x_5^2 S}{\mu_0} + B \cdot \frac{x_6^2 S}{\mu_0} + C \cdot f_1 + D \cdot f_2 + g \\ x_4 \\ B \cdot \frac{x_5^2 S}{\mu_0} + A \cdot \frac{x_6^2 S}{\mu_0} + D \cdot f_1 + C \cdot f_2 + g \\ -\frac{R \cdot (3x_1 + x_3)}{2\mu_0 S N^2} x_5 \\ -\frac{R \cdot (x_1 + 3x_3)}{2\mu_0 S N^2} x_6 \end{bmatrix}, \tag{12}$$

$$g(x) = \begin{bmatrix} g_1(x) & g_2(x) \end{bmatrix} = \begin{bmatrix} 0 & 0 \\ 0 & 0 \\ 0 & 0 \\ 0 & 0 \\ \frac{1}{NS} & 0 \\ 0 & \frac{1}{NS} \end{bmatrix},$$

$$h(x) = \begin{bmatrix} h_1(x) \\ h_2(x) \end{bmatrix} = \begin{bmatrix} x_1 \\ x_3 \end{bmatrix}, \qquad u = \begin{bmatrix} u_1 \\ u_2 \end{bmatrix}.$$

Theorem 1. *MIMO affine nonlinear system is described as*

$$\dot{x} = f(x) + \sum_{i=1}^{m} g_i(x) u_i, \quad y = h(x). \tag{13}$$

Set the equilibrium point $x_0 \in X$. The problem of MIMO nonlinear system feedback linearization will have a solution, if and only if a neighborhood U near x_0 and m functions $h_1(x), \ldots, h_m(x)$ exists. The solution makes relative degree $(r_1, \ldots, r_m)$ of nonlinear system (11) satisfy $r_1 + r_2 + \cdots + r_m = n$, where n denotes the dimension of system and m stands for input dimension.

2.3. Demonstration of Definition 1. Define the equilibrium point of maglev system $x_0 = [z_{10}, \dot{z}_{10}, z_{20}, \dot{z}_{20}, i_{10}, i_{20}]^T$, and suppose state variables $x_1, x_2, x_5, x_6 \in R_+$, according to real maglev system construction. Compute the vector field as follows, by $f(x)$ and $g(x)$:

$$\begin{aligned} &L_{g_1} L_f^0 h_1(x) = 0, && L_{g_2} L_f^0 h_1(x) = 0, \\ &L_{g_1} L_f h_1(x) = 0, && L_{g_2} L_f h_1(x) = 0, \\ &L_{g_1} L_f^2 h_1(x) = \frac{2A \cdot x_5}{\mu_0 N} \neq 0, && L_{g_2} L_f^2 h_1(x) = \frac{2B \cdot x_6}{\mu_0 N} \neq 0, \\ &L_{g_1} L_f^0 h_2(x) = 0, && L_{g_2} L_f^0 h_2(x) = 0, \\ &L_{g_1} L_f h_2(x) = 0, && L_{g_2} L_f h_2(x) = 0, \\ &L_{g_1} L_f^2 h_2(x) = \frac{2B \cdot x_5}{\mu_0 N} \neq 0, && L_{g_2} L_f^2 h_2(x) = \frac{2A \cdot x_6}{\mu_0 N} \neq 0. \end{aligned} \tag{14}$$

Namely, above equations satisfy

$$\begin{aligned} &L_{g_j} L_f^k h_i(x) = 0, \quad \forall x \in V,\ 1 \le j \le m,\ 1 \le i \le m, \\ &\qquad\qquad 0 \le k \le r_i - 2, \\ &L_{g_j} L_f^{r_i} h_i(x) \neq 0, \end{aligned} \tag{15}$$

where

$$H(x) = \begin{bmatrix} L_{g_1} L_f^{r_1 - 1} h_1(x) & L_{g_2} L_f^{r_1 - 1} h_1(x) \\ L_{g_1} L_f^{r_2 - 1} h_2(x) & L_{g_2} L_f^{r_2 - 1} h_2(x) \end{bmatrix} = \begin{bmatrix} \frac{2A \cdot x_5}{\mu_0 N} & \frac{2B \cdot x_6}{\mu_0 N} \\ \frac{2B \cdot x_5}{\mu_0 N} & \frac{2A \cdot x_6}{\mu_0 N} \end{bmatrix}. \tag{16}$$

At the moment, det $H(x) = 4(A^2 - B^2) \cdot x_5 x_6 / \mu_0^2 N^2 \neq 0$; then $E(x)$ is not singular. Relative degree of maglev system is $(r_1, r_2) = (3, 3)$, and

$$r = \sum_{i=1}^{2} r_i = 6 = n. \tag{17}$$

Therefore, Theorem 1 has been proved.

2.4. MIMO Double Suspension Points Model with Feedback Linearization. Compute the vector field generated by $f(x)$ and $h(x)$ as follows:

$$\begin{aligned} L_f^3 h_1(x) &= \frac{1}{\mu_0^2 N^2} \left[-AR(3x_1 + x_3) x_5^2 - BR(x_1 + 3x_3) x_6^2 \right], \\ L_f^3 h_2(x) &= \frac{1}{\mu_0^2 N^2} \left[-BR(3x_1 + x_3) x_5^2 - AR(x_1 + 3x_3) x_6^2 \right]. \end{aligned} \tag{18}$$

Design feedback control value $u = H^{-1}(x)[-b(x) + v]$, where the terms $u = [u_1 \ \ u_2]^T$, $v = [v_1 \ \ v_2]^T$, and

$$H^{-1}(x) = \frac{\mu_0 N}{2(A^2 - B^2)} \cdot \begin{bmatrix} \frac{A}{x_5} & -\frac{B}{x_6} \\ -\frac{B}{x_5} & \frac{A}{x_6} \end{bmatrix},$$

$$b(x) = \left[L_f^{r_1} h_1(x) \quad L_f^{r_2} h_2(x) \right]^T = \left[L_f^3 h_1(x) \quad L_f^3 h_2(x) \right]^T. \tag{19}$$

We can obtain the control value after linearization from (19):

$$u = H^{-1}(x)[-b(x)+v]$$
$$= \begin{cases} \dfrac{\mu_0 N}{2(A^2-B^2)} \cdot \left[-\dfrac{A}{x_5} \cdot L_f^3 h_1(x) + \dfrac{B}{x_6} \cdot L_f^3 h_2(x) + \dfrac{A}{x_5} \cdot v_1 - \dfrac{B}{x_6} \cdot v_2\right] \\ \dfrac{\mu_0 N}{2(A^2-B^2)} \cdot \left[\dfrac{B}{x_5} \cdot L_f^3 h_1(x) - \dfrac{A}{x_6} \cdot L_f^3 h_2(x) - \dfrac{B}{x_5} \cdot v_1 + \dfrac{A}{x_6} \cdot v_2\right]. \end{cases} \quad (20)$$

The diffeomorphic mapping matrix is

$$\Phi(x) = \begin{bmatrix} h_1(x) \\ L_f h_1(x) \\ L_f^2 h_1(x) \\ h_2(x) \\ L_f h_2(x) \\ L_f^2 h_2(x) \end{bmatrix} \quad (21)$$
$$= \begin{bmatrix} x_1 \\ x_2 \\ A \cdot \dfrac{x_5^2 S}{\mu_0} + B \cdot \dfrac{x_6^2 S}{\mu_0} + C \cdot f_1 + D \cdot f_2 + g \\ x_3 \\ x_4 \\ B \cdot \dfrac{x_5^2 S}{\mu_0} + A \cdot \dfrac{x_6^2 S}{\mu_0} + D \cdot f_1 + C \cdot f_2 + g \end{bmatrix}.$$

Choose the coordinates of transformation by matrix $\Phi(x)$:

$$\delta = \left[h_1(x) \;\; L_f h_1(x) \;\; L_f^2 h_1(x) \;\; h_2(x) \;\; L_f h_2(x) \;\; L_f^2 h_2(x)\right]^T. \quad (22)$$

In sum, maglev control system model after linearization is showed as follows:

$$\dot{\delta} = \Psi\delta + Tv,$$
$$y = \Omega\delta, \quad (23)$$

where

$$\Psi = \begin{bmatrix} 0&1&0&0&0&0 \\ 0&0&1&0&0&0 \\ 0&0&0&0&0&0 \\ 0&0&0&0&1&0 \\ 0&0&0&0&0&1 \\ 0&0&0&0&0&0 \end{bmatrix}, \qquad T = \begin{bmatrix} 0&0 \\ 0&0 \\ 1&0 \\ 0&0 \\ 0&0 \\ 0&1 \end{bmatrix}, \quad (24)$$
$$\Omega = \begin{bmatrix} 1&0&0&0&0&0 \\ 0&0&0&1&0&0 \end{bmatrix}.$$

Maglev system after feedback linearization is expressed by two-level integral subsystems:

$$G(s) = \begin{bmatrix} \dfrac{1}{s^3} & 0 \\ 0 & \dfrac{1}{s^3} \end{bmatrix}. \quad (25)$$

Double suspension points' closed control block diagram with magnetic flux feedback is showed in Figure 3.

3. Design of Suspension Controller

Controlled matrix of system (23) is

$$M_c = \left[T \;\; \Psi T \;\; \Psi^2 T \;\; \Psi^3 T \;\; \Psi^4 T \;\; \Psi^5 T\right]$$
$$= \begin{bmatrix} 0&0&0&0&1&0&0&0&0&0&0&0 \\ 0&0&1&0&0&0&0&0&0&0&0&0 \\ 1&0&0&0&0&0&0&0&0&0&0&0 \\ 0&0&0&0&0&1&0&0&0&0&0&0 \\ 0&0&0&1&0&0&0&0&0&0&0&0 \\ 0&1&0&0&0&0&0&0&0&0&0&0 \end{bmatrix}. \quad (26)$$

Because of rank $(M_c) = 6$, maglev system after linearization is controlled completely. So we design double suspension controller for regulating control properties of suspension system:

$$v_1 = K_{p1} \cdot (d_1 - z_0) + K_{d1} \cdot \dot{d}_1 + K_{i1} \cdot \int (d_1 - z_{10}) \cdot dt,$$
$$v_2 = K_{p2} \cdot (d_2 - z_0) + K_{d2} \cdot \dot{d}_2 + K_{i2} \cdot \int (d_2 - z_{20}) \cdot dt. \quad (27)$$

The geometrical relationship between measured positions and real physical positions in (28) is as follows:

$$d_1 = \frac{3}{4} z_1 + \frac{1}{4} z_2,$$
$$d_2 = \frac{1}{4} z_1 + \frac{3}{4} z_2. \quad (28)$$

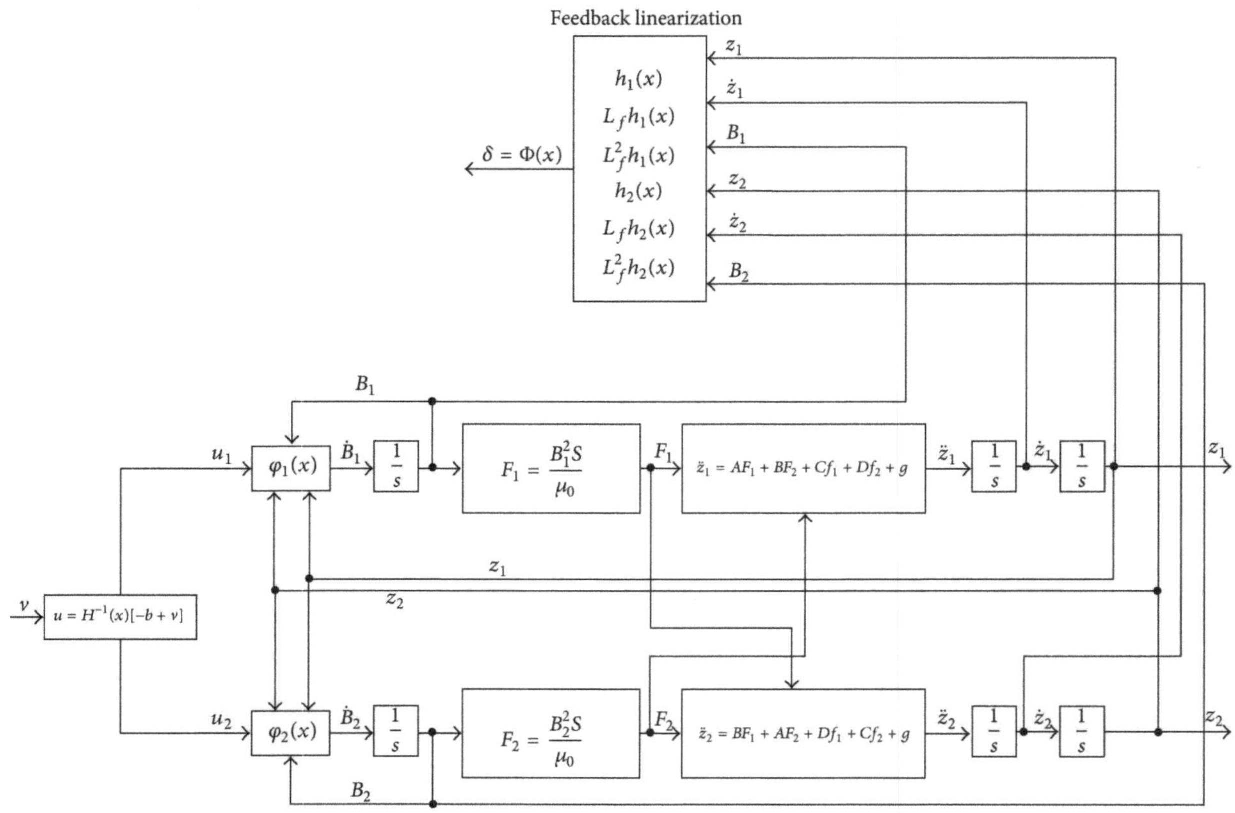

FIGURE 3: Double suspension points' control construction diagram with feedback linearization.

We can get

$$v_1 = K_{p1} \cdot \left(\frac{3}{4}z_1 + \frac{1}{4}z_2 - z_{10}\right) + K_{d1} \cdot \dot{z}_1 + K_{i1} \cdot \int \left(\frac{3}{4}z_1 + \frac{1}{4}z_2 - z_{10}\right) \cdot dt,$$

$$v_2 = K_{p2} \cdot \left(\frac{1}{4}z_1 + \frac{3}{4}z_2 - z_0\right) + K_{d2} \cdot \dot{z}_2 + K_{i2} \cdot \int \left(\frac{1}{4}z_1 + \frac{3}{4}z_2 - z_{20}\right) \cdot dt. \quad (29)$$

By now, double suspension points' controller based on magnetic flux feedback was completed. Construction of the controller is described in Figure 4.

4. Experiments

Some experiments have been completed on CMS04 suspension bogie designed by NUDT (National University of Defense Technology), which is showed in Figure 5. These experiments conclude experiment S1 and experiment S2. Algorithm S1: traditional position-current double cascade suspension PID control method. Algorithm S2: magnetic flux feedback control algorithm based on MIMO feedback linearization. Initial suspension gap of maglev system is 25 mm, and the set value is 10 mm. Decoupling control experiment has been implemented on CMS04 maglev control experiment platform.

TABLE 2: Standard maglev bogie parameters of CMS04 maglev vehicle.

Parameters	Values
m [kg]	653
N [integer]	324
A [m^2]	0.0235
R [Ω]	0.5
z_0/m	0.008
B_0/T	23
g	9.8
μ_0	$4\pi \times 10^{-7}$

Standard maglev bogie's parameters are given in Table 2.

Decoupling property test comes true on one suspension point of maglev bogie. Set expected position 10 mm.

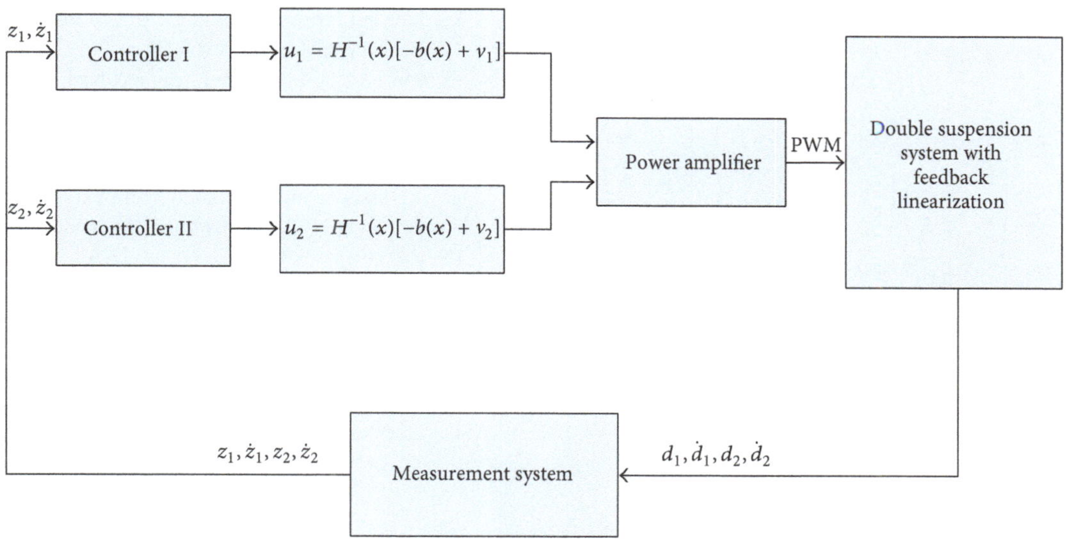

FIGURE 4: Block diagram of double suspension points system.

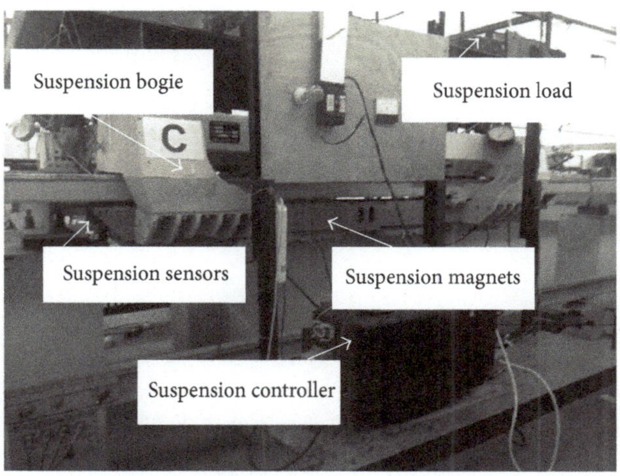

FIGURE 5: CMS04 maglev control experiment platform based on magnetic flux feedback.

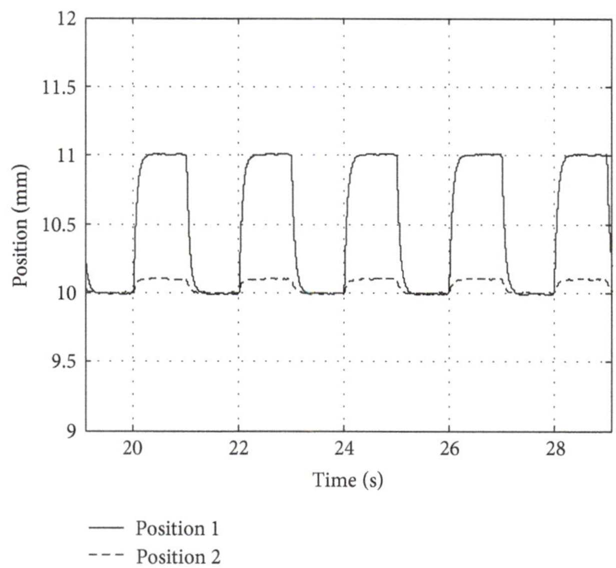

FIGURE 6: Position curves of algorithm S1.

When maglev system becomes stable, we add additional square signal with amplitude 0.5 mm and period 4 s. Observe suspension signals with two control algorithms S1 and S2. Experiment results are showed in Figures 6 and 7.

Experiment results illustrate that when double suspension control system applies algorithm S2, the dynamic decoupling problem has been worked out. S2 raises the robustness and stability properties of maglev control system, but method S1 has no decoupling effect.

5. Conclusions

Double suspension control model has been founded with magnetic flux signal based on MIMO feedback linearization. The feedback linearization algorithm enables accurate linearizing model to keep all properties of original nonlinear system which overcome the disadvantages of Taylor's expansion linearization method. In order to work out dynamic coupling problem and external interference problem of EMS mid-low-speed maglev vehicle, we take magnetic flux signal back to maglev control system and design double suspension compensable controller. Some experiments about new algorithm have been done in maglev vehicle CMS04 designed by NUDT. Experiment results demonstrate that double suspension module is precise based on MIMO state feedback linearization theory. With magnetic flux feedback, the maglev control system has better robustness and adaptability than traditional algorithm.

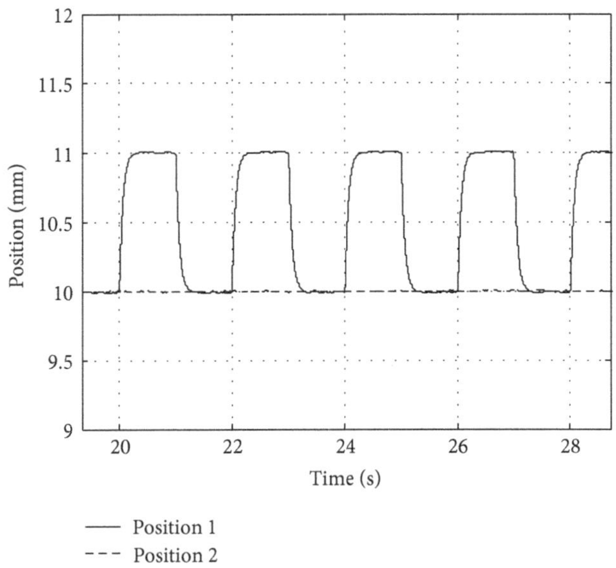

FIGURE 7: Position curves of algorithm S2.

Acknowledgments

This work was financially supported by National Nature and Science Foundation of China (NNNSFC, no. 11202230) and the Twelfth Five-Year National Science and Technology Support Plan (2012BAC07B01).

References

[1] P. Gáspár, Z. Szabó, G. Szederkényi, and J. Bokor, "Design of a two-level controller for an active suspension system," *Asian Journal of Control*, vol. 14, no. 3, pp. 664–678, 2012.

[2] Y. Lin, Y. Shi, and R. Burton, "Modeling and robust discrete-time sliding-mode control design for a fluid power electrohydraulic actuator (EHA) system," *IEEE Transactions on Mechatronics*, vol. 18, no. 1, pp. 1–10, 2013.

[3] P. H. A. Ngoc, "Stability of positive differential systems with delay," *IEEE Transaction on Automatic Control*, vol. 58, no. 1, pp. 203–209, 2013.

[4] F. Mazenc, S.-I. Niculescu, and M. Bekaik, "Stabilization of time-varying nonlinear systems with distributed input delay by feedback of plant's state," *IEEE Transactions on Automatic Control*, vol. 58, no. 1, pp. 264–269, 2013.

[5] W.-H. Chen, G. Yuan, and W. X. Zheng, "Robust stability of singularly perturbed impulsive systems under nonlinear perturbation," *IEEE Transactions on Automatic Control*, vol. 58, no. 1, pp. 168–174, 2013.

[6] Z. Yang, S. Hua, Q. Hongzhuan, and L. Chengrui, "Control reconfigurability of nonlinear system based on control redundancy," in *Proceedings of the 10th IEEE International Conference on Industrial Informatics (INDIN '12)*, pp. 815–820, 2012.

[7] C.-C. Chiang and C.-H. Wu, "Output tracking control for uncertain underactuated systems based on fuzzy sliding mode control approach," in *Proceedings of the IEEE International Conference on Fuzzy Systems (FUZZ-IEEE '12)*, pp. 1–8, 2012.

[8] C. Naichao and M. Shengwei, "Nonlinear stabilizing of multi-machine power systems based on feedback domination method," in *Proceedings of the 27th Chinese Control Conference (CCC '08)*, pp. 412–416, July 2008.

[9] H. Ren and D. Liu, "Nonlinear feedback control of chaos in permanent magnet synchronous motor," *IEEE Transactions on Circuits and Systems II*, vol. 53, no. 1, pp. 45–50, 2006.

[10] B. Roy and H. H. Asada, "Nonlinear feedback control of a gravity-assisted underactuated manipulator with application to aircraft assembly," *IEEE Transactions on Robotics*, vol. 25, no. 5, pp. 1125–1133, 2009.

[11] S. O. Cannizzaro, G. Palumbo, and S. Pennisi, "Effects of nonlinear feedback in the frequency domain," *IEEE Transactions on Circuits and Systems I*, vol. 53, no. 2, pp. 225–234, 2006.

[12] W.-Q. Zhang, *Study and realization on suspension control based on magnetic flux feedback [M.S. thesis]*, National University of Defense Technology, 2009.

Numerical Simulation of Pipeline Deformation Caused by Rockfall Impact

Jie Zhang, Zheng Liang, and Chuanjun Han

School of Mechatronic Engineering, Southwest Petroleum University, Chengdu 610500, China

Correspondence should be addressed to Zheng Liang; liangz_2242@126.com

Academic Editor: Huamin Zhou

Rockfall impact is one of the fatal hazards in pipeline transportation of oil and gas. The deformation of oil and gas pipeline caused by rockfall impact was investigated using the finite element method in this paper. Pipeline deformations under radial impact, longitudinal inclined impact, transverse inclined impact, and lateral eccentric impact of spherical and cube rockfalls were discussed, respectively. The effects of impact angle and eccentricity on the plastic strain of pipeline were analyzed. The results show that the crater depth on pipeline caused by spherical rockfall impact is deeper than by cube rockfall impact with the same volume. In the inclined impact condition, the maximum plastic strain of crater caused by spherical rockfall impact appears when incidence angle α is 45°. The pipeline is prone to rupture under the cube rockfall impact when α is small. The plastic strain distribution of impact crater is more uneven with the increasing of impact angle. In the eccentric impact condition, plastic strain zone of pipeline decreases with the increasing of eccentricity k.

1. Introduction

Long-distance oil and gas pipelines are usually laid underground, while the pipelines are laid on the ground when they go across valleys, rivers, swamps, deserts, permafrost, and other regions [1]. The pipelines on the ground have little effect on the soil environment, and the replacement or maintenance of pipelines on the ground is much more convenient. However, pipelines are vulnerable to external shock loads and wind loads because they are exposed at the ground's surface (see Figure 1). For instance, in April 2005, a pipeline's impact deformation caused by dangerous rockfall took place in Chongqing [2]. Since the pipelines are put into production in 2002, some incidents happened that rockfall impacted the pipelines along Lanzhou—Chengdu—Chongqing Products Pipeline. The rockfall impact has become one of the most serious geological disasters to endanger the safety of the pipeline [3]. The rockfall impact is subject to cause pipeline partial concave or rupture, which easily leads to leakage accident of oil or gas. Therefore, research on pipeline deformation caused by rockfall impact is important for the design, protection, and safety assessment of pipelines.

Currently, researchers often use two kinds of simplified models to research the rockfall impact problems. One is to simplify the rockfall to spherical structure. He et al. [4] and Wang et al. [5] studied the dynamic response problem of the shed-tunnel structure under the rockfall impact. Another is to simplify the rockfall to cube structure. As such, Xiong et al. [2] and Deng et al. [6] analyzed the dynamic response of buried pipes when the cube rockfall impacted the ground by numerical simulation. Based on the actual working conditions of pipeline under rockfall impact, this paper simulated the impact processes of pipelines under spherical and cube rockfall impacting. Also the deformation and plastic strain of pipeline were discussed.

2. Elastic-Plastic Failure Criterion

If oil and gas pipelines are elastic-plastic materials, high strain and strain rate effect should be taken into consideration as these may be significant during rockfall impact condition [7]. Currently, the material constitutive models used in impact problem include Cowper-Symonds model, Johnson-Cook

FIGURE 1: Rockfalls impact pipelines.

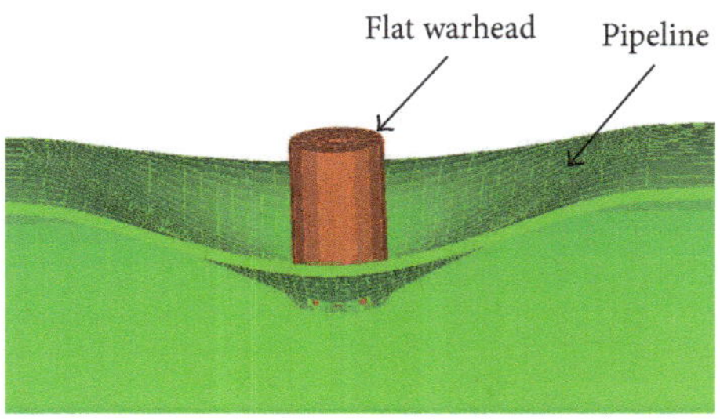

FIGURE 2: The simulation results of example.

model, and Zerilli-Armstrong model. It not only considers the role of strain and strain rate, but also takes the material softening caused by temperature changes produced by plastic expansion into consideration in the latter two models. According to the velocity range of rockfall impact of oil and gas pipeline, the paper chose the Cowper-Symonds model as the constitutive model of pipeline, which ignores the influence of temperature on material properties. Therefore, the model is applicable to analysis of high strain rate impact. The following expression is the dynamic yield stress of the model [8]:

$$\sigma_y = \left[1 + \left(\frac{\dot{\varepsilon}}{C}\right)^{1/P}\right]\left(\sigma_0 + \beta E_P \varepsilon^{\text{eff}}\right), \quad (1)$$

where $\dot{\varepsilon}$ is the strain rate, σ_0 is the initial yield stress, ε^{eff} is the equivalent plastic strain, β is the hardening parameter, E_p is the plastic hardening modulus, and C and P are the coefficients of strain rate, respectively. The plastic reinforcement modulus is

$$E_P = \frac{EE_{\tan}}{E + E_{\tan}}, \quad (2)$$

where E is the elasticity modulus and $E_{\tan}$ is the tangent modulus.

As the impact is a highly nonlinear problem [9], the failure destruction is defined with the rupture strain failure criteria, which means that when the plastic strain of finite element mesh is bigger than the maximum effective plastic strain of the material, the pipeline will have a fracture failure.

3. The Finite Element Model

3.1. Examples Validation. The experimental model of a flat projectile transverse impact pipeline was established in [10], and valuable results were obtained with system test. The same impact model was established by nonlinear finite element program ADINA. The explicit integral algorithm was used to simulate the impacting process. The length of the flat nose is 30 mm, the diameter is 15 mm, and the quality is 41 g. The pipeline's thickness is 3.46 mm, and the outside diameter is 115 mm. The Cowper-Symonds material model and the fracture strain failure criteria were used. Penalty contact algorithm was selected; the friction coefficient between rock and pipelines is 0.3 [2].

Mesh size has a great effect on the simulation results. Many attempts have been made to find the most suitable mesh size. When the mesh size of the pipeline is 2.5 mm, the critical state of the pipeline rupture is obtained as shown in Figure 2. A large pit and partial rupture appear on the impact zone. The critical rupture velocity V is 186 m/s, while the experimental test velocity is 181 m/s, so the error is 2.76%. The critical rupture internal energy E ($E = 1/2mV^2$) obtained by simulation is 709.2 J, which is bigger than the experimental value 671.6 J; the error is 6%. Obviously, the numerical simulation results are in good agreement with

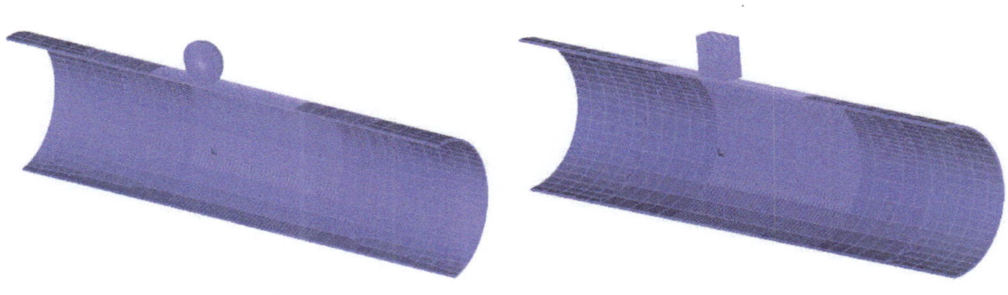

FIGURE 3: Three-dimensional finite element models.

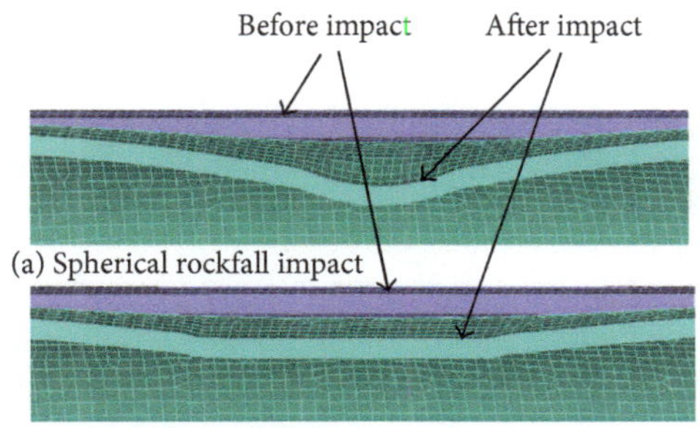

FIGURE 4: Pipeline deformation before and after impact.

the test dates, which proved that the finite element model and mesh size are reliable.

3.2. The Finite Element Model. In the calculation, the spherical and cube rockfalls have the same volume and homogeneous material [11]. Spherical rockfall's radius is 100 mm, the cube rockfall is regular hexahedron, side length is 160 mm, the density of rockfall is 2700 kg/m^3, the elasticity modulus is 55.8 GPa, and Poisson ratio is 0.25. It is assumed that the material of pipeline is the ideal elastic-plastic material, the outer diameter is 599 mm, the thickness is 10 mm, the elastic modulus is 210 GPa, the density is 7800 kg/m^3, Poisson ratio is 0.3, the material's limit fracture strain is 0.74 [9], the limit yield stress is 443 Mpa, and elongation is 0.37 [9]. In this example, the internal pressure of the pipeline is 5 MPa, and span length is 2.6 m.

Considering symmetry of the structure model, the 1/2 entire model was selected to analyze. The calculation model after meshing is shown in Figure 3. The mesh size of pipeline is 2.5 mm, which is the same as the example in Figure 2. Symmetry planes of the model were separately imposed on symmetry constraint; one end of the pipeline is fixed, and the other one is constrained in all directions except the axial direction. Then, the contact relationship between rockfall and pipeline with a penalty function was established, gravity load was applied to the whole model, and the acceleration due to gravity is 9.8 m/s^2.

4. Results and Discussions

4.1. Deformation Analysis under Normal Impact. The pipeline deformation caused by the rockfall impact along the normal direction is shown in Figure 4. Spherical rock causes a spherical cone concave at the bottom of the indentation, while the cube rock causes a rectangular concave. So, the shape of impact crater is related to the shape of rockfall. In addition to the impact contact place, plastic deformation is observed at near site. The further the observation site from the contact place, the smaller the deformation.

The section view of craters impacted by different rocks in the longitudinal plane is shown in Figure 5. With the same volume of rockfalls, the impact crater caused by spherical rock is deeper. There is a small difference between the no-impact zones of the two curves.

As shown in Figure 6, the deformations of pipelines are radial distribution. The farther away from the impact zone, the smaller the deformation. Spherical rockfall can make an oval concave, while the cube rockfall makes a rectangle concave.

Internal pressure has a great effect on the shock stiffness of the pipeline. The largest impact crater depth values under different pipeline pressure are shown in Figure 7. As the internal pressure increases gradually, the largest impact crater depth decreases. When the pressure is 15 MPa, the biggest impact depths are 82.0% and 80.2%, respectively, under spherical and cube rockfalls. So, it is necessary to take internal pressure into consideration when studying the pipeline deformation caused by rockfall impact.

Figure 8 shows the largest impact crater depth under different impact velocity. With the increasing of impact velocity, both impact depth values increase gradually. The smaller the impact velocity, the closer the two impact depths. With the same velocity change range, the change of impact depth of spherical rockfall is bigger than cube rockfall.

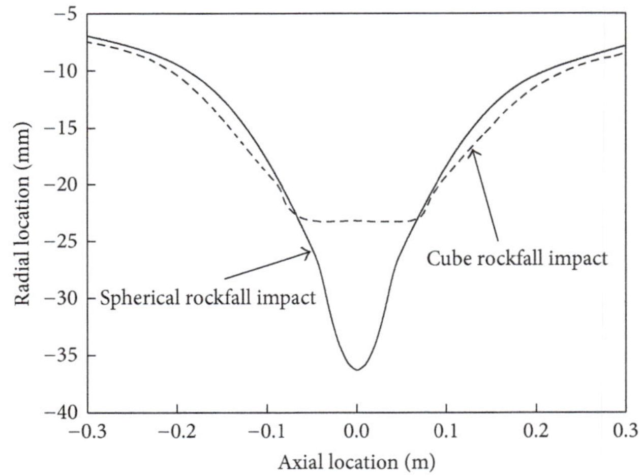

FIGURE 5: Shape caves of the two impact craters in the longitudinal plane.

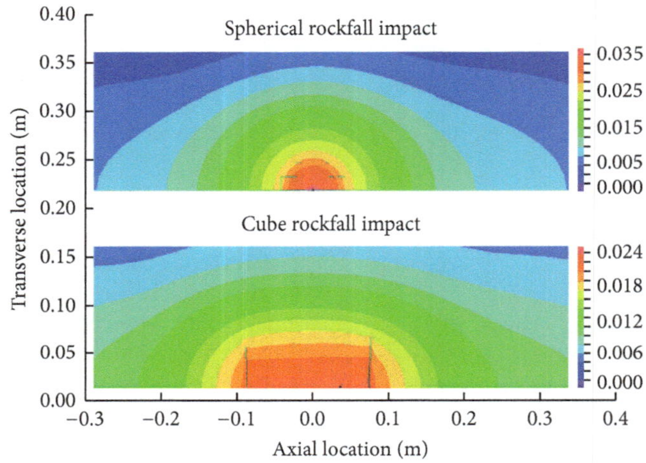

FIGURE 6: Deformation of the two impact craters.

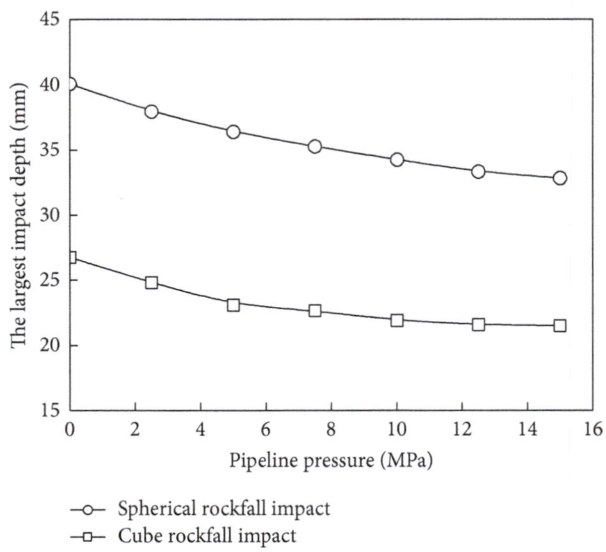

FIGURE 7: Effects of pressure on impact crater depth.

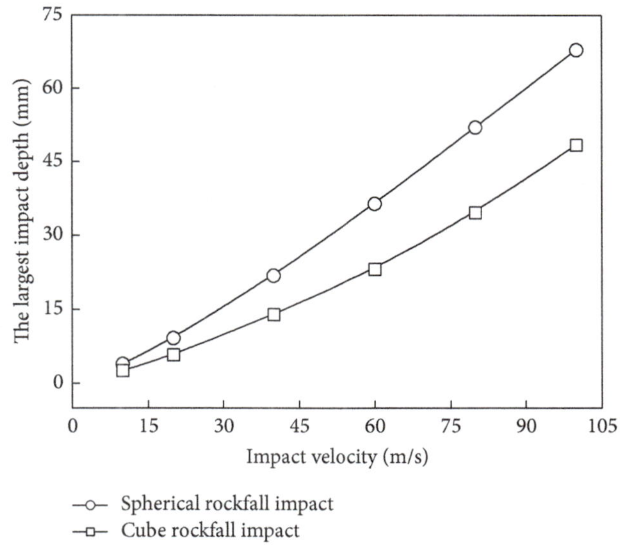

FIGURE 8: Effects of impact velocity on impact depth.

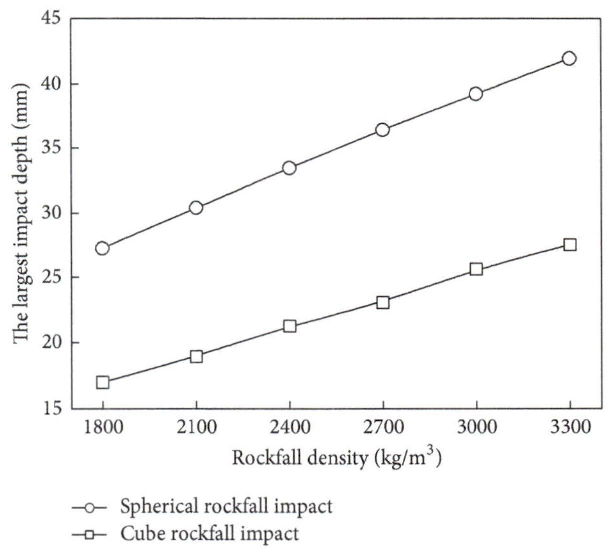

FIGURE 9: Effect of rockfall density on impact depth.

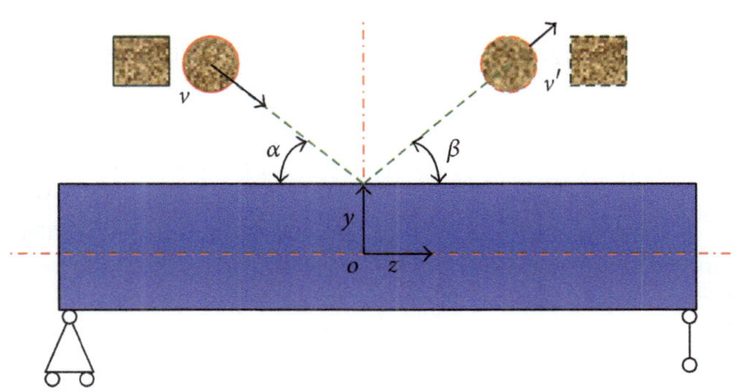

FIGURE 10: Schematic diagram of longitudinal inclined impact.

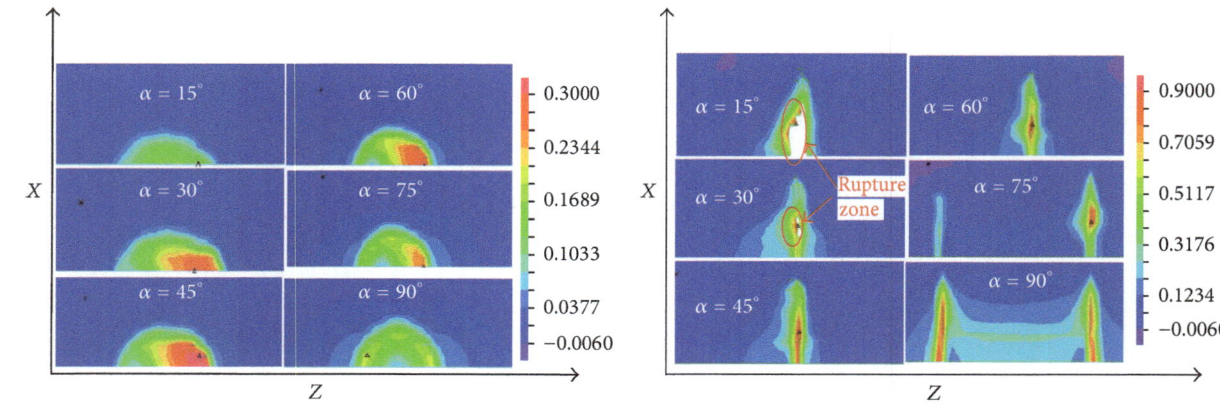

(a) Spherical rockfall impact (b) Cube rockfall impact

FIGURE 11: Plastic strain nephogram under longitudinal inclined impact.

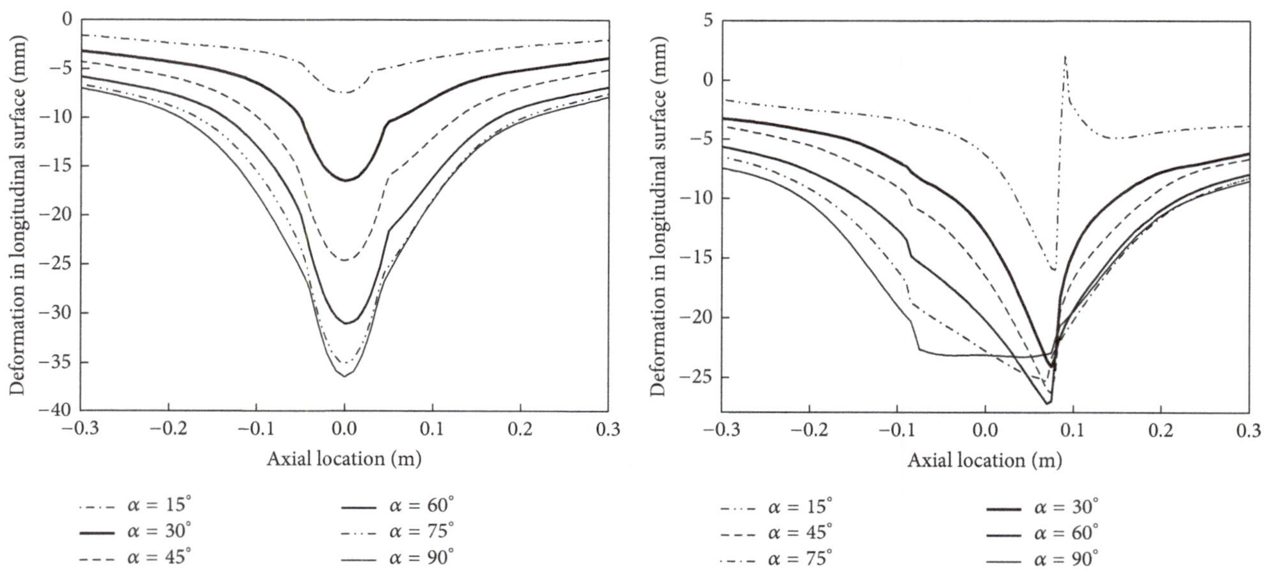

(a) (b)

FIGURE 12: Pipeline deformation in longitudinal surface.

It means that spherical rockfall impact crater is more sensitive to impact velocity.

The density of rock affects its quality and then affects the impulse. The largest impact crater depth values under different rock density are shown in Figure 9. The greater rockfall density, the more serious pipelines' impact deformation. So, the more effective protective measures are needed for the rockfall with large density.

4.2. Deformation Analysis under Longitudinal Inclined Impact. Rockfall may impact pipeline along different angles; therefore, the effect of impact angle on the pipeline deformation should be discussed. The schematic diagram of rockfalls impact pipeline under longitudinal inclined direction is shown in Figure 10. The incidence angle of the rockfall is α, the incidence velocity is v, the rebounding angle is β, the reflection velocity is v', and the initial impact velocity is $v = 60$ m/s.

The plastic strains of pipelines that were impacted by two kinds of rockfalls from different angles are shown in Figure 11. Figure 11(a) shows that the plastic strain of impact zone is uneven when the impact angle is less than 90°. And the smaller the incidence angle, the more obvious this nonuniformity. When $\alpha = 90°$ (the radial impact mentioned above), the impact crater is a symmetric distribution on the x-axis. When α changes from 0° to 90°, the maximum plastic strain of the impact crater increases at first and then decreases, and it reaches the maximum value when $\alpha = 45°$. As shown in Figure 11(b), when the incidence angle is smaller, only one cube edge of rockfall is in contact with the pipeline. As α increases gradually, the second edge gradually comes into contact with the pipeline. When α is 15° and 30°, the pipeline ruptures appear, and the smaller the incidence angle,

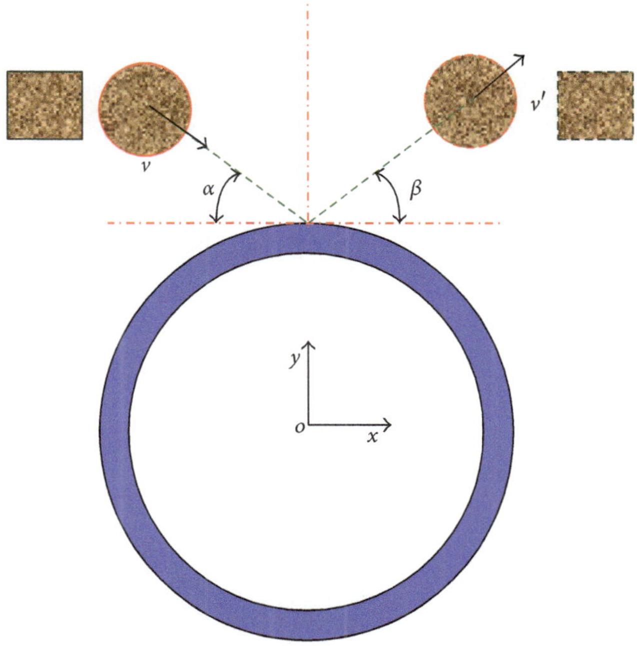

FIGURE 13: Schematic diagram of transverse inclined impact.

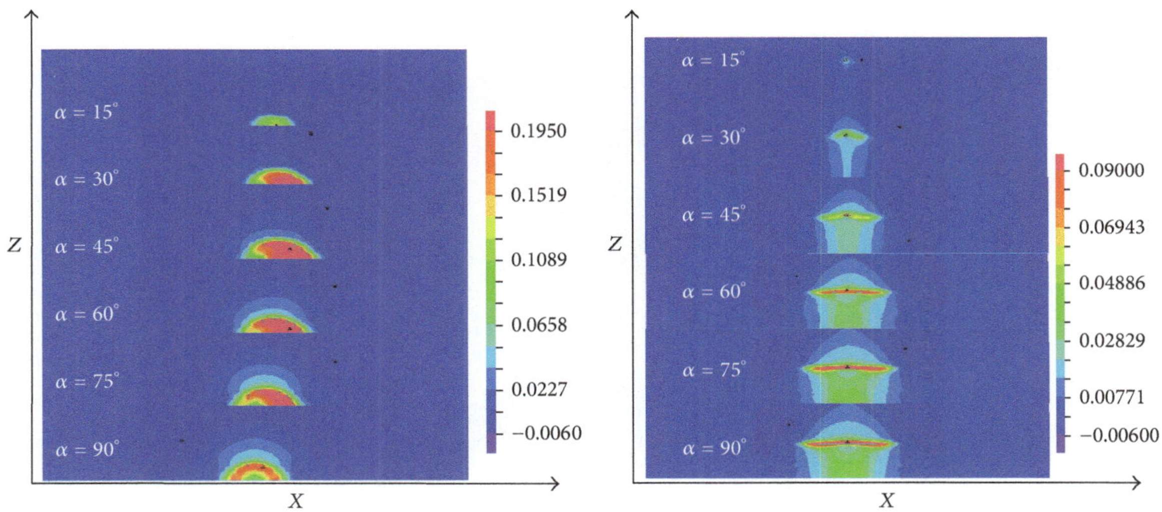

(a) Spherical rockfall impact

(b) Cube rockfall impact

FIGURE 14: Plastic strain nephogram under transverse inclined impact.

the larger the rupture range. The largest plastic strain is the contact zone between rockfall edges and the pipeline.

The deformations of pipeline in longitudinal surface under different incidence angles are shown in Figure 12. Figure 12(a) shows that the smaller the incidence angle, the more inhomogeneous the deformation on both sides of the impact crater. The smaller α, the smaller the pipeline impact crater. As the increasing of α, the largest impact crater depth increases. In Figure 12(b), the impact crater caused by cube rockfall is a triangle when $\alpha \leq 90^\circ$. When $\alpha = 15^\circ$, the mutation of the pipeline deformation curve appears, which means that the rupture occurred here. When $\alpha = 30^\circ$, the deformation curve of longitudinal cross section has no mutation, which indicates that the centre of the impact crater has no rupture.

4.3. Deformation Analysis under Transverse Inclined Impact. The schematic diagram of rockfalls impact pipeline under transverse inclined impact is shown in Figure 13. As the cube rockfall is not axisymmetric structure, this paper only considered the contact between the center of cube rockfall lower surface and the pipeline.

The plastic strains of pipelines under different impact angles are shown in Figure 14. Figure 14(a) shows that impact angle makes an uneven plastic strain distribution of impact

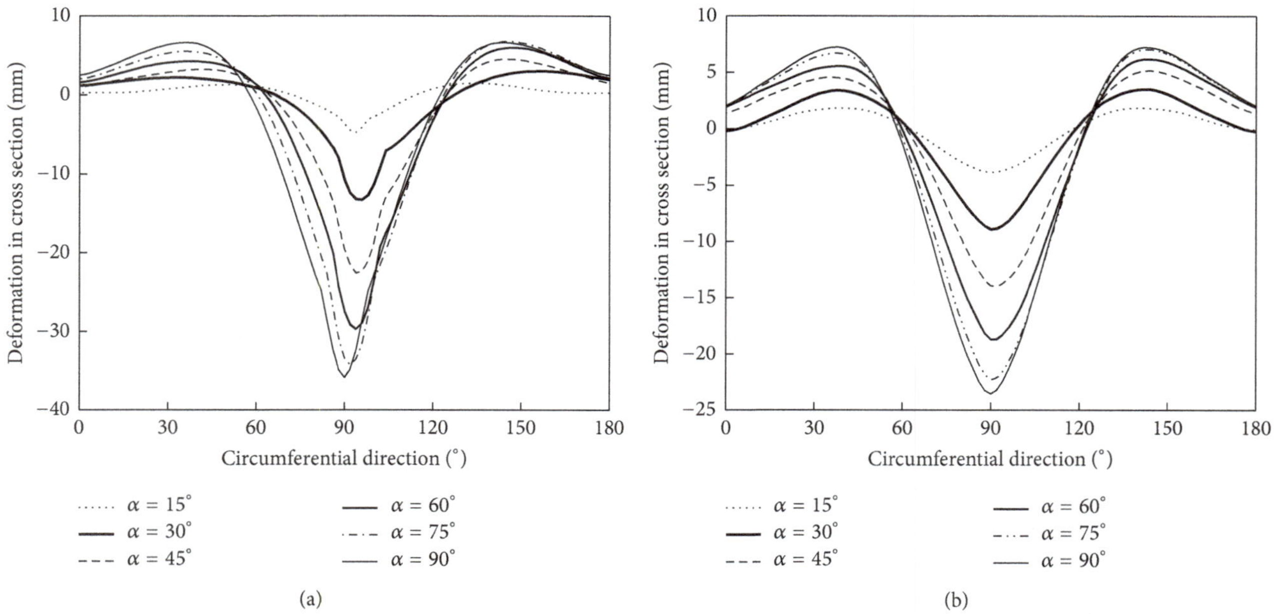

FIGURE 15: Pipeline deformation in cross surface.

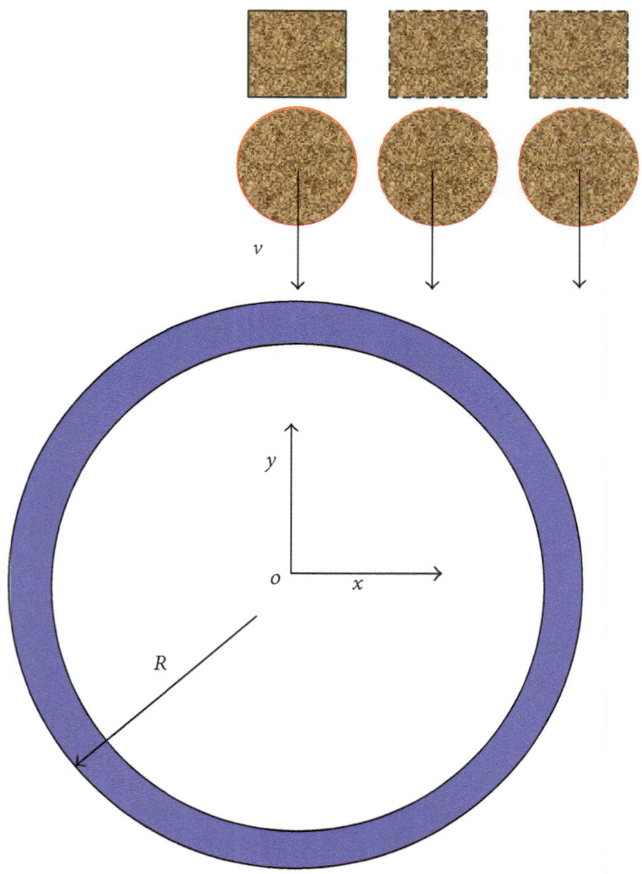

FIGURE 16: Schematic diagram of lateral eccentric impact.

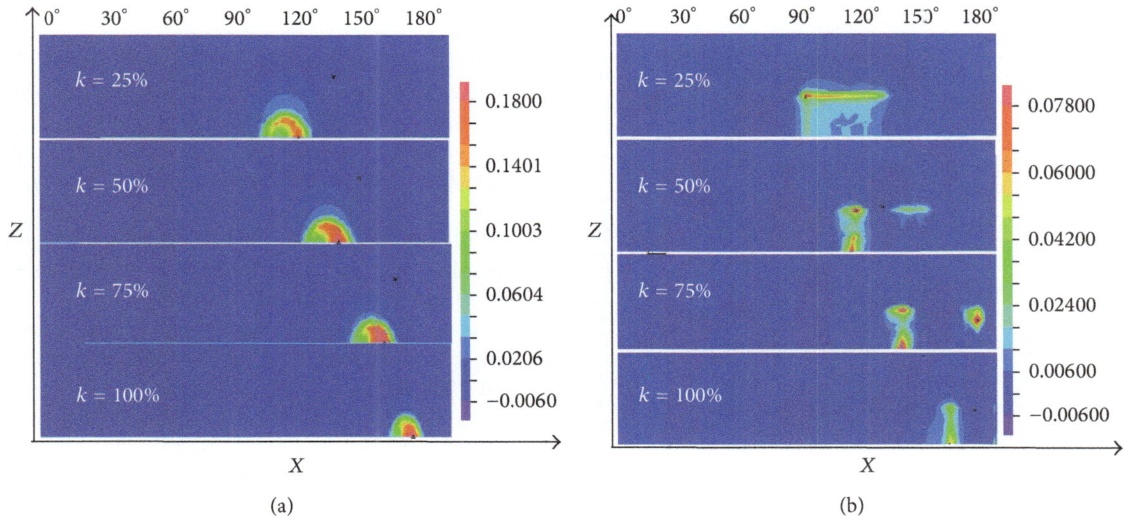

FIGURE 17: Plastic strain nephogram under lateral eccentric impact.

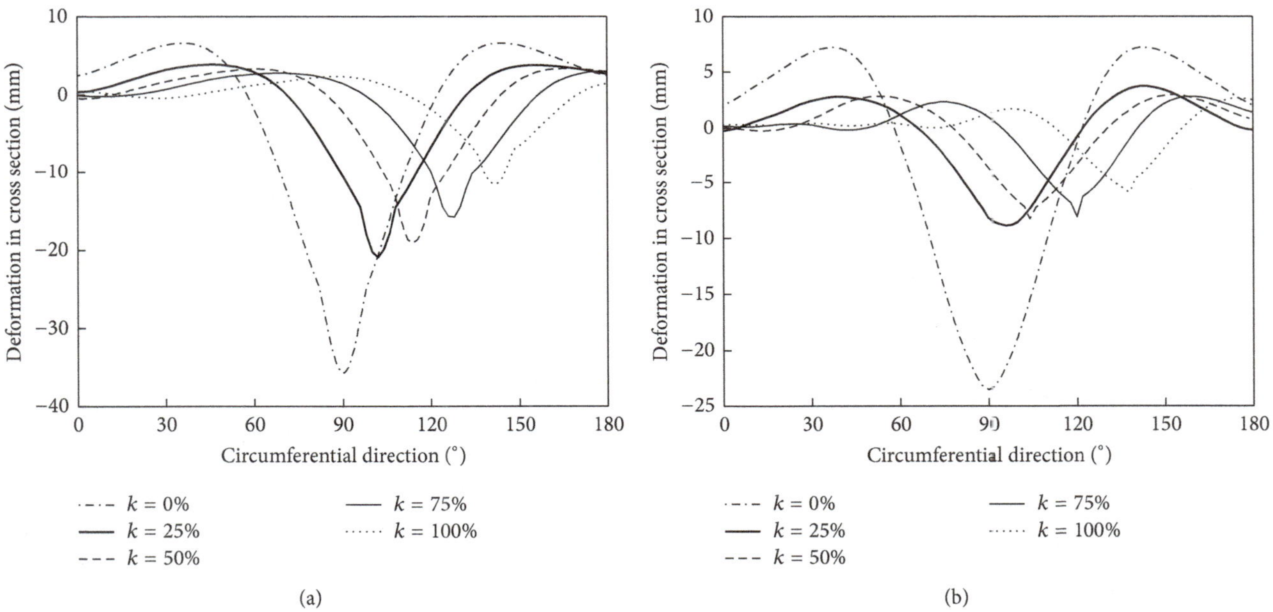

FIGURE 18: Pipeline deformation in cross section.

crater. When $\alpha = 90^\circ$, the elliptical ring appears; the largest plastic strain zone is not in the center of the crater. As the incidence angle increases, the large plastic strain zone increases at first and then decreases. When $\alpha = 45^\circ$, the plastic strain reaches the maximum value. As shown in Figure 14(b), the whole plastic strain increases with the increasing of the incidence angle, and the large plastic strain zone also increases, which is mainly concentrated on the contact zone between cube edges and pipeline.

Figure 15 shows the deformation curves of pipeline in cross section under different impact angles. The pipeline wall shows concave deformation along the circumferential direction 60° ~120°, while the wall shows outer concave deformation along the circumferential direction 60° ~120° and 120°~180°. As α increases, the depth of the impact crater increases. Figure 15(a) shows that the pipeline deformation along the 90° direction is not a symmetrical distribution. As shown in Figure 15(b), the problem of the cube rockfall impact pipeline can be considered as the collision contact problem between a plane and the pipeline.

4.4. Deformation Analysis under Lateral Eccentric Impact. Figure 16 shows the rockfalls impact pipeline from different lateral positions. In order to carry out quantitative description, it is defined as eccentricity $k = x'/R$, where x' is the distance between rockfall center and pipeline center along the x direction and R is the outer radius of the pipeline.

Plastic strains of pipelines impacted under different eccentricities are shown in Figure 17. Figure 17(a) shows that pipelines' plastic strain zone gradually decreases with the increasing of the eccentricity k, and the plastic strain zone also gradually deviates from 90° center position with the increasing of k. As shown in Figure 17(b), the plastic strain zone of the pipeline also gradually decreases with the increasing of k and deviates from center position. The large plastic strain zone is mainly concentrated on the contact region of the cube edges and the pipeline.

Figure 18 shows the deformation curves of the pipelines in the cross section under different eccentricities. With the increasing of k, the depth of the impact crater gradually decreases. And the change rate of impact crater depth caused by spherical rockfall is larger than cube rockfall.

5. Conclusions

The FE models of oil and gas pipelines impacted by spherical and cube rockfalls were established in this paper. The pipeline deformations under radial impact, longitudinal inclined impact, transverse inclined impact, and lateral eccentric impact were analyzed. The conclusions are summarized as follows.

(1) Under radial impact, the crater impacted by spherical rockfall on the pipeline was deeper than by cube rockfall with the same volume.

(2) When α changes from 0° to 90°, the maximum plastic strain of the pipeline caused by spherical rockfall increases at first and then decreases under longitudinal inclined impact. When α is 45°, the plastic strain reaches the maximum value. Specially, when α is small, cube rockfall can probably lead to pipeline's rupture.

(3) The plastic strain distribution of pipeline is extremely uneven under longitudinal inclined impact. As α increases gradually, the plastic zone increases, and large plastic strain zone is mainly concentrated on contact zone between the cube rockfall edges and the pipeline.

(4) Under lateral eccentric impact, the pipeline plastic strain zone decreases with the increasing of k. And the large plastic strain zone deviates from 90° center position with the increasing of k.

Conflict of Interests

The authors declare that there is no conflict of interests regarding the publication of this paper.

Acknowledgment

This research work was supported by the National Natural Science Foundation of China (51004083).

References

[1] J. Shuai and G. J. Yu, *Strength Design of Pipeline and Storage Tank*, Petroleum Industry Press, Beijing, China, 2006.

[2] J. Xiong, Q. L. Deng, and H. L. Zhang, "Safety assessment on the response of buried pipeline caused by rockfall impact load," *Safety and Environmental Engineering*, vol. 20, no. 1, pp. 108–114, 2013.

[3] X. W. Shi, Q. L. Deng, and G. L. Dong, "The hazards of landslides and rockslides to pipeline," *Oil & Gas Storage and Transportation*, vol. 23, no. 3, pp. 295–299, 2013.

[4] S.-M. He, J. Shen, and Y. Wu, "Rock shed dynamic response to impact of rock-fall," *Rock and Soil Mechanics*, vol. 32, no. 3, pp. 781–788, 2011.

[5] D. P. Wang, S. M. He, and C. J. Ouyang, "Study of dynamic response of shed reinforced concrete slab to impact load of rock-fall," *Rock and Soil Mechanics*, vol. 34, no. 3, pp. 881–886, 2013.

[6] X.-J. Deng, S.-F. Xue, and X.-H. Tong, "Numerical simulation on response of buried pipeline induced by rock-fall transverse impaction," *Journal of China University of Petroleum (Edition of Natural Science)*, vol. 33, no. 6, pp. 111–115, 2009.

[7] H. L. Lu, P. X. Yao, and H. Y. Liu, "Simulation of controlled shot peening process based on ANSYS/LS- DYNA," *Machinery Design & Manufacture*, vol. 47, no. 2, pp. 214–216, 2009.

[8] X. Y. Lu, *Dynamic constitutive relationship of TC4 titanium alloy [Ph.D. thesis]*, Nanjing University of Aeronautics and Astronautics, Jiangsu Province, China, 2010.

[9] J.-L. Yu and L. Qin, "Rupture failure of an elbow subjected to internal impact," *Journal of Vibration and Shock*, vol. 29, no. 10, pp. 228–231, 2010.

[10] T. G. Zhang and W. J. Stronge, "Rupture of thin ductile tubes by oblique impact of blunt missiles: experiments," *International Journal of Impact Engineering*, vol. 21, no. 7, pp. 571–587, 1998.

[11] G. Zhang, X. Xiang, and H. Tang, "Field test and numerical calculation of restitution coefficient of rockfall collision," *Chinese Journal of Rock Mechanics and Engineering*, vol. 30, no. 6, pp. 1266–1273, 2011.

12

An Alternate Method to Springback Compensation for Sheet Metal Forming

Waluyo Adi Siswanto,[1] Agus Dwi Anggono,[2] Badrul Omar,[3] and Kamaruzaman Jusoff[4]

[1] *Department of Engineering Mechanics, Universiti Tun Hussein Onn Malaysia, Parit Raja, Batu Pahat, 86400 Johor, Malaysia*
[2] *Department of Mechanical Engineering, Muhammadiyah University of Surakarta, Pabelan, Kartasura 57162, Indonesia*
[3] *Department of Materials Engineering and Design, Universiti Tun Hussein Onn Malaysia, Parit Raja, 86400 Johor, Malaysia*
[4] *Faculty of Forestry, Universiti Putra Malaysia, Serdang, 43400 Selangor, Malaysia*

Correspondence should be addressed to Waluyo Adi Siswanto; waluyo@uthm.edu.my

Academic Editor: Hailiang Yu

The aim of this work is to improve the accuracy of cold stamping product by accommodating springback. This is a numerical approach to improve the accuracy of springback analysis and die compensation process combining the displacement adjustment (DA) method and the spring forward (SF) algorithm. This alternate hybrid method (HM) is conducted by firstly employing DA method followed by the SF method instead of either DA or SF method individually. The springback shape and the target part are used to optimize the die surfaces compensating springback. The hybrid method (HM) algorithm has been coded in Fortran and tested in two- and three-dimensional models. By implementing the HM, the springback error can be decreased and the dimensional deviation falls in the predefined tolerance range.

1. Introduction

The Sheet metal forming (SMF) is one of forming processes performed on metal sheets, strips, and coils. Press working is the term often applied to sheet metal operations because the machines used to perform these operations are press machines. A part produced in a sheet metal operation is often called a stamping product. SMF process consists of stamping, forming, bending, stretching, and trimming. The terms refer to various processes used to convert a sheet metal into different shapes for a large variety of finished useful products. Since SMF process produces the same products, it has become one of the most important manufacturing processes in industry, particularly in the automotive and steel industries [1].

Every process of sheet metal forming or stamping process involves elastic forming and then followed by permanent plastic deformation. Since the presence of elastic properties of metals, after the unloading phase, the elastic springback phenomenon always occurs resulting in an off-target formed shape. Springback is dimensional deviation due to the elasticity of a metal sheet during unloading and following forming [2]. Although it is impossible to eliminate the springback phenomenon, minimizing springback can be done by adopting three approaches. The first one is based on controlling the blank holder force (BHF). The blank holding force is not fixed, but, as a controlled variable, the blank holding force is known as variable blank holder force (VBHF) so that the springback can be minimized. The second approach is by using a hot forming process and the third one is by optimizing die surfaces (die compensation) to accommodate the springback.

The VBHF was initially proposed by Liu et al. [3] to reduce the springback error of U-bend model of NUMISHEET'93. The VBHF was determined by calculating the value of low blank holder force (BHFL) and high blank holder force

(BHFS). In the report [3], the VBHF approach successfully reduced the springback error compared with that from constant blank holder force (CBHF). The application of VBHF has been extended and published by researchers recently: Li et al. [4] investigated the subsection of VBHF in rectangular parts, and then Kitayama et al. [5] presented specifically the optimization of VBHF to reduce the springback. A precision binder force control during forming is required, making this process sensitive to any variations in manufacturing conditions such as punch speed, time control, and friction.

The study of springback reduction by the temperature value of tool in hot forming has been investigated [6]. It is found that the cold punch combined with the hot die can reduce the springback up to 20% when compared to conventional room temperature bending test in the sheet forming process of U-bend aluminum 1050. The experimental investigations in hot forming process can be found in the latest publications [7–9]. Temperature in the springback prediction is becoming a topic that could be improved in the future investigation [2]. The springback value before and after splitting process is different. This has been studied using splitring tests for aluminum alloy [10].

The major advantage of the first strategy to minimize springback is that it is not necessary to modify the tooling surface. However, there are several implementation difficulties such as implementing the force control or force sensor in the application of VBHF being highly cost sensitive. Similarly, in the hot forming process, it requires additional equipment to control the temperatures of the tools and the blank sheet so that the forming process becomes more complicated and expensive.

In the third method (die compensation), it requires many steps of works on tooling design stage, but its potential to compensate springback completely is faster and cheaper even for complex model. Die face adjustment to compensate the springback was done manually in the past, by doing extensive measurements on the prototype or even production tools and refining the tool surface geometry by hand polishing which is time consuming [11]. Wagoner et al. [2] categorized the current research of springback in five topics, that is, plastic constitutive equations, variable Young's modulus, through-thickness integration of stress, magnesium, and advanced high strength steels (AHSS).

Finite element (FE) and numerical and process simulation have gained popularity in the stamping industry due to its speed and low cost, and it has been proven to be effective and efficient in the prediction of form ability and springback behavior [12, 13]. The accuracy of springback simulation is not only related to springback analysis itself but also strongly dependent on the accuracy of forming processes. Any calculation error obtained from every simulation step of forming processes will be accumulated. As a result, the accumulated error influence the accuracy of the springback prediction analysis at the last step of the simulation. Ling et al. [13] tried to reduce springback in L-bending by using the optimization of die radius, clearance, step height, and step distance. The results show a good reduction in springback, but this method is applied and tested in the L-bending only. The U-bending model is analyzed by Slota et al. [14] in both numerical and experimental analyses. The experimental analysis is conducted by using specimens of steel DC06, UHSS TRIP RAK 40/70, and HSS H220PD in orientations 0°, 45°, and 90°. The results show that the simulations are in a good agreement with the experiments. The springback in sheet forming can be simulated accurately by using finite element method, but the big problem is to apply the results to produce stamped parts in accurate dimension.

In order to accommodate the springback and to optimize the die based on surface modification, there are two methodologies available called displacement adjustment (DA) compensation and spring forward (SF) compensation. The concept of DA method [15] is to translate the nodes in the direction opposite to the springback which is adopted from the real springback investigation, whereas the SF method is based on stress state during forming, which is multiplied by negative factor and then loaded to the deformed shape to realize the compensated surface [16]. The compensation method based on calculated springback by implementing the DA method can be done in a computer program [17]. The modification of die surface for springback compensation can be done faster. There are many researchers who reported the implementation of these two methods in various springback accommodation problems and are dealing with minimizing springback errors [18–20].

One of the advantages of DA method is that it can converge rapidly because of the algorithm being based on the real springback measurements, not virtual springback calculation, which may develop instability during the numerical solution [15]. Unfortunately, the original DA method is mostly applicable to compensate surfaces in the direction of the punch travel. The wall area of the dies, therefore, could not be compensated [21].

On the other hand, the SF method could compensate die surface in almost all directions because of the deformation being caused by the stresses of the deformed part in the opposite direction [22]. In the original concept of spring forward [16], only inverted bending stress will be used to create the spring forward shape. The inverted membrane stresses must be eliminated since they will contribute to the unstable buckling condition during the numerical calculation. Another disadvantage of SF method is that the approach is not based on the reality of physical occurrence but an inverted springbok. The complexity of inverted bending stress distribution in the deformed shape is very difficult to apply in an experiment, even impossible [21]. In order to solve the problem, finite element simulation analysis can be used to explain the influence of residual stress in springback forming as done by Brabie et al. [23]. It shows that residual stress is influenced by the BHF and coefficient of friction in both cylindrical and conical forming. They analyze the residual stress during forming and after springback but do not explain how to minimize the springback related to the residual stress.

Force descriptor method (FDM) proposed by Karafillis and Boyce [22] is SF method based. This method has two parameter vectors, the original part shape and the internal forces. The theoretical base of the FDM is that after being formed to the original tooling shape, the part would springback to the original unloaded shape, recovering from

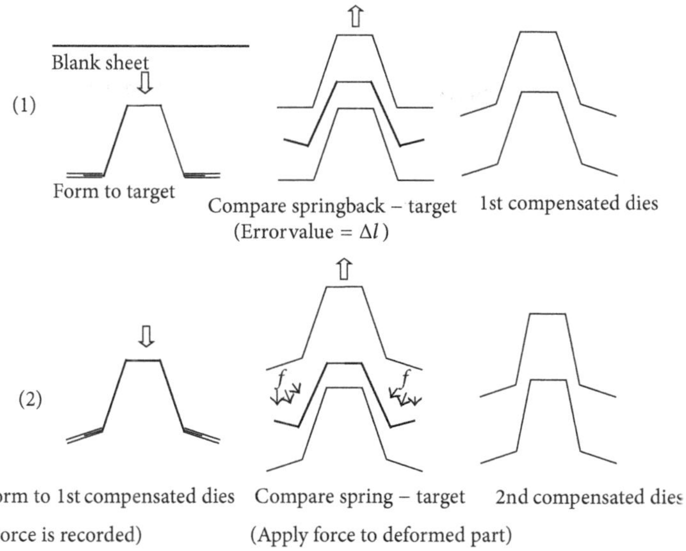

FIGURE 1: Illustration of compensation procedures in hybrid method.

the internal force. By spring forward, the stress free original shape with the reversed internal force, a tooling shape close to the original shape, could be obtained. Considering a simple bending case, the internal force in the first iteration is lower than the final force, which means that the FDM needed more iterations. Cheng et al. [21] modified this approach in the first iteration by replacing the internal force with the internal force obtained from the manual calculation to accelerate the compensation.

Wagoner et al. [24] proposed the displacement adjustment method. At first forming, a flat sheet is deformed to the original die shape. After springback, the shape is compared with its target to obtain the shape error. In the next step, the amount of shape error is added to the current die shape to obtain the new compensated shape. During the next iteration, a blank sheet is deformed to the new compensated shape. If the shape error is not in the range of tolerance, another iteration will be conducted. Wagoner and coworkers claimed this method is effective and converges more rapidly than the FDM.

In the present research, a new approach in accommodating springback is the main objective. This research aims at the development of a compensation procedure that can perform the optimization process, using the combination of fast convergence displacement adjustment (DA) and the flexible spring forward (SF) methods so that the combination method will be fast and applicable for all die surfaces. The method guides the die surface modification process to compensate for springback following the displacement adjustment; then it continues to compensate the springback error in forward direction (spring forward). The new combined method is then called hybrid method (HM) die compensation.

2. Proposed Compensation Method

The hybrid method (HM) is a combination of DA and SF method. The procedures of HM are illustrated in Figure 1. The basic process of the hybrid method can be explained as a simple forming process; a flat metal sheet is bent downwards beyond its elastic region.

The part is deformed into the initial reference geometry as indicated by C^i_{tool}. After the springback, the formed shape distorted to an error value ΔL as seen in Figure 1(1). The reference target shape is then adjusted to $\Delta y = -\Delta L$. The new shape modification field C^{i+1}_{tool} is adjusted directly from the reference product C^i_{tool}, producing the first compensated die geometry $C^{i+1}_{\text{tool}} = C^i_{\text{tool}} + \Delta y^i$.

This process is basically following the DA method. The new die shape is then used for the next forming as seen in Figure 1(2).

If the springback after the unloading process falls out of tolerance at the first compensation, then the part is bent downwards to the new geometry by applying the inverted internal force F^i_{internal} calculated from residual stress (SF algorithm), read as

$$C^{i+1}_{\text{tool}} = C^i_{\text{tool}} + F^i_{\text{internal}} \tag{1}$$

and then sprung forward to the new shape. The deviation should be lower than before and much closer to the reference geometry. If the error is still out of the acceptable tolerance, the compensation process is then continued to new iteration cycle by considering the deviation field (Δy^{i+1}) to create a new die surface. The shape deviation field is calculated between the reference geometry to the last springback geometry. The resulting shape modification field needs to be applied following (1) to the last compensation geometry.

The iteration process continues until the formed shape after the springback falls withn the allowable tolerance of the desired shape by implementing an alternate hybrid algorithm of DA and SF. The algorithms are coded in Fortran to do the iteration processes. The internal residual stresses of the formed shape are obtained from the fully formed simulation results. The internal bending stresses are then converted into finite elemental forces as the input load to the HM spring forward phase. The forces are then applied to the finite elements of the springback shape. The result is compared to the original shape to check the shape deviation error.

To validate the accuracy of the developed hybrid algorithm for die compensation, a compensated die surface generated by hybrid method was tested in Autoform. The springback was then simulated in this software and then compared with the reference shape. To use the die surface of HM in Autoform, the finite elements of the die surfaces should be modified to complete die parts, that is, binder and punch. For this purpose, CATIA was selected to modify and create a surface for springback simulation in Autoform. The HM springback result is then compared with that from Autoform build-in springback compensation module. The tool surfaces are automatically adapted to the new compensated geometry. On the compensation stage, the sprung back shape and the reference geometry are needed. The overall procedure is shown in Figure 2.

The flow of simulations starts from input data of surface die tools consisting of die, punch, binder, and reference surface. The part after springback is compared to the reference geometry for a deviation gap (error) checking. If the error is larger than the allowable tolerance, the die surface needs to be compensated. The generation of compensating surface data is delivered by the software based on DA algorithm. Another simulation is conducted with the new compensated surface until the error is within the tolerance [25].

In this presented research work, the results of HM are compared with those from the stand alone of DA and SF. The comparison results show some critical issues: the maximum error at every iteration cycle, the number of iterations to compensate the die surface, and the total CPU time to converge the compensated die surfaces.

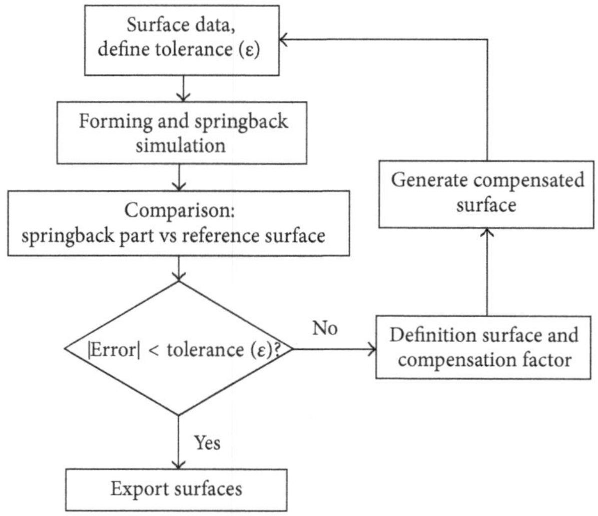

FIGURE 2: Procedure of die compensation.

TABLE 1: Properties of mesh refinement.

Accuracy	Radius penetration (mm)	Maximum element angle (deg)	Maximum displacement angle (deg)
Rough	0.32	45	3.2
Standard	0.22	30	2.2
Fine	0.16	22.5	1.6

3. Numerical Simulation Results

The usage of the hybrid method (HM) algorithm to compensate springback by combining two methods of DA and SF is presented in this section. The results are compared against those from application software Autoform with springback compensation module.

Two models representing two- and three-dimensional problems are used. The two-dimensional (2D) springback model is taken from the U-bending springback problem in NUMISHEET 1993 [26], while the three-dimensional (3D) springback problem is the S-rail benchmark model of NUMISHEET 2008 [27].

3.1. Springback of 2D Model. The U-bending problem consists of a set of punch, holder, die, and blank sheet [26]. The thickness of the blank sheet is 0.8 mm with the maximum punch drawing 70 mm at constant speed of 4.3 m/s. The blank holder force varies between 1, 5, 10, 15, 20, and 25 kN. The sheet material is mild steel DC04 having a Young's modulus 210 GPa and Poisson's ratio 0.3. An initial yield stress is 167.9 MPa, the reference stress value K is equal to 550 MPa, and the work hardening exponent $n = 0.223$.

The mesh quality in Autoform depends on the sets value of parameters radius penetration, maximum element angle, and maximum displacement angle. The mesh refinement varies from rough, standard, and fine, as shown in Table 1.

The results of springback in various BHF are shown in Figure 3. The lowest springback deviation recorded in fine mesh is 1.32 mm under the highest BHF of 25 kN. The higher the BHF, the less the springback. Kergen and Jodogne [28] have shown similar BHF trends against springback in determining the minimum BHF for steels. Hishida and Wagoner [29] determined that a quality defect of products formed by stamping processes, such as fracture, wrinkles, and surface distortion, can be suppressed by the selection of the optimum BHF values. For the present 2D model here, the lowest springback can be achieved by applying higher BHF.

Investigation of the BHF effect continued with increasing the value to reach the failure in forming simulation and to see the thinning history. The holding forces are extended to 26, 27.2, and 27.5 kN in terms of finding the optimum BHF. Figure 3 and Table 2 show clearly that the high springback deviations (higher than 3 mm) occur when the applied holding forces are at 1, 5, 10, and 15 kN. The risks of failure, thinning, and high strain arise under excessive holding forces at 25, 27, and 27.2 kN. In the range from 25 to 27.5 kN, the simulations fail due to the high risk of splitting and failure. This can be seen from the forming limit diagram shown in Figure 4, as indicated in bright color. The result variable maximum failure is defined as the ratio between the

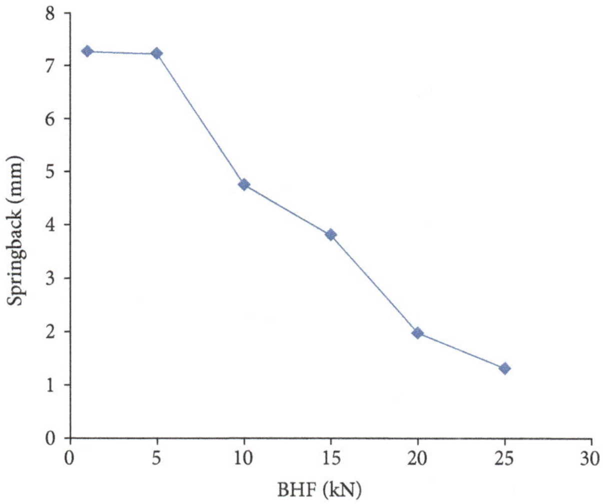

FIGURE 3: Principal result of springback prediction.

TABLE 2: BHF effect in springback, thinning, strain, and failure.

BHF (kN)	Springback (mm)	Thinning (%)	Plastic strain	Failure
1	7.27	0.02	0.07	0.08
5	7.23	0.02	0.07	0.07
10	4.75	0.04	0.08	0.13
15	3.81	0.04	0.09	0.19
20	1.98	0.06	0.12	0.28
25	1.32	0.08	0.16	0.4
26	1.31	0.08	0.20	0.44
27	1.31	0.11	0.30	0.6
27.2	1.31	0.14	0.40	0.7
27.5	1.36	0.20	0.60	1.0

maximum major strain of an element and the major strain on the forming limit curve (FLC) for the same minor strain. Based on the results, the holding force in the range from 20 to 24 kN is suggested as the optimum range of BHF that gives less springback without having a failure risk on the formed sheet.

The adaptive refinement mesh type in Autoform affects the results of springback. Figure 5 shows how element quality affects the springback in the selected holding force 20 kN. The fine meshing shows the closest springback to the experimental result [26]. The highest springback, however, occurs at the flange position. The deviations of other positions of the workpiece are closer to the experimental results.

3.2. Compensation of 2D Model. For the compensation testing, the case BHF 20 kN is selected with the fine mesh quality and the friction coefficient 0.3. This coefficient value is for the general lubricated condition in the forming process [11]. In the sheet metal forming, boundary lubrication is the most widely encountered. It is defined as a condition where the solid surfaces are so close. The reference and springback nodes generated by Autoform are modified to the coordinate x, y, and z under text format to suit the HM input format. There are 601 common points being selected at the reference and springback used for the calculation of error and nodes translations.

Forming simulation for the second cycle is conducted by using new compensated surfaces which were generated under compensation factor 1.0. The compensation types available in Autoform, direct, fixed, and rigid body, are tried. It can be seen in Table 3 that springback results are higher than the initial springback error 1.98 mm (see Table 2) for rigid body types. Figure 6 shows the comparison of springback history results for all compensation types. The fixed draft type converges into the smallest springback error 1.53 mm after the fifth iteration.

In rigid body, there are three adjustments to compensate: automatic, automatic z-direction, and manual. The entire area defined as a rigid body is compensated relative to the average vector. From the result, this is only effective for a number of iterations not more than two. At the second iteration, the rigid body can reach the lowest springback value 1.58 mm and then deviates more on the next iterations. Direct compensation type is influenced by compensation factors only. The compensation factor is 1.0 in each iteration. This compensation makes translation in every direction (x, y, z) become based on the reference of a springback shape. By using fixed draft type, it allows the compensation of surface regions while maintaining their angle towards working direction. This fixed type gives better result than others and the springback deviation is only 1.53 mm.

The sectional view of compensated die and springback position can be seen in Figure 7. At the first iteration, the springback position falls below the reference part (Figure 7(a)), whereas the fifth iteration, the unloaded part, stays at the top of reference surface (Figure 7(b)).

Since the compensation algorithm in Autoform adopts the DA method which considers punch-axis direction, therefore the shape distortion in the sidewall is still significant as is clearly seen in Figure 8.

The springback comparison between HM and Autoform for 2D model is shown in Figure 9. It can be seen clearly that HM performs better than Autoform. At the first iteration, the springback can be decreased significantly to 1.26 mm. The error continues to decrease in the second and third iterations, but then it fluctuates in the fourth iteration showing a higher value than the previous one. The springback reduction trend continues to the last (fifth) iteration. The second and third iteration results are getting higher due to the application of SF method. In the first iteration, it gives a higher residual stress than that from the third iteration that will be used in the SF method in the second and fourth iterations. Therefore, the high residual stress has delivered high compensation value too, as shown at the second iteration for 1.17 mm and 0.97 mm at the fourth iteration. The proposed alternate hybrid method of DA and SF has successfully decreased the springback deviation to 0.67 mm (reducing 66% from

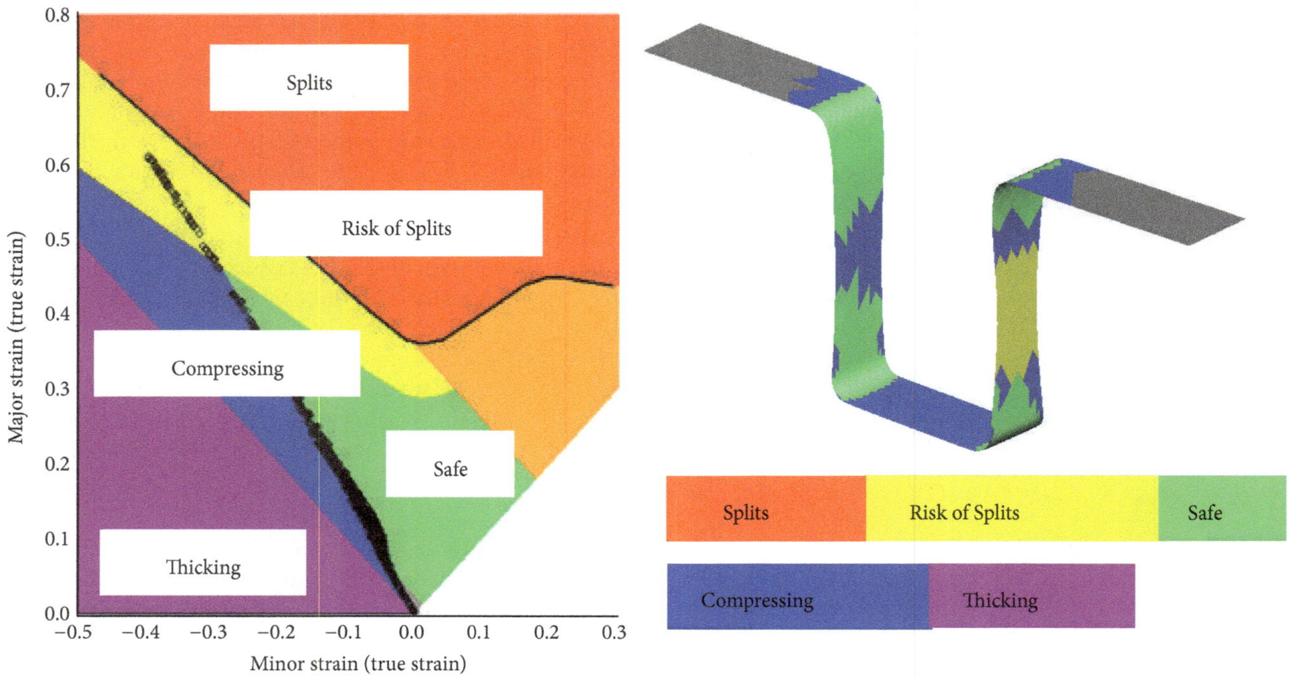

FIGURE 4: Forming limit diagram under BHF of 27.5 kN.

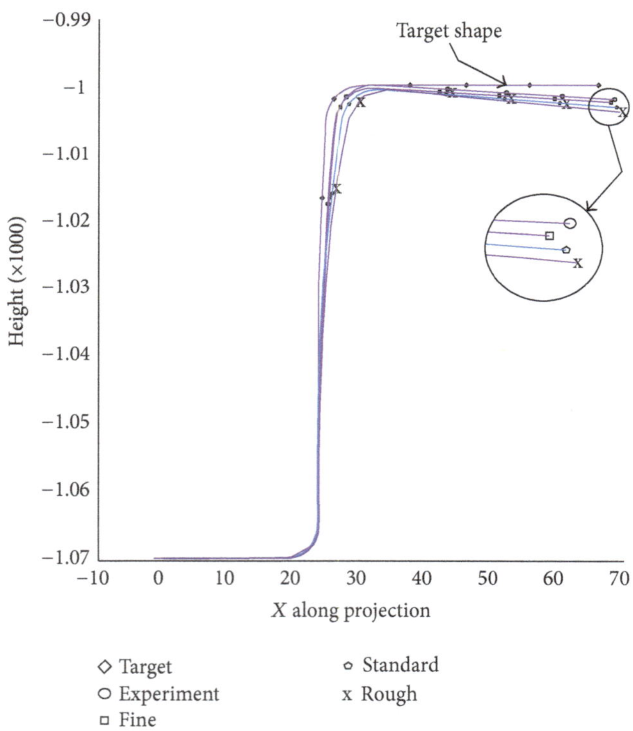

FIGURE 5: The influence of mesh refinement, BHF 20 kN.

the initial error), while Autoform decreased the springback to 1.49 mm (reducing 25% from the initial error).

The alternate HM method also improves the accuracy of the formed product in the sidewall. The maximum deviation

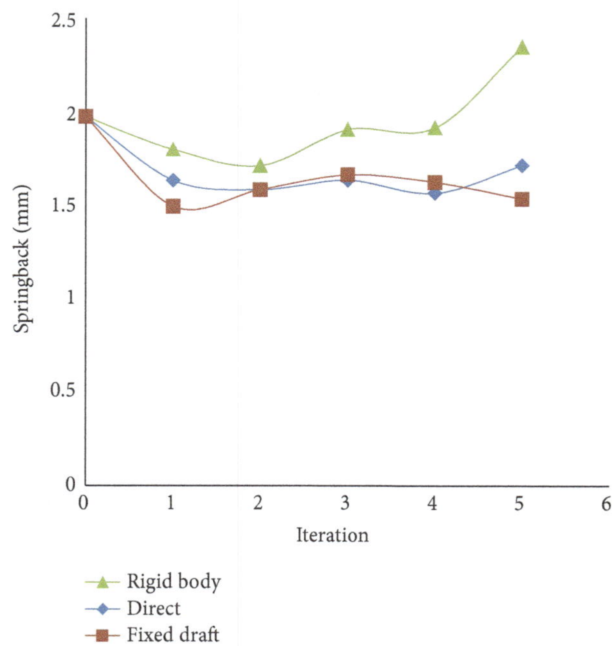

FIGURE 6: Springback error in every cycle of compensation.

after the fifth iteration is 0.935 mm, which is lower than that from Autoform result 1.533 mm.

3.3. Springback of 3D Model. The benchmark of NUMISHEET 2008 has included an additional problem which is related to springback. In this analysis, the accuracy of springback prediction is presented for three different models, that is,

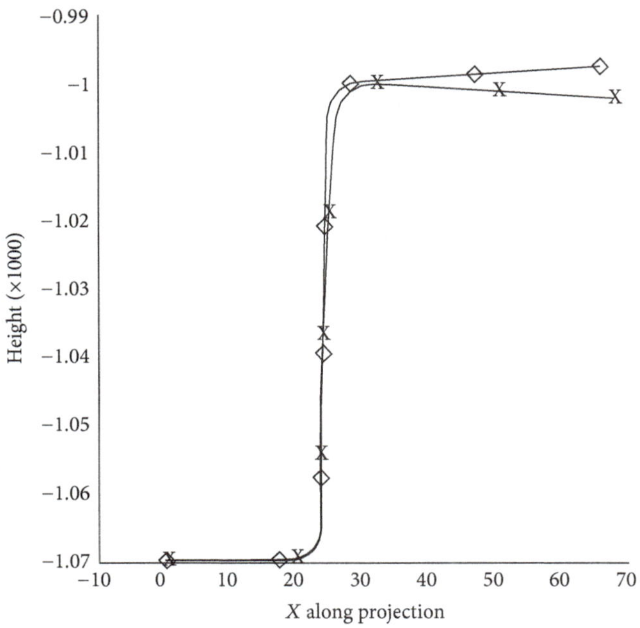

(a) Result from the first iteration

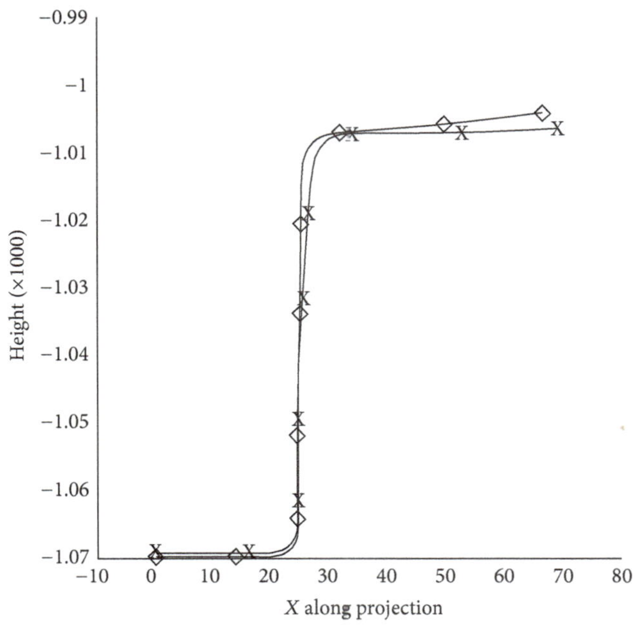

(b) Result from the fifth iteration

FIGURE 7: The influence of mesh refinement, BHF 20 kN.

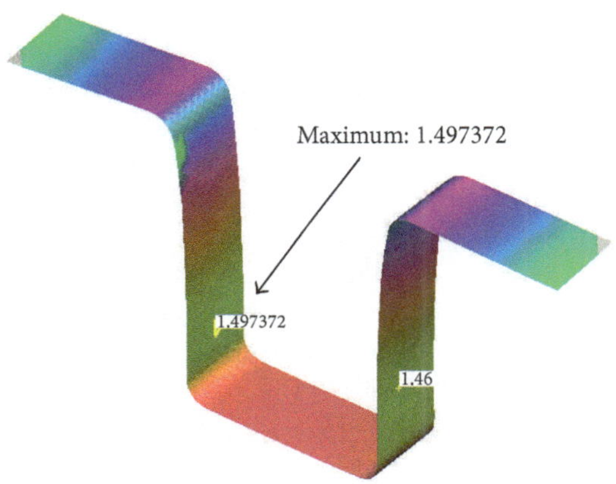

(a) 1st iteration

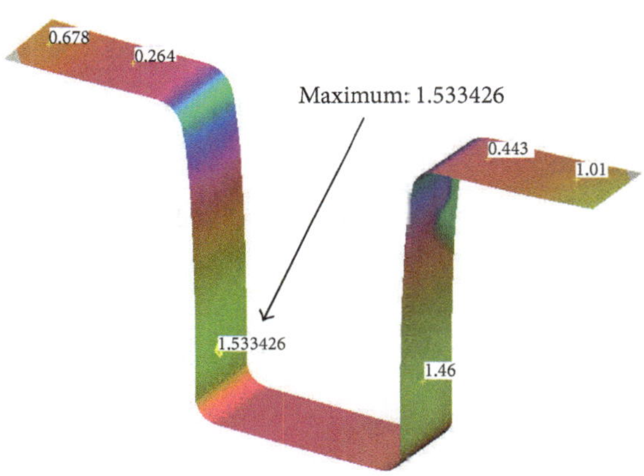

(b) 5th iteration

FIGURE 8: Highest deviation in the sidewall area.

TABLE 3: Springback error after the fifth compensation, BHF 20 kN.

Type	Compensation factor	Working direction	Springback
Direct	1.0	x, y, z	1.71
Fixed draft	1.0	z	1.53
Rigid body	1.0	z	2.35

without draw beads, smooth draw beads, and locking draw beads.

The material HX260LAD type of microalloyed steel grades with high yield strength for cold forming has been selected as the blank sheet material. The summary of mechanical properties of the material is shown in Table 4.

Variations of BHF 90, 120, and 150 kN are applied on the blank holder parallel to punch direction. The comparison results use several reference sections of A, B, C, and D [27].

The difference of draw bead design in the part geometry has a significant effect on springback. The springback deviation in every section does not show consistent trend when different BHFs are applied. The inconsistent springback

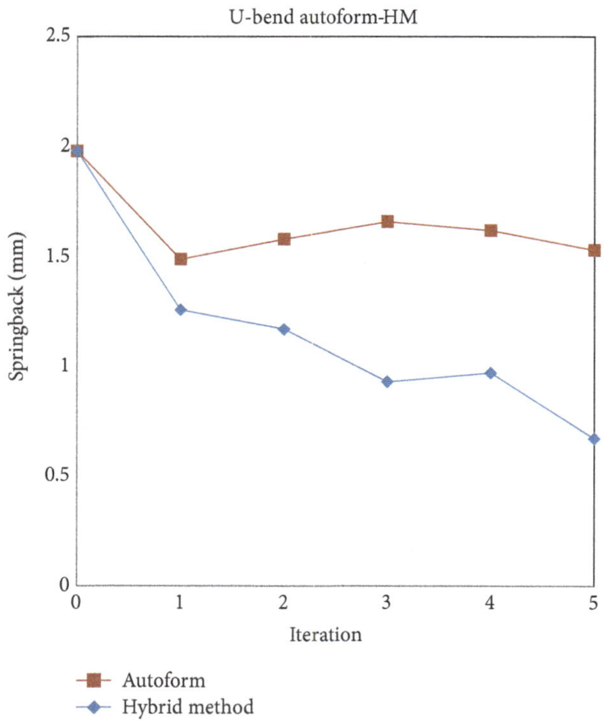

FIGURE 9: Compensation of U-bending results.

TABLE 4: Summary of mechanical properties HX260LAD.

Orient.	Thickness (mm)	Yield stress (MPa)	U.T.S (MPa)	Uniform elongation %	r-value
L	1.00	394.3	463.7	16.4	0.581
T	1.00	427.7	466.0	17.5	1.013
D	1.00	395.3	447.0	17.0	1.166
Mean	1.00	405.8	458.9	16.9	0.981

results are presented in Figures 10(a), 10(b), and 10(c). In the lock bead model, the increase of BHF from 90 to 120 kN gives higher springback of 2.06 mm and 2.15 mm, in sections B and D, and lower springback in sections A and C of 2.12 mm and 1.95 mm, respectively. When the BHF is applied higher at 150 kN, the springback decreased in sections B and C to 1.92 mm and 1.95 mm, respectively. All of these deviations show that the S-rail is in a twisting mode. In a square cup model where twisting does not occur, Demirci et al. [30] showed the trend of higher BHF to reduce the springback error. In this result, the experimental validation with draw beads is not directly performed. The springback experimental validation however has been conducted in different springback benchmark problems [26, 27] and found that Autoform can predict springback in the range of experimental results.

By looking at the average springback in all sections (Figure 10(d)), the trend of springback in different holding forces can be seen. Both smooth and locked beads show that increasing the holding force can reduce the springback deviation. Under the smooth beads, the springback decreases from 1.97 mm to 1.87 mm when the holder force increased from 90 kN to 150 kN. A similar trend is shown for locked beads. It decreases the springback from 2.1 mm to 2.05 mm. When the drawing process does not use any beads, the tendency of the springback error to the increasing holding force cannot be predicted but fluctuated.

In terms of the meshing variation, the recorded mean springback error in standard mesh type (33147 elements) is 1.87 mm, the fined mesh 1.78 mm (56054 elements), and the rough mesh 1.92 mm (17758 elements). The mesh density and the computer CPU (Central Processing Unit) time in Intel i5 processor are shown in Figures 11(a) and 11(b). The total CPU time to finish the simulation for fine mesh type is 26.786 minutes; the standard mesh needs 12.136 minutes and the rough type only needs 5.11 minutes.

The increment number of fines, standard, and rough types are 36, 33, and 29 increments at the end of the simulation.

The optimum result can be achieved with the combination parameters of holder force 150 kN with smooth draw beads in standard mesh type. The forming limit diagram (FLD) using the mentioned combination is shown in Figure 12. The figure shows that there is no risk of splits and excess thinning when the blank sheet has been fully formed.

3.4. Compensation of 3D Model. For die compensation, the S-rail model material HX260LAD, BHF 150 kN with smooth beads in standard mesh type, is selected. The mean springback result of this case is 1.87 mm as depicted in Figure 13(a). The reference and springback model of S-rail is converted as point clouds to be used as the input file of HM. The first iteration is conducted by translating the springback nodes following the DA method and then to the inverse of springback (SF algorithm) to get the first compensated die as shown in Figure 13(b).

For convergence control, a tolerance error 1 mm is used. After the fifth cycle, the mean springback can be decreased as low as 0.83 mm and the iteration process terminated (error < 1 mm). Figure 14 shows the compensated die surface and the springback after fifth iteration.

To use the die surface of HM results in Autoform; the elements must be modified to form the die surface components, that is, binder and punch. CATIA is used to modify and to create surface for springback simulation in Autoform. After springback simulation, the HM result is then compared with the springback from Autoform.

Figure 15 shows the springback compensation history of the alternate HM method compared with Autoform. The first iteration of HM has decreased the springback error to 1.13 mm (39% reduction from initial value). With the alternate DA and SF, after the fifth iteration, the springback deviation can be reduced by 55% from the initial error value from 1.87 mm to 0.83 mm while Autoform can only reduce to 1.23 mm or 34% reduction.

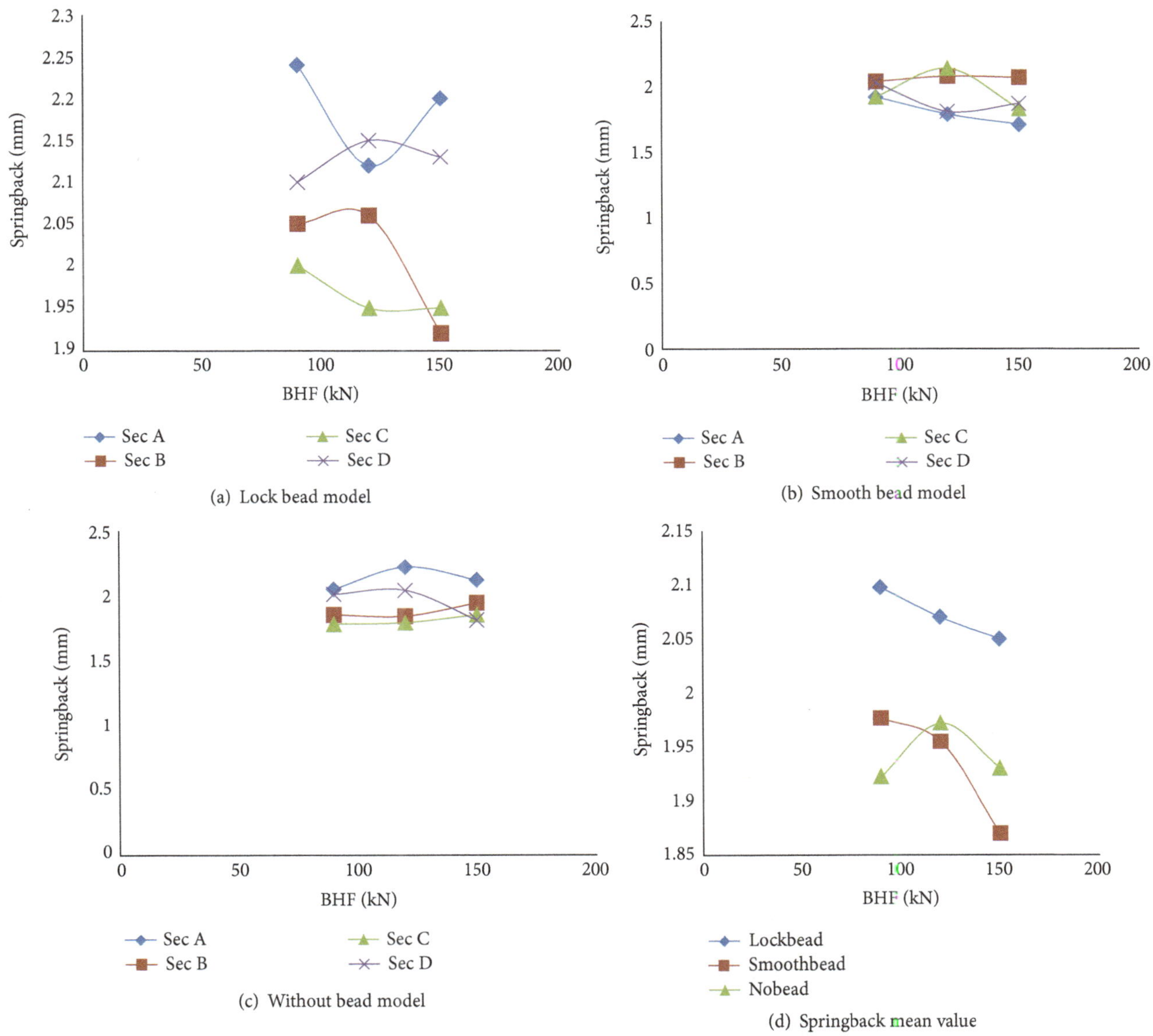

FIGURE 10: Springback error in sections A, B, C, and D.

The fluctuation of HM shown in Figure 15 is due to the switching method from DA to SF in the first time. In the next iteration cycle, the shape is in a stable condition and can converge to reduce the springback deviation.

4. Discussion

In die compensation, the die surface results rely on springback analysis and the deflection history of elements under the bending and unbending processes.

Deformation theory could be used in prediction instead of flow theory. Accurate data and parameters will require more complicated optimization to accommodate deformation history. Therefore, the springback analysis accuracy was influenced by many factors.

The use of appropriate material model in springback analysis is one factor to improve the accuracy because the correct material can provide an accurate stress state at the end of the forming stage. It is useful if the material model is based on the initial yielding and hardening parameters obtained from the average of multiple experiments. The accuracy of springback prediction is also influenced by the coefficient of friction between the blank sheet and the die surfaces.

The use of combination between optimum blank holder force and draw beads will reduce the springback deviation. For complicated three-dimensional model where the springback is not as simple as bending problem but involving twisting mode, finding the optimum drawing parameters is not straight forward, but a simulation series is required to find the best configuration for a minimum springback.

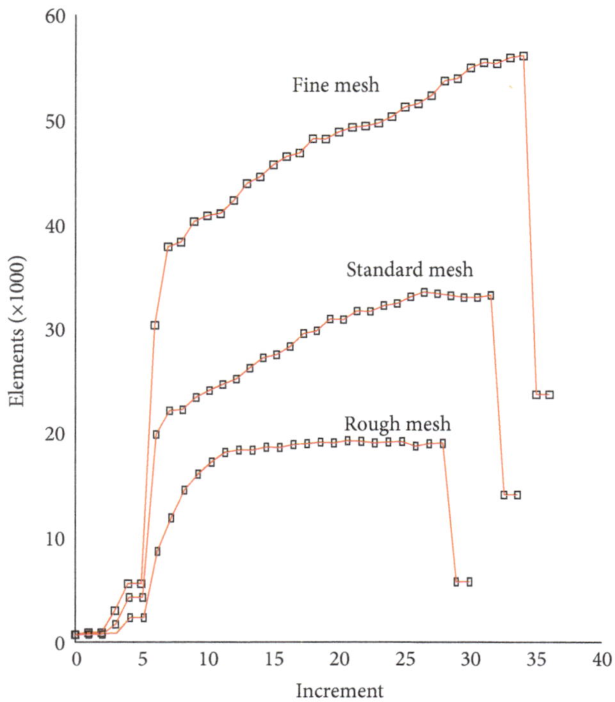

(a) Element mesh increment

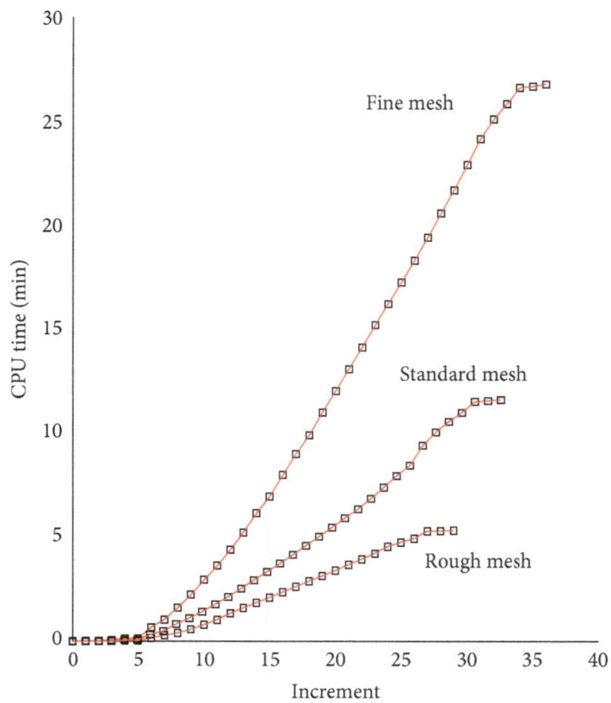

(b) CPU time

FIGURE 11: Springback error in sections A, B, C, and D.

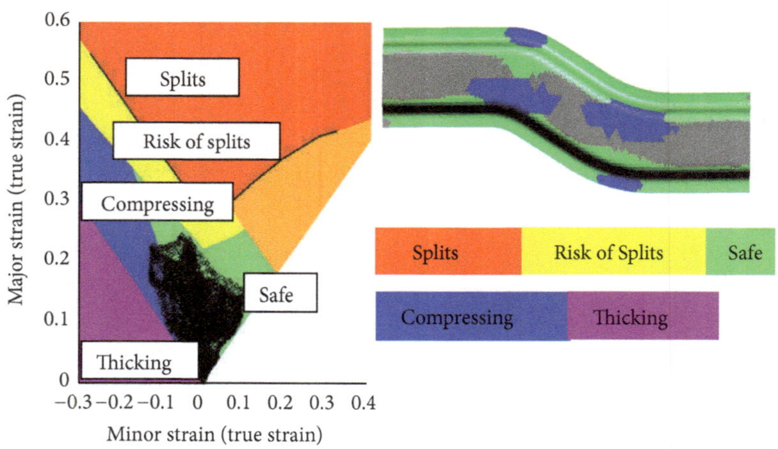

FIGURE 12: Forming limit diagram of the BHF 150 kN.

Springback distortion also depends on the bending moment which in turn depends on the stress distribution through sheet thickness. Shell elements require numerical integration of stress and strain distribution. The largest of integration points will improve the accuracy, but consequently is more time consuming.

In these works, the HM has been coded in Fortran mainly to enable the alternate methods of DA and SF. The displacement adjustment (DA) strategy is intended to generate compensated nodes by translating the nodes in the opposite direction to the springback. The magnitude of the vector translations calculated from the origin and springback position are distributed in all directions Δy, Δx, and Δz corresponding to the axes of x, y, and z.

The strategy to speed up the springback reduction is by using a large compensation factor to the translation vectors which later copied to the die surfaces. This reduces the springback error significantly in the first cycle (see Figure 15).

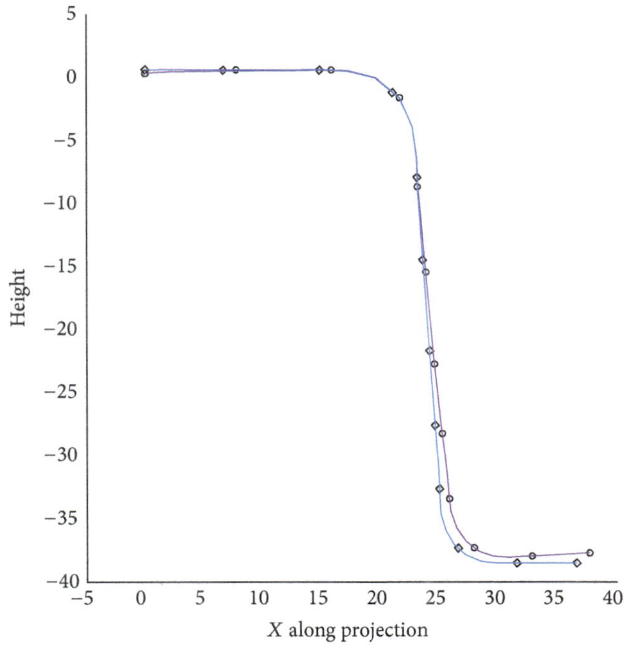

(a) Springback from original die

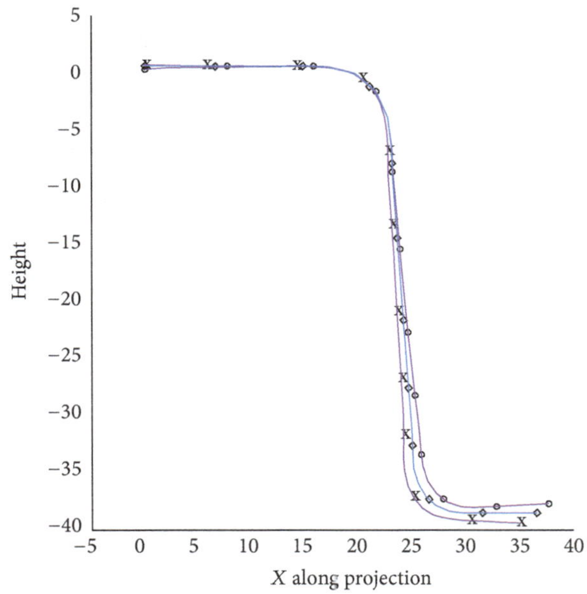

(b) Compensation from the first springback

FIGURE 13: Springback and compensation.

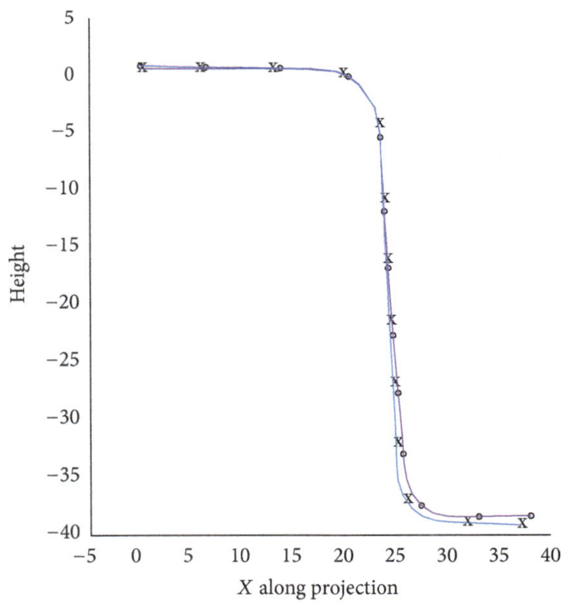

FIGURE 14: HM die compensation after five cycles.

In the subsequent process, the applied load force in spring forward algorithm is obtained from the internal stress at fully loaded phase. The stresses are converted to inverted load forces.

5. Conclusion

In this research, two different methods, displacement adjustment (DA) and spring forward (SF), are joined in alternate manner to compensate the die tools to minimize springback error called hybrid method (HM). When it is used in one algorithm, there are advantages to converge faster and abilities to compensate in all sides. A new approach in springback accommodation using the alternate hybrid method (HM) has been tested in two- and three-dimensional springback problem models. The results show that in two-dimensional models, it can reduce the springback up to 66% in after five iteration cycles and in three-dimensional 55%. In the comparison result with Autoform, the HM method shows better performance while the Autoform can reduce 22% and 35% only for two- and three-dimensional models. The HM is an alternate method to reduce springback error based on die tool compensation. This is applicable for sheet metal forming on stamping process having lower and upper dies.

The implication of this research is that the approach of accommodating springback can be further extended to the implementation of HM in user friendly application software. In this work, the proposed HM approach works on finite element nodes and uses an external CAD program to modify them to point clouds before redrawing them as the die parts. In the future study, an automatic implementation of HM should be ported in a finite element forming simulation program so that the implementation will be easier and user friendly.

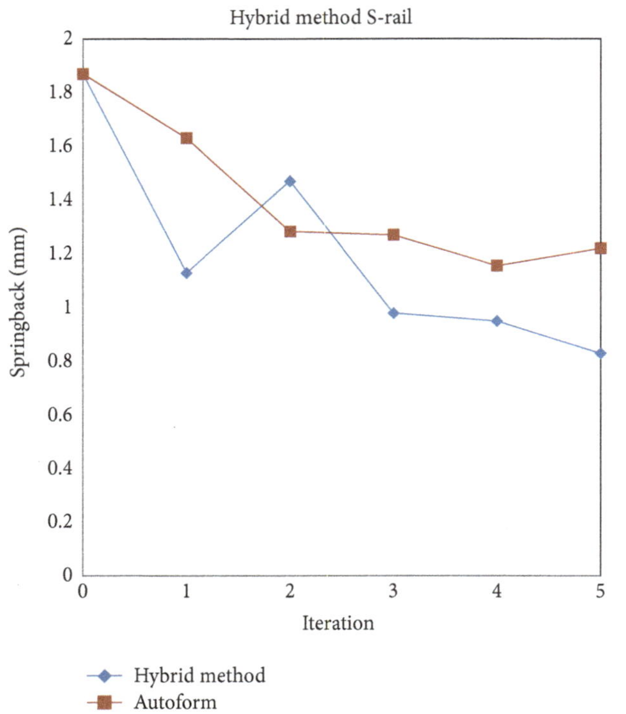

FIGURE 15: Comparisons between Autoform and HM compensation.

Conflict of Interests

The authors declare that there is no conflict of interests regarding the publication of this paper.

Acknowledgments

The authors would like to thank the Universiti Tun Hussein Onn Malaysia (UTHM). This research work has been supported by a Fundamental Research Grant Scheme (FRGS) Vot. no. 0748.

References

[1] M. P. Groover, *Fundamentals of Modern Manufacturing, Materials, Processes, and Systems*, John Wiley & Sons, New York, NY, USA, 2007.

[2] R. H. Wagoner, H. Lim, and M.-G. Lee, "Advanced issues in springback," *International Journal of Plasticity*, vol. 45, pp. 3–20, 2013.

[3] G. Liu, Z. Lin, W. Xu, and Y. Bao, "Variable blankholder force in U-shaped part forming for eliminating springback error," *Journal of Materials Processing Technology*, vol. 120, no. 1-3, pp. 259–264, 2002.

[4] J. Q. Li, S. F. Jiang, H. P. Wu, W. Zhu, and C. D. Lu, "Study on subsection variable blank holder force in deep drawing of rectangular parts," *Materials Research Innovations*, vol. 15, supplement 1, pp. S230–S233, 2011.

[5] S. Kitayama, S. Huang, and K. Yamazaki, "Optimization of variable blank holder force trajectory for springback reduction via sequential approximate optimization with radial basis function network," *Structural and Multidisciplinary Optimization*, vol. 47, no. 2, pp. 289–300, 2013.

[6] Y. H. Moon, S. S. Kang, J. R. Cho, and T. G. Kim, "Effect of tool temperature on the reduction of the springback of aluminum sheets," *Journal of Materials Processing Technology*, vol. 132, no. 1-3, pp. 365–368, 2003.

[7] R. Barea, G. Conejero, N. Candela, and M. Carsí, "Hot forming of a new steel used in stamping dies and tooling," *International Journal of Materials Research*, vol. 104, no. 3, pp. 281–285, 2013.

[8] X. Fan, Z. He, S. Yuan, and K. Zheng, "Experimental investigation on hot forming-quenching integrated process of 6A02 aluminum alloy sheet," *Materials Science and Engineering A*, vol. 573, pp. 154–160, 2013.

[9] S. Hoja, H. Klümper-Westkamp, F. Hofmann, and H.-W. Zoch, "Plasma nitriding of tool steels for hot massive forming," *Journal of Heat Treatment and Materials*, vol. 68, no. 1, pp. 3–12, 2013.

[10] H. Laurent, R. Grèze, M. C. Oliveira, L. F. Menezes, P. Y. Manach, and J. L. Alves, "Numerical study of springback using the split-ring test for an AA5754 aluminum alloy," *Finite Elements in Analysis and Design*, vol. 46, no. 9, pp. 751–759, 2010.

[11] H. Tschaetsch, *Metal Forming Practise: Processes—Machines—Tools*, Springer, New York, NY, USA, 2005.

[12] M. Kadkhodayan and I. Zafarparandeh, "On the relation of equivalent plastic strain and springback in sheet draw bending," *International Journal of Material Forming*, vol. 1, no. 1, pp. 141–144, 2008.

[13] Y. E. Ling, H. P. Lee, and B. T. Cheok, "Finite element analysis of springback in L-bending of sheet metal," *Journal of Materials Processing Technology*, vol. 168, no. 2, pp. 296–302, 2005.

[14] J. Slota, M. Jurcisin, and M. Dvorak, "Experimental and numerical analysis of springback prediction in U-bendings of anisotropic sheet metals," *Zeszyty Naukowe Politechniki Rzeszowskiej. Mechanika*, vol. 85, no. 4–13, pp. 525–533, 2013.

[15] W. Gan and R. H. Wagoner, "Die design method for sheet springback," *International Journal of Mechanical Sciences*, vol. 46, no. 7, pp. 1097–1113, 2004.

[16] A. P. Karafillis and M. C. Boyce, "Tooling design in sheet metal forming using springback calculations," *International Journal of Mechanical Sciences*, vol. 34, no. 2, pp. 113–131, 1992.

[17] Autoform, *Autoform Plus R2 Documentation*, Autoform, Neerach, Switzerland, 2010.

[18] W. A. Siswanto and B. Omar, "Die surface design optimization accommodating springback assisted by an automatic surface generator," *International Journal of Material Forming*, vol. 2, no. 1 supplement, pp. 797–800, 2009.

[19] X. Li, X. Zhan, X. Liu, and B. Deng, "Die surface compensation based on the springback," *Journal of Theoretical and Applied Information Technology*, vol. 45, no. 2, pp. 675–680, 2012.

[20] X.-A. Yang and F. Ruan, "Compensation direction for die-face adjustment based on springback compensation," *Journal of Jilin University (Engineering and Technology Edition)*, vol. 42, no. 1, pp. 103–108, 2012.

[21] H. S. Cheng, J. Cao, and Z. C. Xia, "An accelerated springback compensation method," *International Journal of Mechanical Sciences*, vol. 49, no. 3, pp. 267–279, 2007.

[22] A. P. Karafillis and M. C. Boyce, "Tooling and binder design for sheet metal forming processes compensating springback error," *International Journal of Machine Tools and Manufacture*, vol. 36, no. 4, pp. 503–526, 1996.

[23] G. Brabie, B. Chirita, N. Nanu, and V. Ciubotariu, "Analysis of springback and residual stresses generated by cold plastic forming in drawn round parts made from steel sheets," *Metalurgia International*, vol. 14, no. 12, pp. 21–27, 2009.

[24] R. H. Wagoner, W. Gan, K. Mao, S. Prise, and F. Rasouli, "Design of sheet forming dies for springback compensation," in *Proceedings of the 6th International ESAFORM Conference*, 2003.

[25] L. Wei and Y. Yuying, "Multi-objective optimization of sheet metal forming process using pareto-based genetic algorithm," *Journal of Materials Processing Technology*, vol. 208, no. 1–3, pp. 499–506, 2008.

[26] K. B. Nielsen and N. Bronnberg, "Simulation of the 2D draw bending process: numisheet benchmark," in *Proceedings of the 2nd International Conference (NUMISHEET '93)*, Numisheet, Tokyo, Japan, August-September 1993.

[27] K. Roll, K. Wiegand, and P. Hora, "Benchmark 2—Influence of drawbeads on the springback behaviour," in *Proceeding of the 7th International Conference and Workshop on Numerical Simulation of 3d Sheet Metal Forming Processes*, Interlaken, Switzerland, September 2008.

[28] R. Kergen and P. Jodogne, "Computerized control of the blank holder pressure on deep drawing presses," Technical Paper 920433, Society of Automotive Engineers, 1992.

[29] Y. Hishida and R. H. Wagoner, "Experimental analysis of blank holding force control in sheet forming," Technical Paper 930285, Society of Automotive Engineers, 1993.

[30] H. I. Demirci, C. Esner, and M. Yasar, "Effect of the blank holder force on drawing of aluminum alloy square cup: theoretical and experimental investigation," *Journal of Materials Processing Technology*, vol. 206, no. 1–3, pp. 152–160, 2008.

13

Modelling and Analysis on Biomechanical Dynamic Characteristics of Knee Flexion Movement under Squatting

Jianping Wang,[1] Kun Tao,[2] Huanyi Li,[3] and Chengtao Wang[4]

[1] *School of Mechanical Engineering, Henan Polytechnic University, Jiaozuo 454003, China*
[2] *Department of Orthopedics, Shanghai 10th People's Hospital, Tongji University School of Medicine, Shanghai 200072, China*
[3] *Department of Orthopedics, Zhenjiang Jinshan Hospital, Zhenjiang 212001, China*
[4] *School of Mechanical Engineering, Shanghai Jiao Tong University, Shanghai 200240, China*

Correspondence should be addressed to Chengtao Wang; sjtuerwang@126.com

Academic Editors: S. Hirokawa and B. Mody

The model of three-dimensional (3D) geometric knee was built, which included femoral-tibial, patellofemoral articulations and the bone and soft tissues. Dynamic finite element (FE) model of knee was developed to simulate both the kinematics and the internal stresses during knee flexion. The biomechanical experimental system of knee was built to simulate knee squatting using cadaver knees. The flexion motion and dynamic contact characteristics of knee were analyzed, and verified by comparing with the data from in vitro experiment. The results showed that the established dynamic FE models of knee are capable of predicting kinematics and the contact stresses during flexion, and could be an efficient tool for the analysis of total knee replacement (TKR) and knee prosthesis design.

1. Introduction

All kinds of movements of the knee joint are harmonious in each joint. Due to the complicated structure and large quantity of motion, its noneffective rate in all the joints is on top. The success rate of artificial joint replacement surgery has reached 90% [1]. Even so, there were functional failure, prosthesis loosening or dislocation and the excessive wear of prosthesis, and so forth, postoperatively [2, 3]. The main factors for operation failure, in addition to the pathological reasons, come from operative and prosthetic aspects. However, the disease prevention of human knee, the design of artificial knee prosthesis, and the improvement of surgical technique depended on the research into the movement, stress, and such biomechanical characteristics about natural and artificial knee joint [4].

The internal joint contact stress and distribution of natural and artificial knee joint are directly related to its functional activities. It becomes more important to analyze the biomechanical characteristics of the motion and stress during high flexion activities [5], especially significant for population groups where lifestyle and work activities or religious activities demand deep flexion such as squatting and kneeling. However, the corresponding measurement and prediction are relatively difficult because of the limitation of ethic and measuring devices. So far, there were few effective methods to directly measure the internal stress and distribution for in vivo knee. Therefore, establishing the knee joint calculation model becomes a widely used method for predicting the internal stress and distribution. Among them, the dynamic finite element analysis has been developed into an effective method to predict the internal stress and distribution under dynamic loading conditions [6–9]. In previous models, relatively small range of flexion was conducted, usually not more than 120 degrees; there were less finite element models which can simultaneously proceed with dynamic synchronous prediction for the patellofemoral joint and femorotibial joint [7], and less have represented the whole joint with physiological soft tissue constraint [10].

In this paper, the anatomical model of three-dimensional geometric natural knee was reconstructed. Dynamic finite element (FE) model of natural knee joint, which includes tibiofemoral, patellofemoral articulations and the surrounding soft tissues, was developed in this research, to simulate both the kinematics and internal stress during knee flexion. The biomechanical experimental system of knee flexion motion was set up to simulate human knee squatting using cadaver knees. The flexion motion and dynamic contact characteristics of knee joint were analyzed and were verified by comparison with the data from cadaver in vitro experiment.

2. Materials and Methods

2.1. Dynamic FE Knee Model

2.1.1. The Modeling of Geometric Knee. The knee of a healthy volunteer (height 1.73 m, weight 60 kg, male) had been scanned by CT (computed tomography) and MRI (magnetic resonance imaging), respectively. Then the simulation models of the knee joint bone and soft tissues were reconstructed, respectively. Due to the errors of reconstruction leading to the errors of measurement and calculation of motion, the research of precision of cortical bone reconstruction had been carried out [11]. However, the extraction of CT and MRI data was in different coordinates. It could not be directly obtained for geometric anatomic model of knee, which includes both bone and soft tissue. Bone's point cloud contour with clear geometric feature point can be cached through the MRI, so the principle of three-dimensional image registration [12] was adopted to register the point cloud contour obtained by MRI and the bone tissue profile obtained by CT. Then, a complete geometry simulation model of knee joint was built by registration of soft and bone tissues (Figure 1). The methodology and models used in this work were described thoroughly by Wang et al. (2009) [13].

2.1.2. The FE Model of Knee. Based on the above anatomical model, the mesh model of natural knee was established, which includes both bone and soft tissues, such as cartilage, meniscus, anterior and posterior cruciate ligament, medial and lateral collateral ligaments, and patellar tendon (Figure 2). The hexahedron units were adopted in all bone and soft tissues of knee FE model to reduce calculation cost. The methodology and models used in this work were described thoroughly by Wang et al. (2009) [14].

There are different features of each tissue's material properties. The material properties of the different tissues derived from literatures facilitate comparison with them. For MCL and LCL, it is an average constant from Gardiner et al. (2001) [15]. For ACL, it is single axial stress strain curve from Butler et al. (1980) [16]. For cruciate ligament and patellar tendon, it was from Suggs et al. (2008) [17]. The meniscus can be regarded as elastic and isotropic in axial, radial, and circumferential direction, respectively [18]. In this paper, the meniscus material parameters were from LeRoux and Setton (2002) [19].

Load and boundary conditions, for the FE model of natural knee, were consistent with the experimental conditions. The value of 400 N was applied to the quadriceps, which was parallel to the femur shaft and directed to the starting point of quadriceps. The value of 300 N force was applied along with the knee joint force line to simulate the weight of body [20]. Under the control of the applied muscles force, the femur move is relative to the tibia with full freedom. The movement of tibia was not active and it was *determined* by the loading conditions distributed in knee model. Nine surface contact pairs were defined for the femur, tibia, patella, and other soft tissue.

2.2. Experimental Verification. Six male volunteers' lower limb specimens were mounted into the specially designed loading and connecting device, which were connected with the standard material testing machine (CSS-44010, Changchun Research Institute for Testing machines Co., Ltd, China). This measurement system was adopted to measure both the dynamic movement and contact stress of both the tibiofemoral and the patellofemoral joints (Figure 3).

2.2.1. Experimental Devices. To realize simulation of squat movement and force loading, loading and connecting mechanisms were designed connected with the tension/compression testing machine, based on the biomechanical and boundary conditions of FEA. Accompanying squatting, three sets of measuring system were used to measure *synchronously* the relative motion and the stress of tibiofemoral and patellofemoral joint, realizing force control and loading. Experimental devices were composed of devices of loading and connective devices and the tension/compression testing machine. The loading and connective devices were designed to connect the lower limbs of cadavers and tension/compression testing machine, loading the gravity and quadriceps force to realize joint flexion, which were composed of the upper and the lower connection device (Figure 4). The cadaver knee's dynamic translation and rotation (and lock) in coronal, sagittal, and cross sectional planes can be implemented, simulating living knee movement.

Optical tracking system (Polaris hybrid optical tracking system, NDI, Calgary, AB, Canada) was used to measure the trajectory of the femur, tibia, and patella under squatting for analysis of knee relative motion by coordinate transformation [21] (Figure 5). Coordinate system was established by measuring knee's bone marker points, and knee joint rotations were defined according to the clinical joint coordinate system (JCS) [22, 23]. The same method was used for FE model.

Tecscan measurement system (Tescan Inc., Boston, MA, USA) was used to implement contact measurement (Figure 6). The measurement system is comprised of I-scan sensor, data conversion handle, data analysis, and calibration software. The special sensor I-scan 4000 was used for tibiofemoral joint. This sensor consists of two pieces of separate sensors, whose specifications are of 33 mm * 28 mm, 0.1 mm thick. The I-scan 5051 sensor was used for patellofemoral joint's contact measurement [24]. In the experiments, the articular capsule was opened, and then the sensing piece of I-scan sensor was, respectively, put into the joint gaps of tibiofemoral and patellofemoral joint (Figure 7).

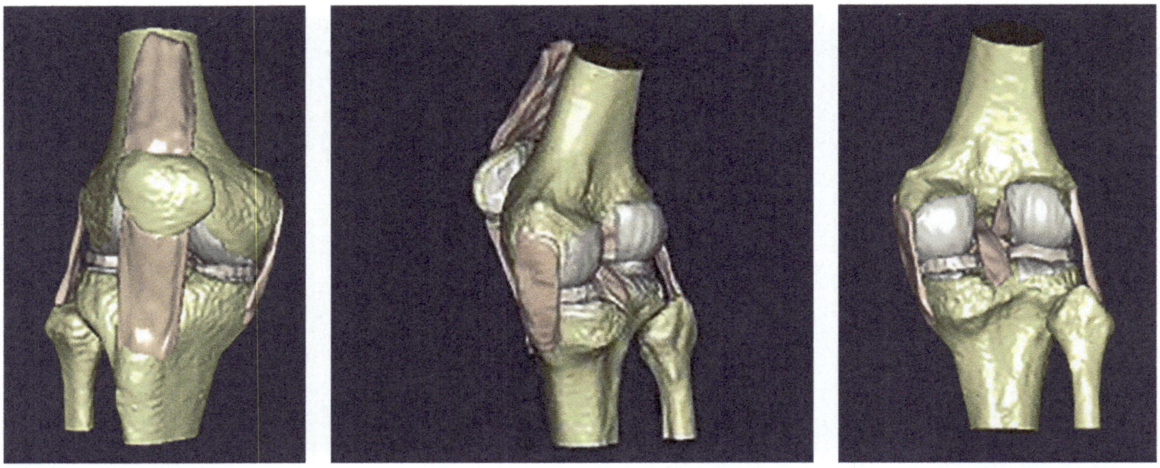

(a) (b) (c)

FIGURE 1: The geometric anatomy model of human knee.

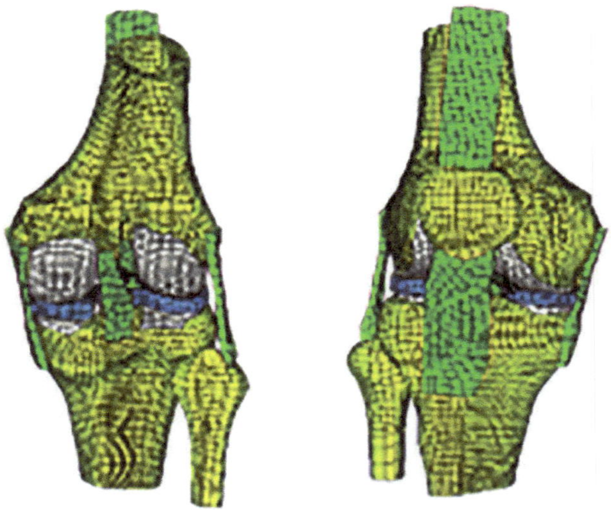

(a) (b)

FIGURE 2: The FE model of knee.

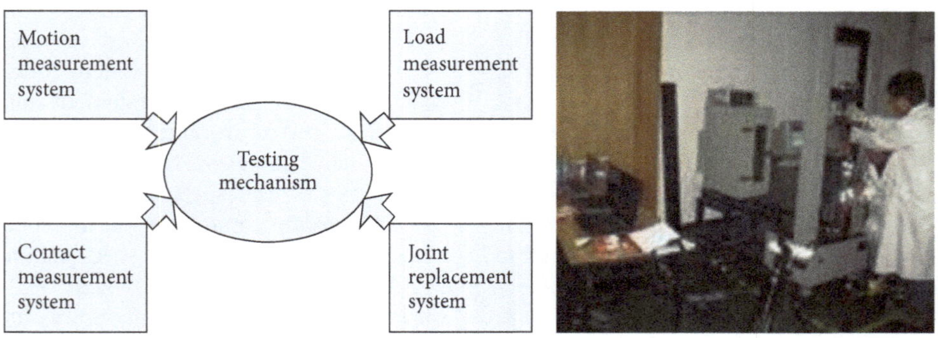

(a) (b)

FIGURE 3: The flowchart and the real of knee experiment system.

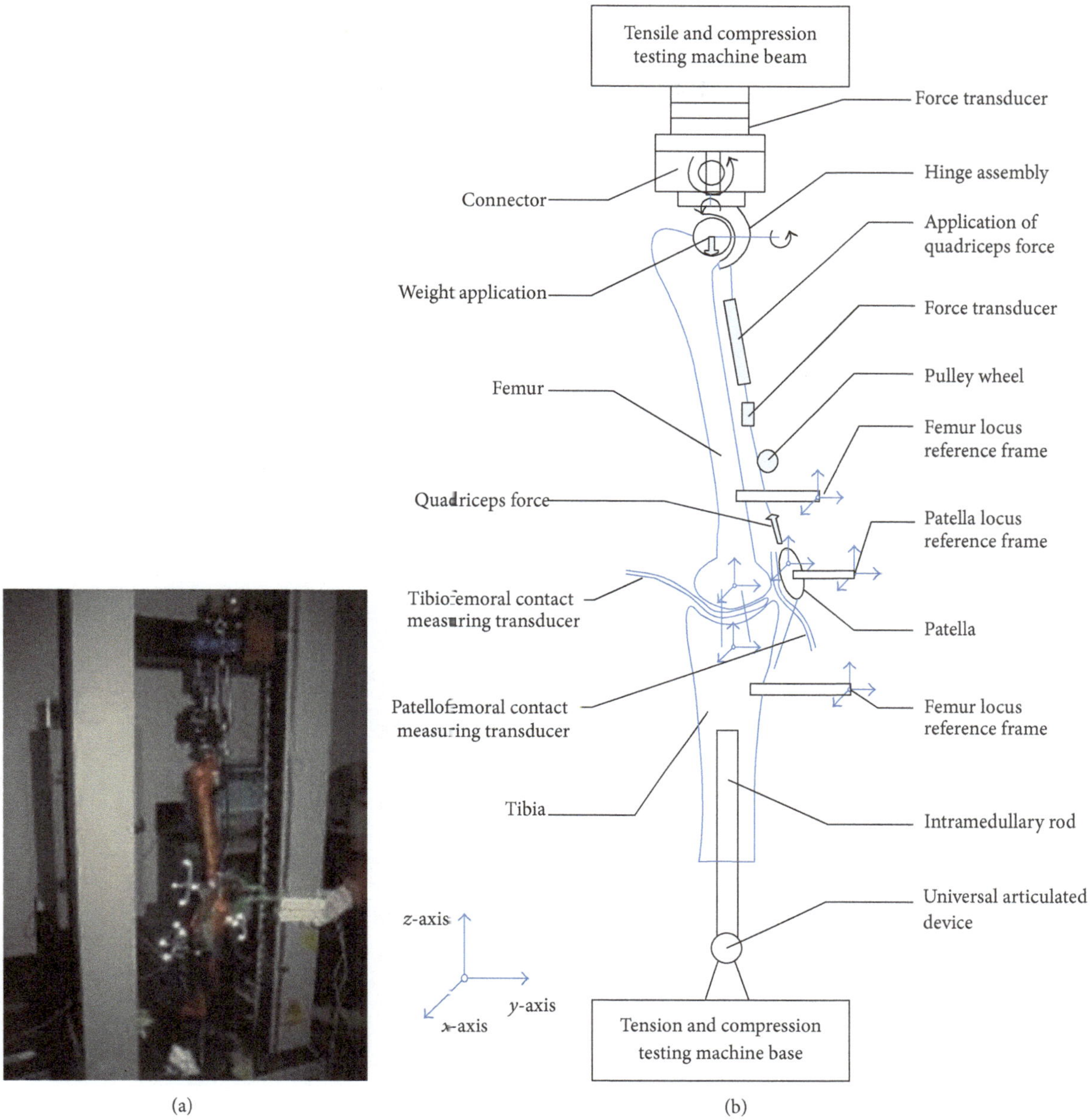

Figure 4: Setup of knee experiment.

The gravity loading of the femoral head was controlled and *measured* by the sensor on the mobile beam of tension/compression testing machine. The quadriceps force and connection test device was connected to the force sensor CFBLS-25 (25 Kg maximum load, 0.03% FS, Shanghai Yichuan Instrument Factory) and the amplifier VM641 (±0.1% accuracy, Guangzhou Huamao Sensor Co., Ltd.). The sensor, pulley, and screw were connected to form loading and measuring device, which was used to measure and control the tension of quadriceps. The output signal of the amplifier was collected by the DAQ Card (data acquisition card) (Yanhua PCI-1710L, Advantech Co., Ltd.). And the measured results were outputted by the commercial software Labview (Laboratory Virtual Instrumentation Engineering Workbench, United States National Instruments Company).

2.2.2. Experimental Loading and Measurement. The I-scan 4000 and the I-scan 5051 pressure test pieces were placed, respectively, in tibiofemoral and patellofemoral joint, and the articular capsule was sutured. The specimen was fixed into the experimental platform. The gravity loading on the femoral head was controlled and measured by sensor of tension/compression testing machine. The tension of quadriceps tendon was outputted by the DAQ Card and the commercial software Labview. Simultaneously, the contact pressure of tibiofemoral joint was recorded by the I-scan contact measurement system. For dynamic movement measurement, three reference frames consisting of 14 mm diameter markers were fixed in tibia, femur, and patella, respectively. The data of markers were captured and the relative motions of patella,

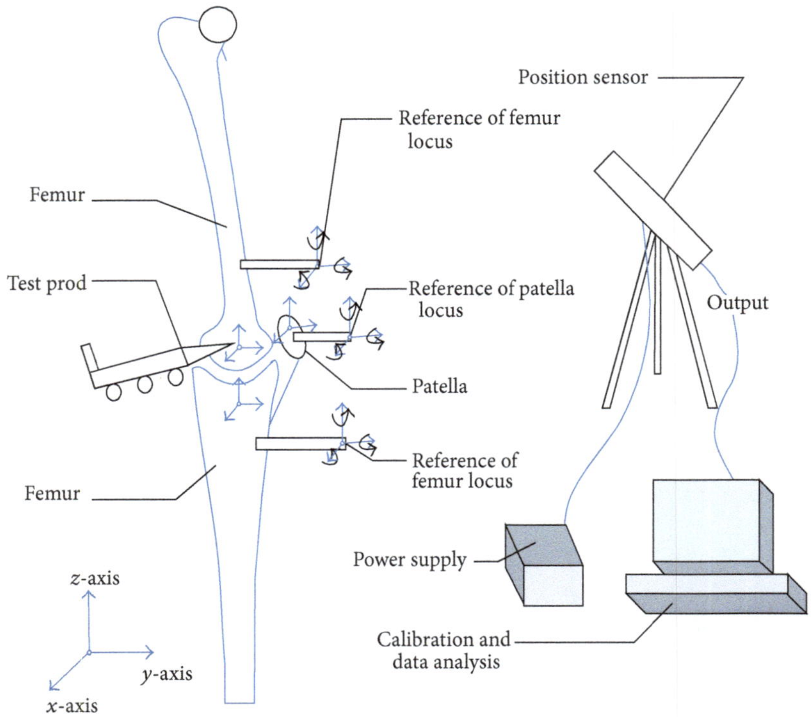

FIGURE 5: The measurement system of movement.

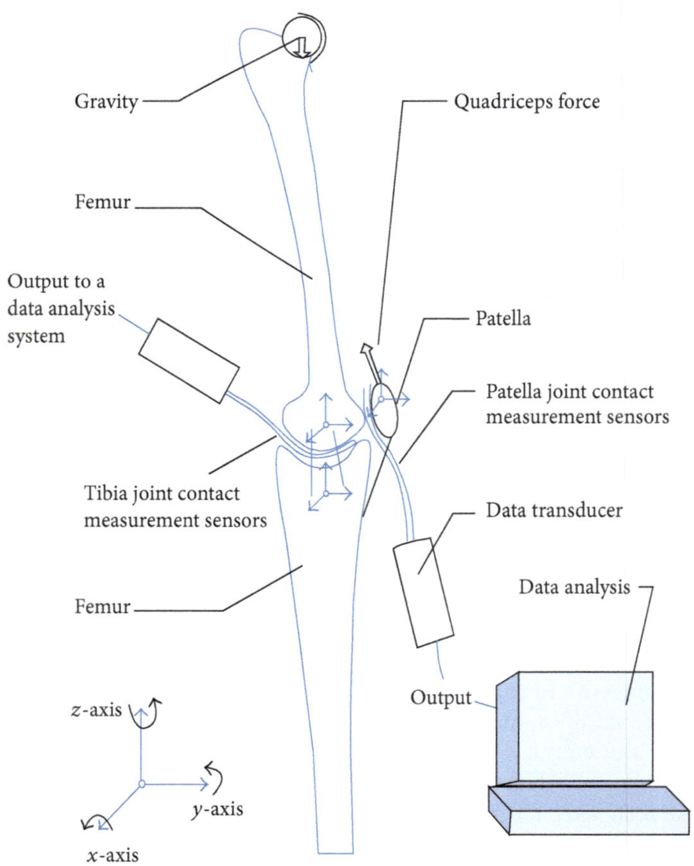

FIGURE 6: The measurement system of contact.

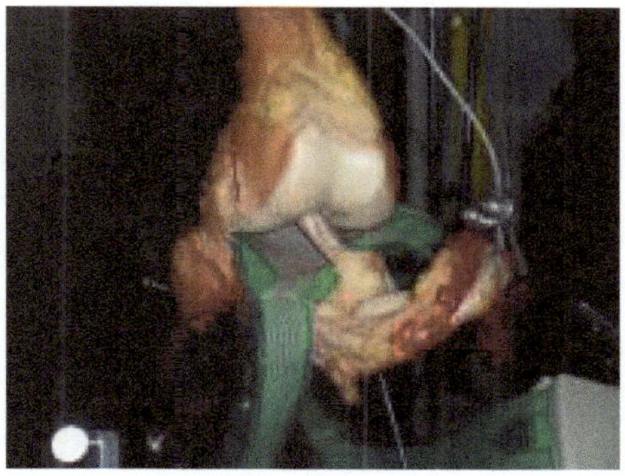

FIGURE 7: The location of Tecscan sensor.

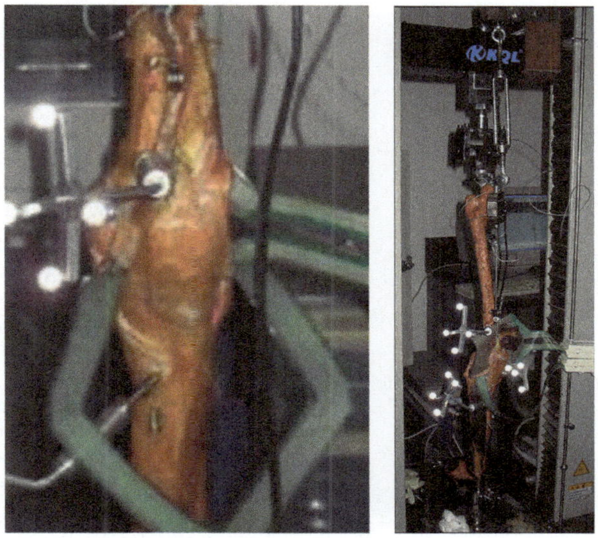

(a) (b)

FIGURE 8: The device after installation.

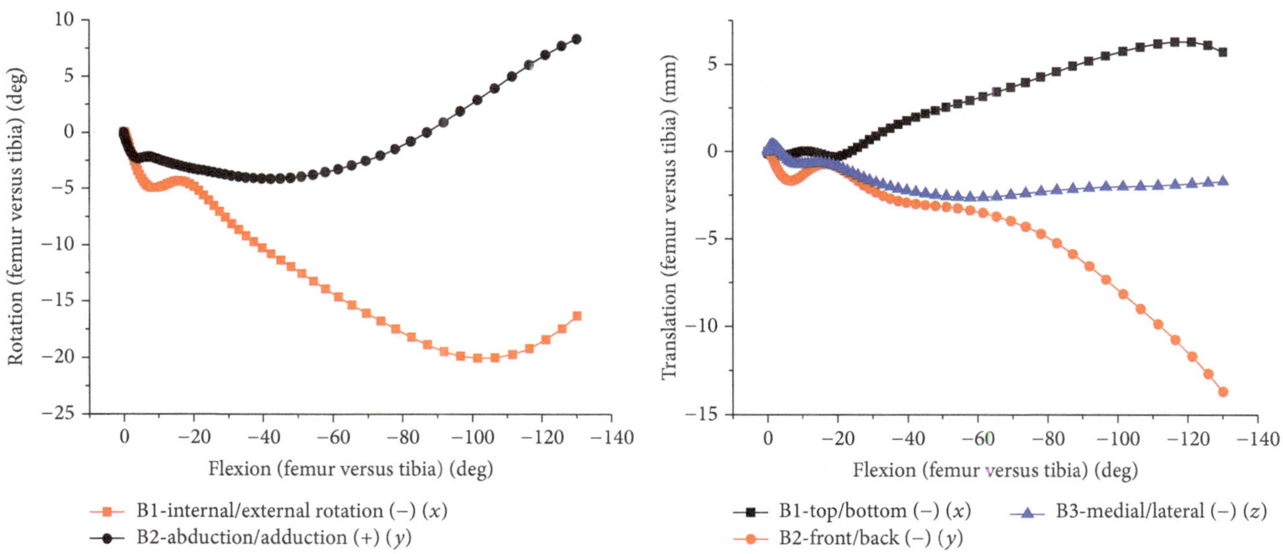

(a) (b)

FIGURE 9: The relative movement of natural tibiofemoral joint. (a) Internal/external and abduction/adduction of tibiofemoral joint. (b) Translation of tibiofemoral joint.

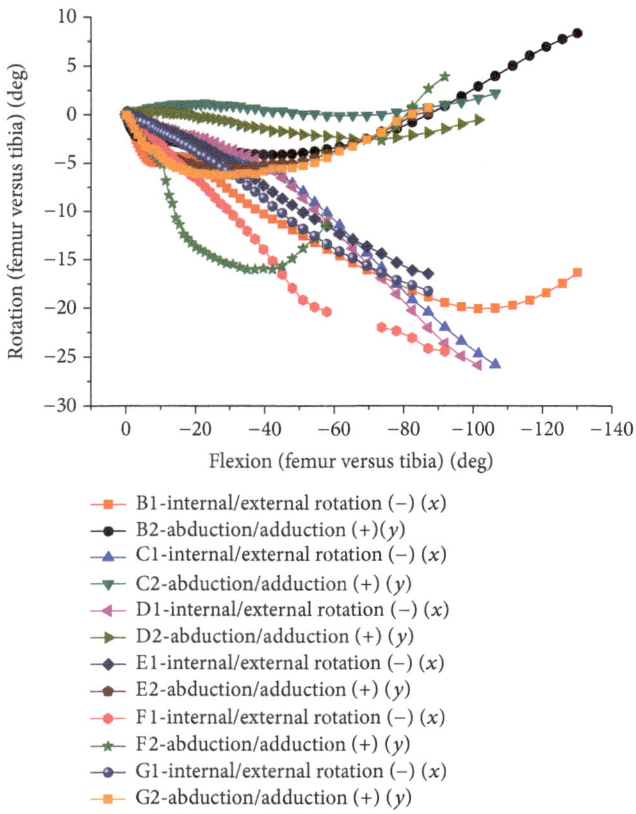

FIGURE 10: Internal/external and abduction/adduction of tibiofemoral joint relative movements in both test and FEA.

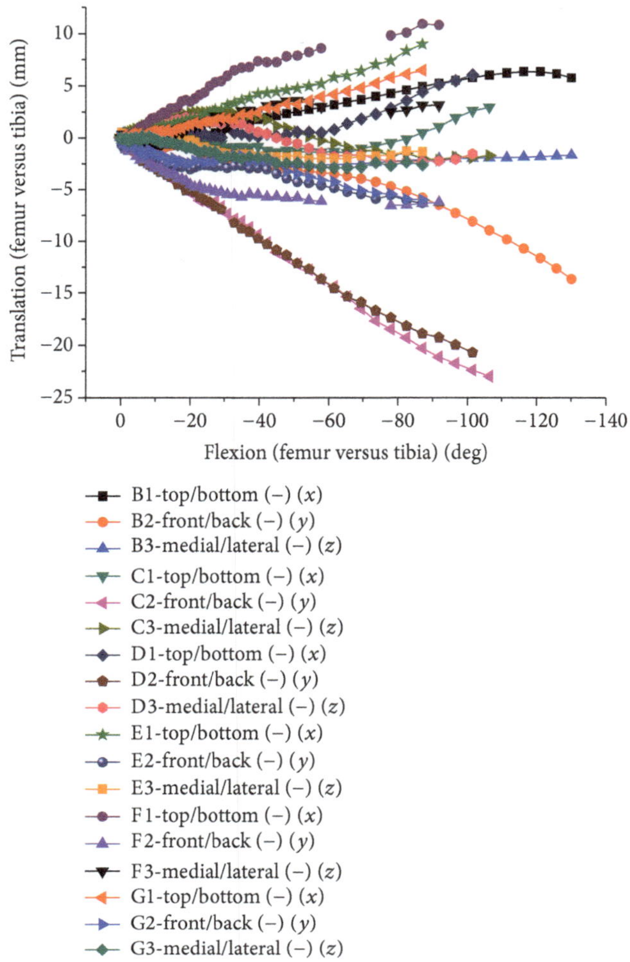

FIGURE 11: The relative translation of tibiofemoral joint in both test and FEA.

tibia, and femur were measured by the Polaris optical tracking system (Figure 8).

3. Results

3.1. The FE Calculation and Experimental Results of Knee Joint. The FE model was developed into the software of ABAQUS-6.5.1 (HKS, Pawtucket, RI) to analyze the articular contact and the relative motion of the patellofemoral joint and tibiofemoral joint synchronously. By calculating the dynamic FEA of natural knee, the movement and joint contact stress of healthy knee could be obtained. The flexion motion and dynamic contact characteristics of knee joint were analyzed and verified by comparison with the data from cadaver in vitro experiment.

3.1.1. The FE Results and Verification Analysis of Tibiofemoral Joint. The tibial internal rotation increases as knee flexes, and femoral relative external rotation decreases after about 90° flexion, accompanying 9° adduction of femur (Figure 9). The femur backward translation increased along with knee flexing. In high flexion, femoral condyle lifted off tibial surface a few millimeters and contacted with posterior meniscus. Smaller medial-lateral translation happened in the entire flexion. The cadaver experiment and FE results were compared in contrast diagram (Figures 10 and 11).

During 0–90-degree knees flexion, femur external rotation was average of 20 degrees, and femur abduction turned to average of 2 degrees. From 90–120-degree flexion, the tibial internal rotation increased while femur external rotation decreased. Synchronously, the femoral adduction increased. There was unusual abduction for one of the specimens during earlier flexion stage. The femur translation trends of the experiment and FE results were basically consistent (Figure 11) in the direction of medial-lateral, up-down, and front-back, respectively. With flexion growth, femoral relative tibial translated upward, inward and backward, respectively. Two cases translated downward within 20–70-degree flexion. The backward translation was relatively small, which was approximately 7 mm at 90-degree flexion, being similar to simulation results. There was a large backward translation for one case, approximately 23 mm around 110-degree flexion. There was bigger difference for translation result, which may be caused by the error of selecting the osseous marker points and the individual difference.

With knee flexion from 0 to 130 degrees, the average peak stress of tibiofemoral joint was 10 MPa at 0 degrees and 6 MPa at 30~90 degrees, and it increased to 21 MPa

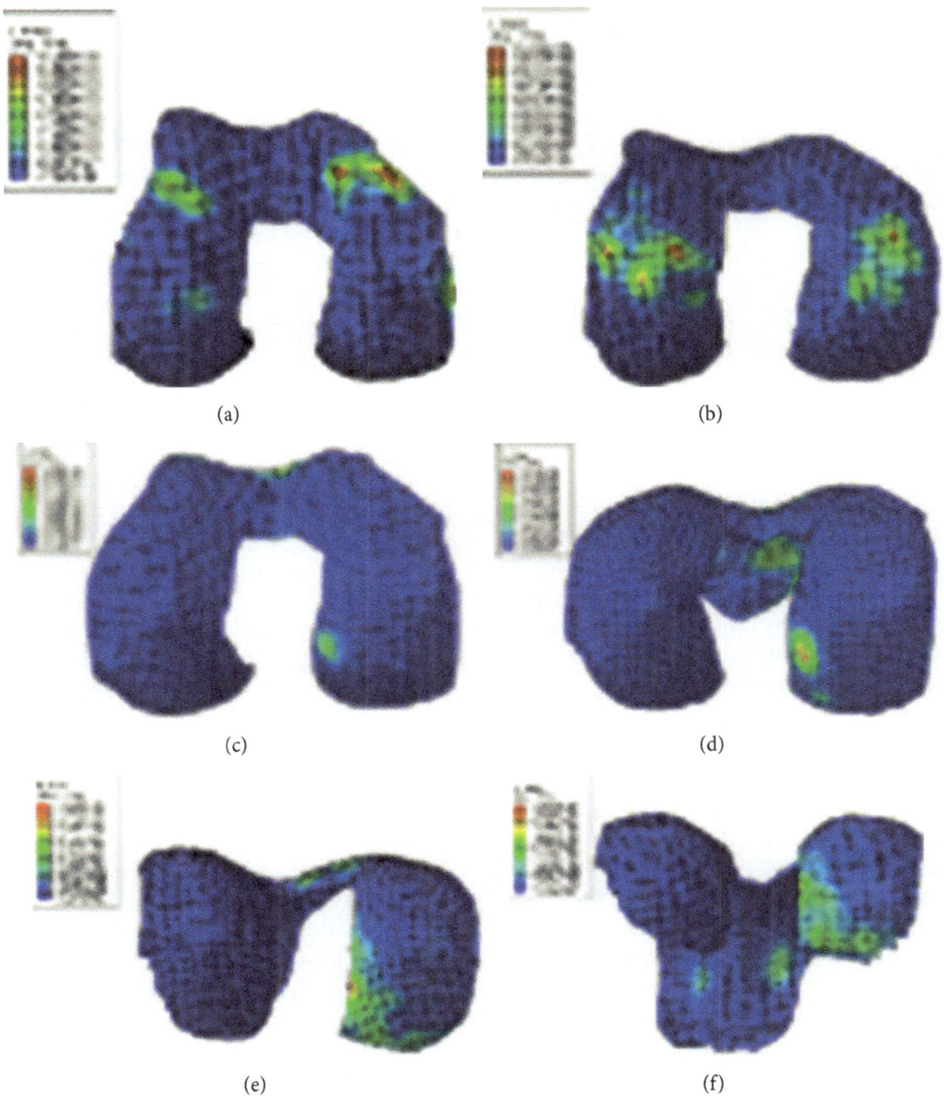

FIGURE 12: Contact stress of 0-30-60-90-130 degree flexion of tibiofemoral joint in FEA. (a) Stress distribution of 0° flexion. (b) Stress distribution of 30° flexion. (c) Stress distribution of 60° flexion. (d) Stress distribution of 90° flexion. (e) Stress distribution of 120° flexion. (f) Stress distribution of 130° flexion.

from 90 to 130 degrees. There were differences between the medial and the lateral tibiofemoral joint. From 0- to 60-degree flexion, medial and lateral contact stresses were close for FE results, but the lateral contact stress was higher than the medial contact stress for in vitro test. Starting from 90-degree flexion, larger contact stress was in the medial tibiofemoral joint for both of FE and in vitro test. Simultaneously, in the lateral tibiofemoral joint, there was little contact for FE results and about 4 MPa small contact stress for in vitro test. In 0–30-degree flexion, the contact of tibiofemoral joint mainly occurred in the front of tibia, and contact area was relatively small. Within 30~60-degree flexion, contact was mainly in the central tibia; contact area is increasing and causing smaller contact stress. With flexion deepening, the backward translation of femur increased, contacting with the posterior meniscus, and lateral femoral condyle lift off tibial surface a few millimeters in higher flexion (Figures 12 and 13). Accordingly, the total contact area of tibiofemoral joint relatively decreased, causing larger contact stress in the medial tibiofemoral joint.

3.1.2. The FE Results and Verification Analysis of Patellofemoral Joint. With flexion of tibiofemoral joint, the flexion change of patellofemoral joint was basically linear (Figure 14). Accompanying patella's small angle external rotation and medial tilting relative to femur, there were small medial, backward, and upward translations. Patella was up to about 90 degrees flexion, external rotation of 3.7 degrees, medial tilting of 10 degrees, backward translation of 64 mm, the maximum downward translation of 49 mm at about 90 degrees knee flexion, and the maximum lateral translation of 6 mm at about 80 degrees knee flexion.

The change tendency was basically consistent in simulation and test (Figures 15 and 16), being slightly different.

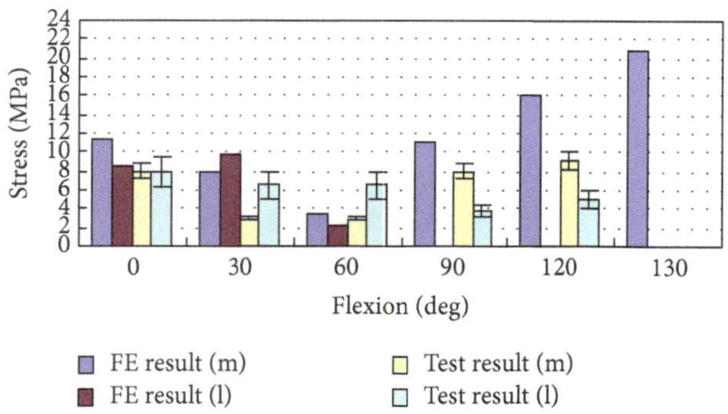

FIGURE 13: Comparison between experimentally measured stress applied to tibiofemoral joint and the FE results, where m refers to the medial femur and l refers to the lateral femur.

The patella internal rotation (20 degrees maximum) was bigger than simulation results. There was little difference for patella relative medial tilting, excepting lateral tilting at 60 degrees flexion in one case. It was basically consistent in relative flexion of patellas, being linear with tibiofemoral flexion. The fluctuations of translation were larger than the rotation. The maximum medial translation in the experiment occurred within 30 degrees flexion earlier than the simulation of about 70 degrees flexion. However, the upward and backward translations were larger than the simulation. This difference may be caused by soft tissue relaxation in the cadaver experiment.

With knee flexion, the contact place of patellofemoral joint gradually moved from the inferior patella to the superior patella and then turned slightly downward after 120 degrees. Within 30 degrees of flexion, the contact places of patellofemoral joint were unsteady. Within 30–90 degrees, the contact places of patellofemoral joint gradually drifted to the medial and lateral patella; the contact places were mainly on the medial patella ridge. Starting from 90-degree flexion, contact position was obviously distributed in medial and lateral margin patella. From 0- to 90-degree flexion, the contact stress was of about 9 MPa for both the simulation and the test results. The medial contact stress was significantly greater than the lateral one except within 30-degree flexion. The contact stress increased after 90°, reaching 22 MPa at 130-degree flexion. The margin patella contacted with femoral epicondylus in higher flexion and decreased contact area causing higher contact stress (Figures 17 and 18).

4. Discussion

4.1. The Experimental System Analysis. To fully understand the function of knee joint, not only the synchronal measurement of both tibiofemoral and patellofemoral articulations but also the synchronal measurement of both kinematics and contact can be implemented in this experiment system and simulation. Because of the difficulty of accurately measuring internal joint contact stress in rational functional conditions, there was seldom contact research about the tibiofemoral joint. Hsu et al. (1997) [25] used unidirectional sensor embedded in patellar prosthesis to measure patellofemoral joint contact force. And it was limited that the force and movement were measured separately. Powers et al. (1998) [26] used electromagnetic tracking equipment to measure the patella movement and used pressure sensitive film to measure patellofemoral joint contact pressure and area. However, tibiofemoral joint was not measured. Halloran et al. (2010) [9] developed a whole joint model, and this included fixture components only. Prior work with cadaveric specimens included displacement control of TF kinematics and only PF soft tissue representations [27].

The in vitro experiment and simulation results were compared in this study, and there were differences between them. The accuracy of test was inevitably influenced by some of the characteristics of experiment system. Although the thickness of sensing piece is only 0.1 mm, the joint surface shape was slightly changed, and the results accuracy of contact area and contact pressure peak in joint was affected. The difference of reference point selection caused the difference of coordinate system, which led to measurement error of movement. Due to the limitation of measurement mechanism itself, it is difficult to perfectly simulate the human knee joint squat movement. In vitro test research showed that quadriceps force size and loading direction influence tibiofemoral joint rotation, especially within 0~90 degrees flexion; the force and angle (Q angle) reduction of quadriceps caused tibial internal rotation reduction [28, 29]. Therefore, the difference between simulation and living quadriceps activities may be one of the reasons of the tibial rotation difference between in vitro and living tests. The signal interference of sensor, measurement errors of force transducer, contact stress sensor, reference frame, and calibration error can also cause measurement error. Because of losing activity, the muscles and joints of specimens will be stiff, and decreased joint fluid lubrication in articular capsule increased joint contact friction. And the mechanical property of bone and soft tissue changed accompanied with relaxing and drying of specimen organization itself. These conditions may all affect the experimental results. So there are differences between the simulation and the results of the experiment, as well as between the specimen and living body. In addition, individual differences can also cause difference. Therefore, the experimental method also needed to be continuously developed to improve the biomechanical research of knee.

4.2. Natural Human Knee Flexion Biomechanical Analysis. Along with the knee flexion, natural knee joint femur rolled backward and tibia internally rotated. The results of movement of femur relative tibia were basically consistent with Iwaki (2000) and Johal et al. (2005) [30, 31]. Within 0- to 30-degree flexion, tibial internal rotation increased (3.3–12.8 degrees). Within 30~90-degree flexion, natural knee tibial rotation was bigger (10.6–17.5), maintaining internal rotation, which is approximate to previous results of passive flexion with no loading (25-degree internal rotation at 100-degree passive flexion) [32]. And it was different from Hirokawa et al. (1992) [29] (1-degree external rotation at 120-degree flexion) and Van Kampen and Husikes (1990) [33] (internal rotation began until 120-degree flexion for three of the four specimens, in which fixed femoral test equipment was used). If rotating force was applied on tibia, especially for the unloading knee,

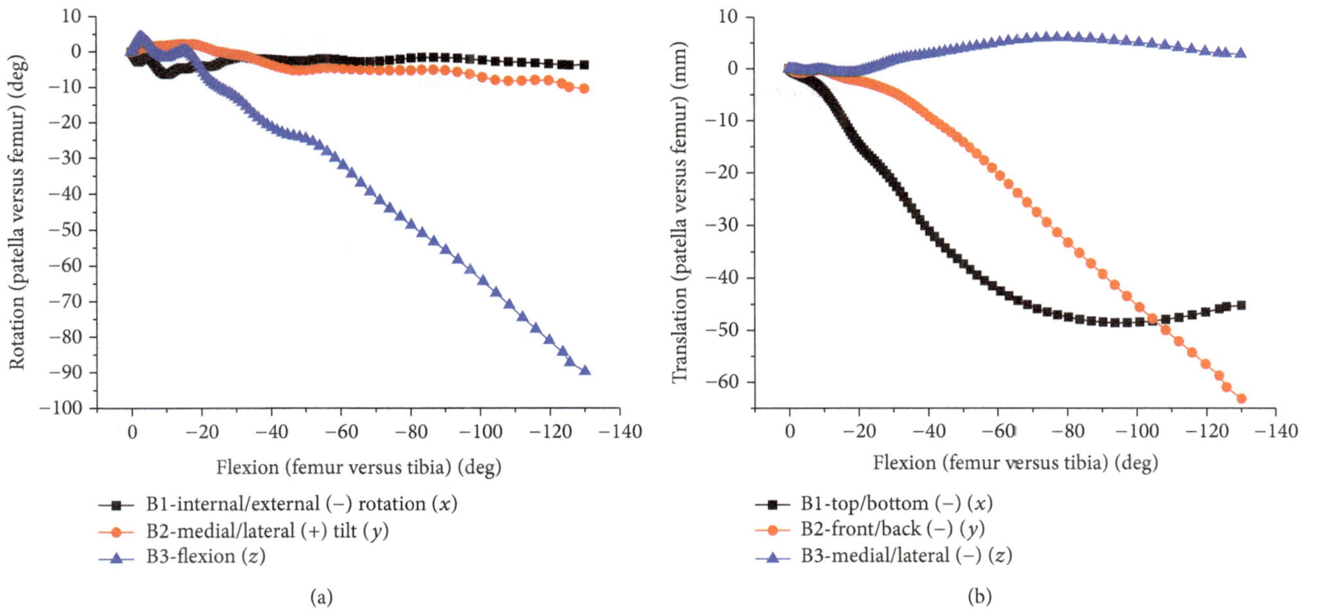

FIGURE 14: The relative movement of patellofemoral joint. (a) Internal/external, abduction/adduction, and flexion of patellofemoral joint. (b) Translation of patellofemoral joint.

femur external rotation at 90 degree is reversible. Therefore, if the tibia flexes 90 degrees at external rotation phase, it is possible to stretch to 20-degree flexion with no lateral condyle backward-forward translation. So, to great extent, the tibia rotation belongs to tibial rotation belong to "combining" not inevitable within 20~90 degrees flexion. However, the 20-degree flexed position was the critical point of tibia in internal rotation. Therefore, tibia internal rotation was inevitable within 0~20 degrees flexion, rotating around the long axis of knee.

The results of knee adduction-abduction about living body were little reported recently. The research in this paper showed that the femoral external rotation was along with the adduction. The reason is that the medial tibia platform is more concave than the lateral one and the projection on coronal plane of medial condyle is lower than that of the lateral condyle [34]. No eversion in knee flexion due to the consistent curvature radius of the medial and lateral femoral posterior condyle, no distal outstanding of medial condyle. The adduction in higher flexion was caused and corresponded with the internal rotation of tibia relative to femur. The research results of Hsieh et al. (1998) [35] (0-degree adduction at 90-degree flexion) and Kurosawa et al. (1985) [36] (2.2-degree adduction at 120-degree flexion) were within the scope of this study.

Within hol-extension to passive flexion (more than 120-degree flexion), there was 10 mm femoral condylar backward rolling. Femoral condyle scrolled up to meniscus posterior horn, almost dislocation to the tibia. This result is consistent with Nakagawa et al. (2000) [37] of no loading deep flexion Japan knee and Abdel-Rahman and Hefzy (1998) [38] ray research results. Wilson et al. (2000) [32] research results were of 24 ± 4 mm backward translation, 13 ± 4 mm distal translation, and 5 ± 3 mm medial translation after 100-degree flexion, which were approximate to this paper. Wretenberg et al. (2002) [39] analyzed 16 nonload right tibiofemoral joint contact at 0-degree, 30-degree, and 60-degree flexion by MRI and observed that the lateral displacement was bigger than the medial. The research of Johal et al. (2005) [31] showed that femoral condyle translated backward within 0~90-degree flexion, accompanied with slipping or rolling.

In this study, the distribution of medial and lateral contact stress was influenced by the joint position. The peak stress of tibiofemoral joint contact was of average 10 MPa at 0-degree flexion and 6 MPa within 30~90-degree flexion, and it grew from 90-degree flexion reaching 21 MPa at 130-degree flexion. In the flexion process, the medial and lateral tibiofemoral joint contact stresses were different. From 0- to 60-degree flexion, medial stress was approximate to the lateral stress for FEA and slightly larger contact stress at lateral joint for in vitro experiment. Within 0~30-degree flexion, contact position was mainly in anterior tibia, and contact area was relatively small. From 30- to 60-degree flexion, the contact mainly occurred in central tibia, increased contact area, and decreased contact stress. Along with deepening flexion, femur backward translation and liftoff increased, which decreased tibiofemoral articular contact area. After 90-degree flexion, larger contact stress occurred in medial joint for FEA and in vitro experiment. The 21 MPa maximum contact stress of simulation analysis results was closed to 25 MPa cartilage cracking limitation [40] (Torzilli et al., 1999), being consistent with Ashvin Thambyah et al. (2005) [5] which is of 14 MPa average stress peak (standard deviation 2.5 MPa) on walking phase. The loaded living research by Iwaki et al. (2000) and Hill et al. (2000) [30, 41] showed that large change occurred in high flexion for tibiofemoral joint, and too much stress (more than 25 MPa) caused cartilage injury, which may be the cause of the continued development

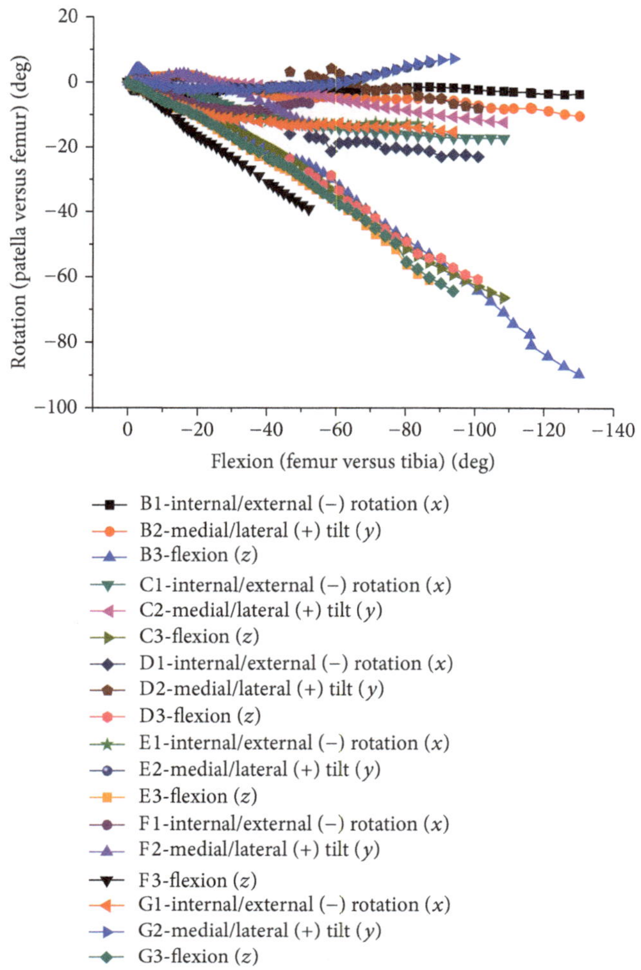

FIGURE 15: Internal/external, abduction/adduction, and flexion of patellofemoral joint relative movements in both test and FEA.

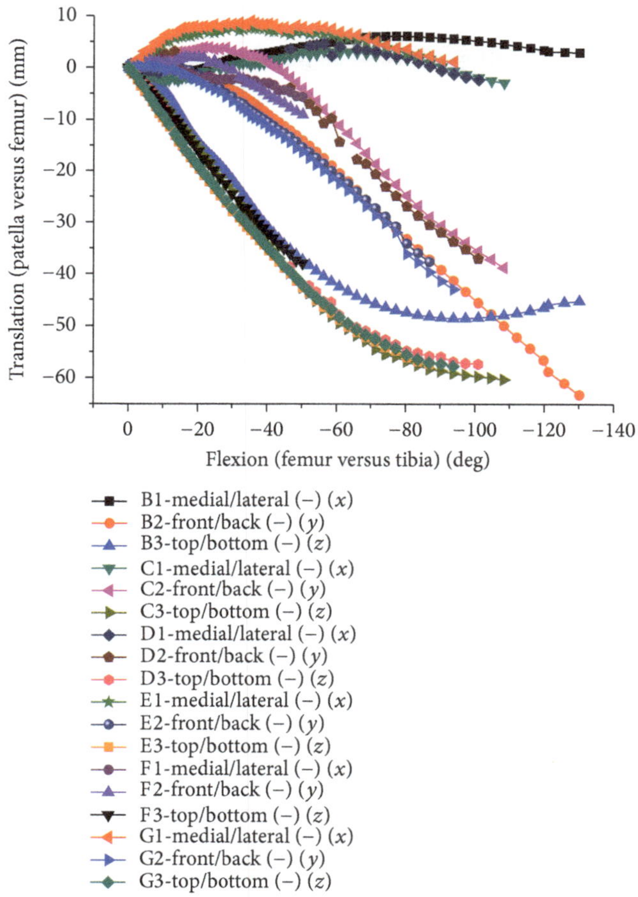

FIGURE 16: Translation of patellofemoral joint relative movement in both test and FEA.

of joint degeneration. Contact area reduction within high flexion was likely the reason of high stress in high flexion, and the cause of high stress was not only from high loading.

In lateral knee joint, considerable backward translation occurred, accompanied femur lifting off the lateral tibia and fall on the posterior horn of meniscus, the femur being subluxation. At the same time medial stress increased, which may be the reason of medial meniscus tear.

As for patellofemoral joint movement, Zavatsky et al. (2004) [42] gave the measurement result, and it is consistent with this research. Even though there are different reference points and reference frames, the calculation and measurement of patellofemoral relative rotation in this paper were within the standard deviation range being consistent with other research [35]. The patellofemoral joint translation motion results (backward, distal, medial, or lateral translation) also were approximate to other results [25, 33, 35]. At the same time, patellofemoral flexion was lagged behind the tibiofemoral flexion, which was consistent with the relative reports [33, 35, 43].

The contact point's upward movement of the healthy patellofemoral joint mainly occurred in early flexion, which was consistent with the previous in vitro [44, 45] and in vivo research [46]. The data of this study showed that, after 60-degree flexion, the contact area was relatively stable in the proximal portion of patella and symmetrically distributed in medial and lateral patellar surface, with no contact near the ridge of the patella. The research of Suzuki et al. (2012) [47] was approximate to this paper. And patellofemoral contact area increased with the knee flexion, of 70% average growth within 30 degrees flexion and of 34% average increase within 30~60 degrees flexion [48]. It was also found in this research that, with the deepening tibiofemoral joint flexion, the patellar tendon resulted in medial tilting so that the patella odd facet contact with the femur reduced the tension of the patellar tendon and quadriceps. At the same time, there was always a contact zone (concave area) at the upper patella and the contact occurred in the intercondylar notch along with the medial and lateral contact, a recess located in the upper patella exactly matching with the trochlear.

In higher flexion, the curvature radius of the femoral condyles was shorter and the lateral collateral ligament and anterior cruciate ligament were slack. Meanwhile, with tibiofemoral joint flexing, the tibiofemoral joint pressure increased, and the tension of the quadriceps tendon and the patellar tendon increased. The study showed that arthritis and cartilage damage was the result of the repeated or high contact stress [49], while the medial tibiofemoral was

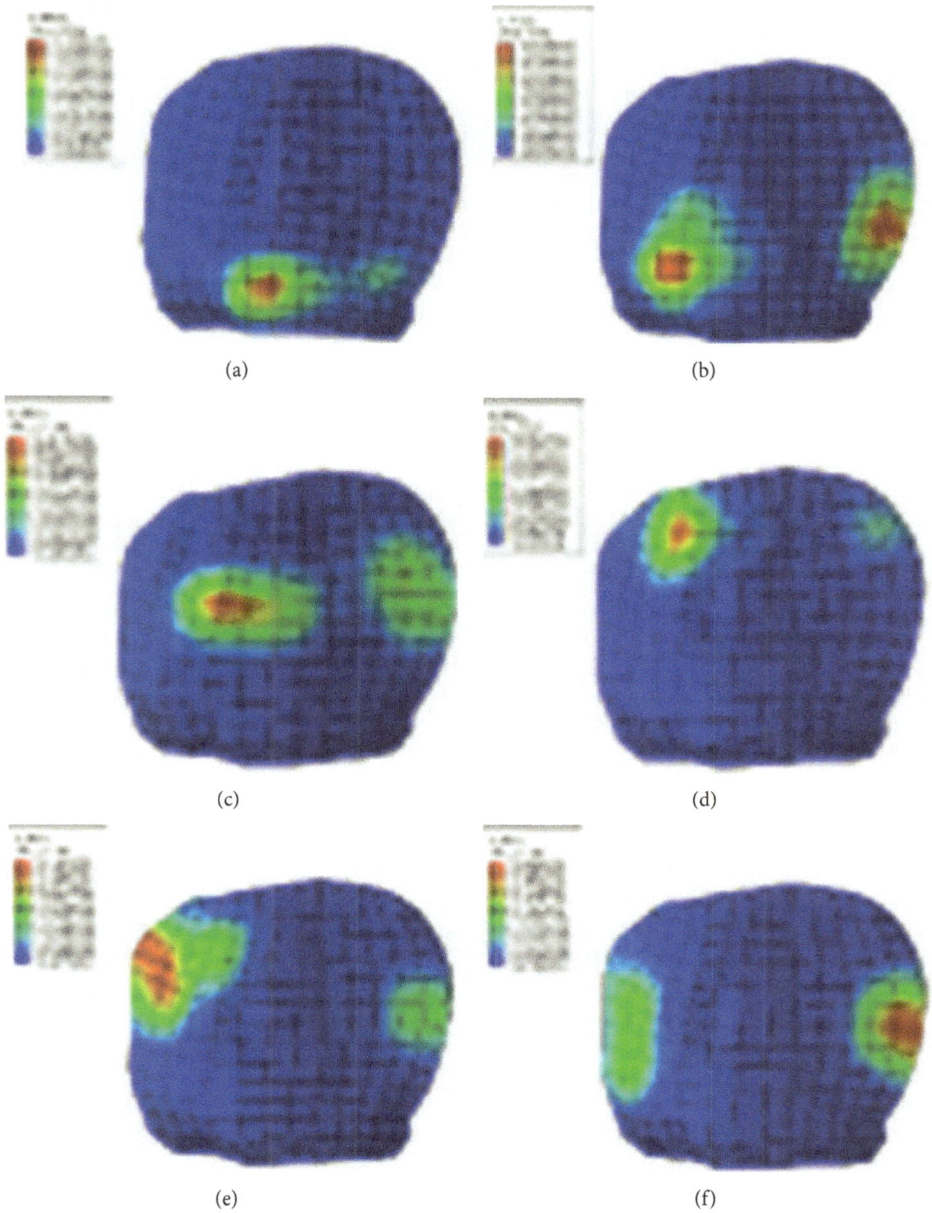

FIGURE 17: Contact stress of 0-30-60-90-130 degree flexion of tibiofemoral joint in FEA. (a) Stress distribution of 0° flexion. (b) Stress distribution of 30° flexion. (c) Stress distribution of 60° flexion. (d) Stress distribution of 90° flexion. (e) Stress distribution of 120° flexion. (f) Stress distribution of 130° flexion.

prone to arthritis, leading to an articulated knee. There was difference between Westerners and Asians, especially in the high flexion activities [50]. Arched knee is more likely to occur in Asians. It could not be ignored for the mechanical factors causing knee disease from the activities or ethnic differences between Asians and Westerners. Especially for Asians, the relationship between mechanics and movement in high flexion should be fully understood.

Conflict of Interests

All authors contributed significantly to the study and paper preparation, and none demonstrates any conflict of interests regarding this submission.

Authors' Contribution

Jianping Wang and Kun Tao are the first authors for this paper.

Acknowledgments

This work was primarily supported by the "Research on the properties of hip and knee joints and basic design of artificial joints for Asia Race" project funded by the National Natural Science Foundation of China (30810103908). And further support was from the "Research on high-flexion stability mechanism of high-flexion artificial knee" project funded by the National Natural Science Foundation of China

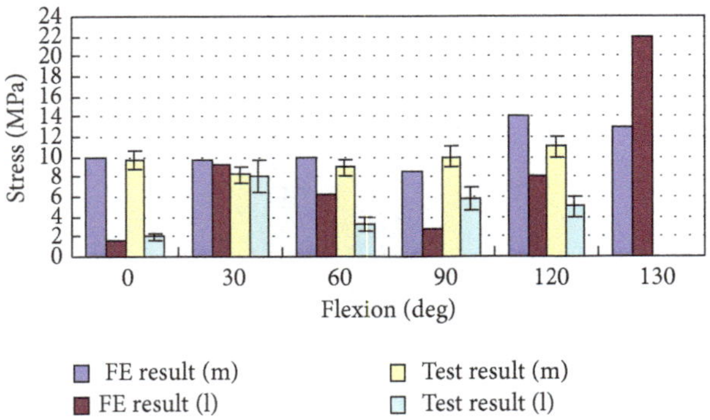

FIGURE 18: Comparison between experimentally measured stress applied to patella cartilage and the FE results, where m refers to the medial patella and l refers to the lateral patella.

(31370999) and National Natural Science Foundation of China (81000808).

References

[1] K. Knutson, S. Lewold, O. Robertsson, and L. Lidgren, "The Swedish knee arthroplasty register: a nation-wide study of 30,003 knees 1976–1992," *Acta Orthopaedica Scandinavica*, vol. 65, no. 4, pp. 375–386, 1994.

[2] E. Luger, S. Sathasivam, and P. S. Walker, "Inherent differences in the laxity and stability between the intact knee and total knee replacements," *Knee*, vol. 4, no. 1, pp. 7–14, 1997.

[3] G. W. Blunn, A. B. Joshi, R. J. Minns et al., "Wear in retrieved condylar knee arthroplasties: a comparison of wear in different designs of 280 retrieved condylar knee prostheses," *Journal of Arthroplasty*, vol. 12, no. 3, pp. 281–290, 1997.

[4] M. S. Hefzy and E. Abdel-Rahman, "Dynamic modeling of the human knee joint: formulation and solution techniques. a review paper," *Biomedical Engineering—Applications, Basis and Communications*, vol. 7, no. 1, pp. 5–21, 1995.

[5] A. Thambyah, J. C. H. Goh, and S. Das De, "Contact stresses in the knee joint in deep flexion," *Medical Engineering and Physics*, vol. 27, no. 4, pp. 329–335, 2005.

[6] A. C. Godest, M. Beaugonin, E. Haug, M. Taylor, and P. J. Gregson, "Simulation of a knee joint replacement during a gait cycle using explicit finite element analysis," *Journal of Biomechanics*, vol. 35, no. 2, pp. 267–275, 2002.

[7] J. P. Halloran, A. J. Petrella, and P. J. Rullkoetter, "Explicit finite element modeling of total knee replacement mechanics," *Journal of Biomechanics*, vol. 38, no. 2, pp. 323–331, 2005.

[8] C. K. Fitzpatrick and P. J. Rullkoetter, "Influence of patellofemoral articular geometry and material on mechanics of the unresurfaced patella," *Journal of Biomechanics*, vol. 45, no. 11, pp. 1909–1915, 2012.

[9] J. P. Halloran, C. W. Clary, L. P. Maletsky, M. Taylor, A. J. Petrella, and P. J. Rullkoetter, "Verification of predicted knee replacement kinematics during simulated gait in the Kansas knee simulator," *Journal of Biomechanical Engineering*, vol. 132, no. 8, Article ID 081010, 2010.

[10] M. A. Baldwin, C. W. Clary, C. K. Fitzpatrick, J. S. Deacy, L. P. Maletsky, and P. J. Rullkoetter, "Dynamic finite element knee simulation for evaluation of knee replacement mechanics," *Journal of Biomechanics*, vol. 45, no. 3, pp. 474–483, 2012.

[11] J. Wang, M. Ye, Z. Liu, and C. Wang, "Precision of cortical bone reconstruction based on 3D CT scans," *Computerized Medical Imaging and Graphics*, vol. 33, no. 3, pp. 235–241, 2009.

[12] P. J. Besl and N. D. McKay, "A method for registration of 3-D shapes," *IEEE Transactions on Pattern Analysis and Machine Intelligence*, vol. 14, no. 2, pp. 239–256, 1992.

[13] J.-P. Wang, L.-W. Qian, and C.-T. Wang, "Simulation of geometric anatomy model of human knee joint," *Journal of System Simulation*, vol. 21, no. 10, pp. 2806–2809, 2009.

[14] J. Wang, H. Wu, and Ch. Wang, "Dynamic finite emement modeling of human knee joint and application in TKR," *Journal of Medical Biomechanics*, vol. 24, no. 5, pp. 333–337, 2009.

[15] J. C. Gardiner, J. A. Weiss, and T. D. Rosenberg, "Strain in the human medial collateral ligament during valgus loading of the knee," *Clinical Orthopaedics and Related Research*, no. 391, pp. 266–274, 2001.

[16] D. L. Butler, F. R. Noyes, and E. S. Grood, "Ligamentous restraints to anterior-posterior drawer in the human knee. A biomechanical study," *Journal of Bone and Joint Surgery A*, vol. 62, no. 2, pp. 259–270, 1980.

[17] J. F. Suggs, G. R. Hanson, S. E. Park, A. L. Moynihan, and G. Li, "Patient function after a posterior stabilizing total knee arthroplasty: cam-post engagement and knee kinematics," *Knee Surgery, Sports Traumatology, Arthroscopy*, vol. 16, no. 3, pp. 290–296, 2008.

[18] F. R. Noyes and E. S. Grood, "The strength of the anterior cruciate ligament in humans and rhesus monkeys: age related and species related changes," *Journal of Bone and Joint Surgery A*, vol. 58, no. 8, pp. 1074–1082, 1976.

[19] M. A. LeRoux and L. A. Setton, "Experimental and biphasic FEM determinations of the material properties and hydraulic permeability of the meniscus in tension," *Journal of Biomechanical Engineering*, vol. 124, no. 3, pp. 315–321, 2002.

[20] G. Li, E. Most, E. Otterberg et al., "Biomechanics of posterior-substituting total knee arthroplasty: an in vitro study," *Clinical Orthopaedics and Related Research*, no. 404, pp. 214–225, 2002.

[21] L. Klimek, U. Ecke, B. Lübben, J. Witte, and W. Mann, "A passive-marker—based optical system for computer-aided surgery in otorhinolaryngology: development and first clinical experiences," *Laryngoscope*, vol. 109, no. 9, pp. 1509–1515, 1999.

[22] E. S. Grood and W. J. Suntay, "A joint coordinate system for the clinical description of three-dimensional motions: application to the knee," *Journal of Biomechanical Engineering*, vol. 105, no. 2, pp. 136–144, 1983.

[23] G. K. Cole, B. M. Nigg, J. L. Ronsky, and M. R. Yeadon, "Application of the joint coordinate system to three-dimensional joint attitude and movement representation: a standardization proposal," *Journal of Biomechanical Engineering*, vol. 115, no. 4, pp. 344–349, 1993.

[24] M. L. Harris, P. Morberg, W. J. M. Bruce, and W. R. Walsh, "An improved method for measuring tibiofemoral contact areas in total knee arthroplasty: a comparison of K-scan sensor and Fuji film," *Journal of Biomechanics*, vol. 32, no. 9, pp. 951–958, 1999.

[25] H.-C. Hsu, Z.-P. Luo, J. A. Rand, and K.-N. An, "Influence of lateral release on patellar tracking and patellofemoral contact characteristics after total knee arthroplasty," *Journal of Arthroplasty*, vol. 12, no. 1, pp. 74–83, 1997.

[26] C. M. Powers, J. C. Lilley, and T. Q. Lee, "The effects of axial and multi-plane loading of the extensor mechanism on the patellofemoral joint," *Clinical Biomechanics*, vol. 13, no. 8, pp. 616–624, 1998.

[27] M. A. Baldwin, C. Clary, L. P. Maletsky, and P. J. Rullkoetter, "Verification of predicted specimen-specific natural and implanted patellofemoral kinematics during simulated deep knee bend," *Journal of Biomechanics*, vol. 42, no. 14, pp. 2341–2348, 2009.

[28] S. C. Shoemaker, D. Adams, D. M. Daniel, and S. L. Woo, "Quadriceps/anterior cruciate graft interaction: an in vitro study of joint kinematics and anterior cruciate ligament graft tension," *Clinical Orthopaedics and Related Research*, no. 294, pp. 379–390, 1993.

[29] S. Hirokawa, M. Solomonow, Y. Lu, Z.-P. Lou, and R. D'Ambrosia, "Anterior-posterior and rotational displacement of the tibia elicited by quadriceps contraction," *The American Journal of Sports Medicine*, vol. 20, no. 3, pp. 299–306, 1992.

[30] H. Iwaki, V. Pinskerova, and M. A. R. Freeman, "Tibiofemoral movement 1: the shape and relative movements of the femur and tibia in the unloaded cadaver knee," *Journal of Bone and Joint Surgery B*, vol. 82, no. 8, pp. 1189–1195, 2000.

[31] P. Johal, A. Williams, P. Wragg, D. Hunt, and W. Gedroyc, "Tibio-femoral movement in the living knee. A study of weight bearing and non-weight bearing knee kinematics using 'interventional'MRI," *Journal of Biomechanics*, vol. 38, no. 2, pp. 269–276, 2005.

[32] D. R. Wilson, J. D. Feikes, A. B. Zavatsky, and J. J. O'Connor, "The components of passive knee movement are coupled to flexion angle," *Journal of Biomechanics*, vol. 33, no. 4, pp. 465–473, 2000.

[33] A. Van Kampen and R. Husikes, "The three—dimensional tracking pattern of the human patella," *Journal of Orthopaedic Research*, vol. 8, no. 3, pp. 372–382, 1990.

[34] R. P. Welsh, "Knee joint structure and function," *Clinical Orthopaedics and Related Research*, vol. 147, pp. 7–14, 1980.

[35] Y.-F. Hsieh, L. F. Draganich, S. H. Ho, and B. Reider, "The effects of removal and reconstruction of the anterior cruciate ligament on patellofemoral kinematics," *The American Journal of Sports Medicine*, vol. 26, no. 2, pp. 201–209, 1998.

[36] H. Kurosawa, P. S. Walker, S. Abe, A. Garg, and T. Hunter, "Geometry and motion of the knee for implant and orthotic design," *Journal of Biomechanics*, vol. 18, no. 7, pp. 487–499, 1985.

[37] S. Nakagawa, Y. Kadoya, S. Todo et al., "Tibiofemoral movement 3: full flexion in the living knee studied by MRI," *Journal of Bone and Joint Surgery B*, vol. 82, no. 8, pp. 1199–1200, 2000.

[38] E. M. Abdel-Rahman and M. S. Hefzy, "Three-dimensional dynamic behaviour of the human knee joint under impact loading," *Medical Engineering and Physics*, vol. 20, no. 4, pp. 276–290, 1998.

[39] P. Wretenberg, D. K. Ramsey, and G. Németh, "Tibiofemoral contact points relative to flexion angle measured with MRI," *Clinical Biomechanics*, vol. 17, no. 6, pp. 477–485, 2002.

[40] P. A. Torzilli, R. Grigiene, J. Borrelli Jr., and D. L. Helfet, "Effect of impact load on articular cartilage: cell metabolism and viability, and matrix water content," *Journal of Biomechanical Engineering*, vol. 121, no. 5, pp. 433–441, 1999.

[41] P. F. Hill, V. Vedi, A. Williams, H. Iwaki, V. Pinskerova, and M. A. R. Freeman, "Tibiofemoral movement 2: the loaded and unloaded living knee studied by MRI," *Journal of Bone and Joint Surgery B*, vol. 82, no. 8, pp. 1196–1198, 2000.

[42] A. B. Zavatsky, P. T. Oppold, and A. J. Price, "Simultaneous in vitro measurement of patellofemoral kinematics and forces," *Journal of Biomechanical Engineering*, vol. 126, no. 3, pp. 351–356, 2004.

[43] T. M. G. J. Van Eijden, W. De Boer, and W. A. Weijs, "The orientation of the distal part of the quadriceps femoris muscle as a function of the knee flexion-extension angle," *Journal of Biomechanics*, vol. 18, no. 10, pp. 803–809, 1985.

[44] M. S. Hefzy, W. T. Jackson, S. R. Saddemi, and Y.-F. Hsieh, "Effects of tibial rotations on patellar tracking and patello-femoral contact areas," *Journal of Biomedical Engineering*, vol. 14, no. 4, pp. 329–343, 1992.

[45] R. Singerman, "Effects of patella alta and patella infera on patellofemoral contact forces," *Journal of Biomechanics*, vol. 27, no. 8, pp. 1059–1065, 1994.

[46] R. D. Komistek, D. A. Dennis, J. A. Mabe, and S. A. Walker, "An in vivo determination of patellofemoral contact positions," *Clinical Biomechanics*, vol. 15, no. 1, pp. 29–36, 2000.

[47] T. Suzuki, A. Hosseini, J.-S. Li, T. J. Gill, and G. Li, "In vivo patellar tracking and patellofemoral cartilage contacts during dynamic stair ascending," *Journal of Biomechanics*, vol. 45, no. 14, pp. 2432–2437, 2012.

[48] G. B. Salsich, S. R. Ward, M. R. Terk, and C. M. Powers, "In vivo assessment of patellofemoral joint contact area in individuals who are pain free," *Clinical Orthopaedics and Related Research*, no. 417, pp. 277–284, 2003.

[49] T. Farquhar, Y. Xia, K. Mann et al., "Swelling and fibronectin accumulation in articular cartilage explants after cyclical impact," *Journal of Orthopaedic Research*, vol. 14, no. 3, pp. 417–423, 1996.

[50] S. J. Mulholland and U. P. Wyss, "Activities of daily living in non-Western cultures: range of motion requirements for hip and knee joint implants," *International Journal of Rehabilitation Research*, vol. 24, no. 3, pp. 191–198, 2001.

14

Calculation of Reaction Forces in the Boiler Supports Using the Method of Equivalent Stiffness of Membrane Wall

Josip Sertić, Dražan Kozak, and Ivan Samardžić

Mechanical Engineering Faculty in Slavonski Brod, Trg Ivane Brlić-Mažuranić 2, 35 000 Slavonski Brod, Croatia

Correspondence should be addressed to Josip Sertić; josip.sertic@sfsb.hr

Academic Editors: S. Kou and K.-L. Ou

The values of reaction forces in the boiler supports are the basis for the dimensioning of bearing steel structure of steam boiler. In this paper, the application of the method of equivalent stiffness of membrane wall is proposed for the calculation of reaction forces. The method of equalizing displacement, as the method of homogenization of membrane wall stiffness, was applied. On the example of "Milano" boiler, using the finite element method, the calculation of reactions in the supports for the real geometry discretized by the shell finite element was made. The second calculation was performed with the assumption of ideal stiffness of membrane walls and the third using the method of equivalent stiffness of membrane wall. In the third case, the membrane walls are approximated by the equivalent orthotropic plate. The approximation of membrane wall stiffness is achieved using the elasticity matrix of equivalent orthotropic plate at the level of finite element. The obtained results were compared, and the advantages of using the method of equivalent stiffness of membrane wall for the calculation of reactions in the boiler supports were emphasized.

1. Introduction

Nowadays, the steam boilers of waste incineration power plants are made specifically according to the requirements of investors. Due to the location where the boiler will be set up, the amount and type of fuel to be burned in a given time, and technical parameters of produced steam, the production of steam boilers is mainly individual. This represents an exceptional effort for the designers. The pipes and sheet metal are the basic structural elements used for making boilers. Larger structural units called the membrane walls are obtained by welding these elements. The membrane walls make the basic structure of boiler and have a double function. The heat exchanger, that is, evaporator, is the primary function, and the secondary is transfer of loading. The steam boiler membrane walls are loaded by the pressure of working fluid (water/steam), their own mass, and the mass of other structural units that rely on the membrane walls, the mass of working fluid, the mass of insulation and refractory, and the fouling mass that is deposited during the combustion process. The loading is transferred across the membrane walls to the boiler supports and then to the bearing steel structure.

Since the water tube boilers are large structures consisting of a large number of pipes connected to the membrane walls, it is very difficult to accurately calculate the reactions in the supports. In practice, there are two approaches to calculate the reactions in the steam boiler supports. The first approach is based on the assumption that the membrane stiffness of the boiler membrane wall is very high, so the boiler is calculated as an ideal rigid body. The second approach takes into account the finite stiffness of the membrane walls using the finite element method. The discretization of real geometry of boiler (membrane walls) requires an extremely large number of finite elements, so this approach is rarely used, eventually for smaller boilers. Most often, the membrane wall stiffness is approximated by the equivalent stiffness of orthotropic plate/shell, which allows a multiple reduction of degrees of freedom of a numerical model.

It is difficult to find more detailed research papers on the topic of homogenization of steam boiler membrane wall. In 2012, Milošević-Mitić [1] with a group of authors proposed a procedure for calculating the strength of water tube boiler using the finite element of orthotropic plate.

The homogenization of membrane wall stiffness is performed numerically, using the method of equalizing displacement.

When calculating the deflection of membrane walls due to the influence of underpressure/overpressure of flue gases, the equivalent wall thickness method is most commonly used in practice. The bending stiffness of membrane wall pipe is equal to the bending stiffness of equivalent wall thickness for the axis perpendicular to the longitudinal axis of membrane wall pipe, and the bending stiffness of membrane wall tape is equal to the bending stiffness of equivalent wall thickness for the axis parallel to the longitudinal axis of membrane wall pipe.

The steam boiler membrane wall represents a structurally orthotropic plate. The stiffened plates that are frequently used, both in building and in the processing/power plants, where it is necessary to achieve the optimal structural solution considering the mass of structure, also belong to this class of structures. The earliest research on the topic of stiffness of stiffened plates was most likely done by Huber [2–4]. His investigations were inspired by the lack of exact expressions for solving the problems of bending and buckling of stiffened structures by using the theory of homogeneous orthotropic plates. Similarly, Flügge [5] published the first paper on the topic of structurally stiffened shell in 1932, and 15 years later, so did van der Neut [6]. Over time, the understanding of the theory of deformation of structurally orthotropic plates and shells has progressed. The terms "equivalent stiffness" and "equivalent wall thickness" became the basis for the analysis of deformation of stiffened or structurally orthotropic plates/shells. Furthermore, Dale et al. [7] applied the equivalent stiffness in the research of buckling of pressure loaded sandwich plates. Smith et al. [8] proposed an improved formulation that takes into consideration the change of position of the neutral surface that is connected with the local interaction between the plate and stiffener. Pflüger [9], Gomza and Seide [10], and Libove and Hubka [11] applied in their research the method of equivalent stiffness and thickness of stiffened plate/shell. In the papers [12–14], the theory of symmetric sandwich plates, with the included transverse shear stiffness, is applied. In 1956, Huffington Jr. [15] analyzed the equivalent stiffness of orthogonally stiffened plate without the stiffener eccentricity. During the last fifty years, the equivalent plate or shell approach was still being used as a first approximation method [16–19]. For the experimental determination of elasticity constants, the method of testing the load-deformation characteristics of repeating structural part of the structurally orthotropic plate was most commonly used [20]. In 1969, Soong [21–23] presented a derivation of the stiffness expressions for orthotropic cylinders based on a deformation energy approach. Throughout the following years, special attention was given to the interaction between the plate and stiffeners [24], for a different layout or grid of stiffeners [25–27] and different combinations of loads. In 1995, Jaunky et al. [28, 29] presented a refined smeared-stiffener theory for grid-stiffened laminated-composite panels, based upon the work presented by Smith et al. [8]. Similarly, Wodesenbet presented an improved smeared-stiffener theory for grid-stiffened laminated-composite cylinders in 2003.

It can be concluded that the current research is mainly based on two systematic analytical methods that take or do not take into account the transverse shear stiffness. First, there is a direct equilibrium-compatibility method. The first method uses equilibrium and compatibility in a direct manner for plates reinforced by stiffeners. The second one is called the basic-cell energy-equivalence method. The second method is based on defining the equivalence between the deformation energy of the basic repeating cell and the deformation energy of the equivalent orthotropic plate. This method is suitable for the determination of the members of elasticity matrix for grid-stiffened plates and sandwich plates.

Unlike the stiffened plates, a steam boiler membrane wall consists of a homogeneous plate with attached stiffeners. Therefore, the two mentioned methods are not entirely suitable for the determination of equivalent stiffness of membrane wall, that is, the equivalent elasticity constants. The process of homogenization of steam boiler membrane wall is presented in the papers [30–32]. The elasticity constants for the membrane and bending stiffness of membrane wall are determined theoretically, by equalizing the deformation of membrane wall with the deformation of equivalent orthotropic plate. In these papers, the transverse shear stiffness is not taken into consideration.

2. Steam Boiler Membrane Wall as the Structurally Orthotropic Plate

The steam boiler membrane wall can be approximated as the equivalent orthotropic plate, that is, the orthotropic plate, which will have the same elastic properties as the real membrane wall. Using Kirchhoff-Love theory of shells [33, 34], the constitutive equations of equivalent orthotropic plate or shell can be written as follows:

$$\boldsymbol{\sigma} = \mathbf{D} \cdot \boldsymbol{\varepsilon}, \tag{1}$$

where $\boldsymbol{\sigma}$ is vector of section forces, $\mathbf{D}$ is matrix of elasticity, and $\boldsymbol{\varepsilon}$ is vector of deformation. According to [35], it can be said that the constitutive equations for the analysis of plates and shells are analog. The matrix expression (1) represents the six constituent equations of membrane wall as the structurally orthotropic plate or its equivalent orthotropic plate, which connect the section forces with the corresponding deformations

$$\begin{bmatrix} N_1 \\ N_2 \\ N_{12} \\ M_1 \\ M_2 \\ M_{12} \end{bmatrix} = \begin{bmatrix} A_{11} & A_{12} & 0 & 0 & 0 & 0 \\ & A_{22} & 0 & 0 & 0 & 0 \\ & & A_{33} & 0 & 0 & 0 \\ & & & D_{11} & D_{12} & 0 \\ & \text{sym.} & & & D_{22} & 0 \\ & & & & & D_{33} \end{bmatrix} \cdot \begin{bmatrix} \varepsilon_1 \\ \varepsilon_2 \\ \varepsilon_{12} \\ \kappa_1 \\ \kappa_2 \\ 2 \cdot \kappa_{12} \end{bmatrix}. \tag{2}$$

N_1, N_2, and N_{12} are the section forces (Figure 1), and, M_1, M_2 and M_{12} are the section moments (Figure 2) in the membrane wall, that is, the equivalent orthotropic plate. A_{11}, A_{12}, A_{22}, and A_{33} are the members of elasticity matrix for membrane loads, and D_{11}, D_{12}, D_{22}, and D_{33} are the members of

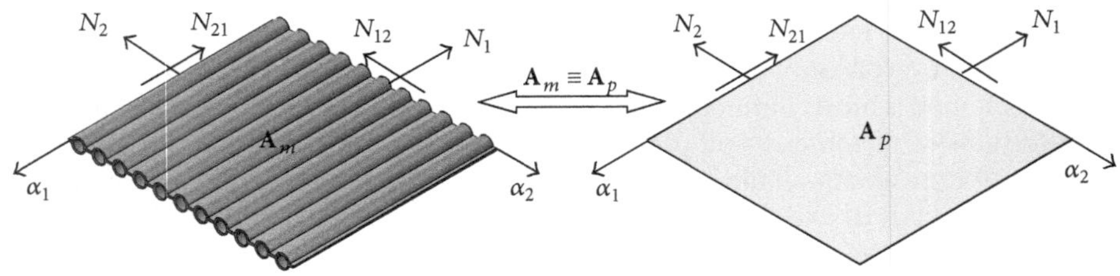

FIGURE 1: Section forces in the membrane wall, that is, the equivalent orthotropic plate, with the membrane stiffness of membrane wall $\mathbf{A}_m$ equivalent to the membrane stiffness of orthotropic plate $\mathbf{A}_p$.

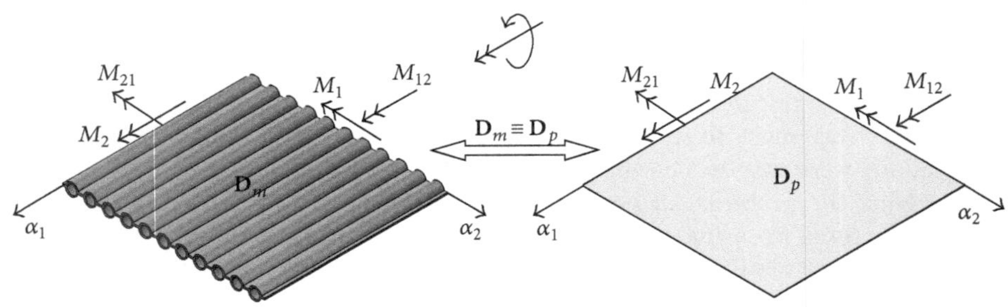

FIGURE 2: Section moments in the membrane wall, that is, the equivalent orthotropic plate, with the bending stiffness of membrane wall $\mathbf{D}_m$ equivalent to the bending stiffness of orthotropic plate $\mathbf{D}_p$.

elasticity matrix for bending loads of equivalent orthotropic plate. ε_1, ε_2, and ε_{12} are the membrane deformations, and κ_1, κ_2, and κ_{12} are the curvatures of neutral surface of structurally orthotropic plate, that is, the equivalent orthotropic plate.

The members of elasticity matrix referring to the membrane stiffness [30, 34] of membrane wall can be calculated using the following expressions:

$$A_{11} = \frac{E_{m,1} \cdot h}{1 - \nu_{m,12} \cdot \nu_{m,21}}, \qquad A_{12} = \frac{E_{m,2} \cdot h \cdot \nu_{m,12}}{1 - \nu_{m,12} \cdot \nu_{m,21}},$$
$$A_{22} = \frac{E_{m,2} \cdot h}{1 - \nu_{m,12} \cdot \nu_{m,21}}, \qquad A_{33} = G_{m,12} \cdot h, \qquad (3)$$

where $E_{m,1}$ is modulus of elasticity for the membrane stiffness in the direction of axis α_1 (the axis parallel to the axis of pipe in the membrane wall), $E_{m,2}$ is modulus of elasticity for the membrane stiffness in the direction of axis α_2 (the axis perpendicular to the axis of pipe in the membrane wall), $\nu_{m,12}$ is Poisson's ratio for the membrane stiffness in the direction of axis α_1, $\nu_{m,21}$ is Poisson's ratio for the membrane stiffness in the direction of axis α_2, and $G_{m,12}$ is shear modulus for the membrane stiffness. The members of elasticity matrix referring to the bending stiffness [29, 33] of membrane wall can be calculated using the following expressions:

$$D_{11} = \frac{E_{b,1} \cdot h^3}{12 \cdot (1 - \nu_{b,12} \cdot \nu_{b,21})},$$
$$D_{12} = \frac{E_{b,2} \cdot h^3 \cdot \nu_{b,12}}{12 \cdot (1 - \nu_{b,12} \cdot \nu_{b,21})},$$
$$D_{22} = \frac{E_{b,2} \cdot h^3}{12 \cdot (1 - \nu_{b,12} \cdot \nu_{b,21})}, \qquad (4)$$
$$D_{33} = \frac{G_{b,12} \cdot h^3}{12},$$

where the elasticity constants for the bending stiffness are indicated by index b. The expression for the wall thickness of equivalent orthotropic plate [30] can be obtained from the equality of moment of inertia of the cross-section of membrane wall, considering the plane perpendicular to the longitudinal axes of the membrane pipes (Figure 3), and the cross-section of equivalent orthotropic plate. Consider

$$h = \sqrt{\frac{3 \cdot \pi}{16 \cdot D \cdot k} \cdot \left(D^4 - (D - 2 \cdot \delta)^4\right) + \frac{(k - D) \cdot t^3}{D \cdot k}}. \qquad (5)$$

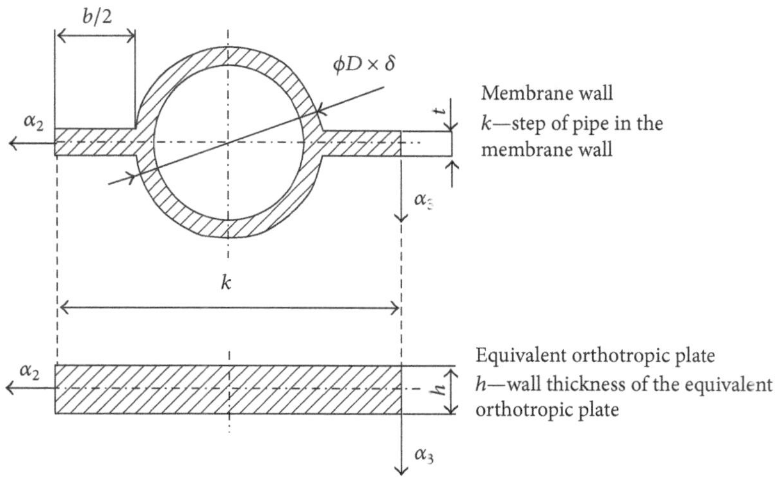

FIGURE 3: Geometry of cross-section of membrane wall and equivalent orthotropic plate.

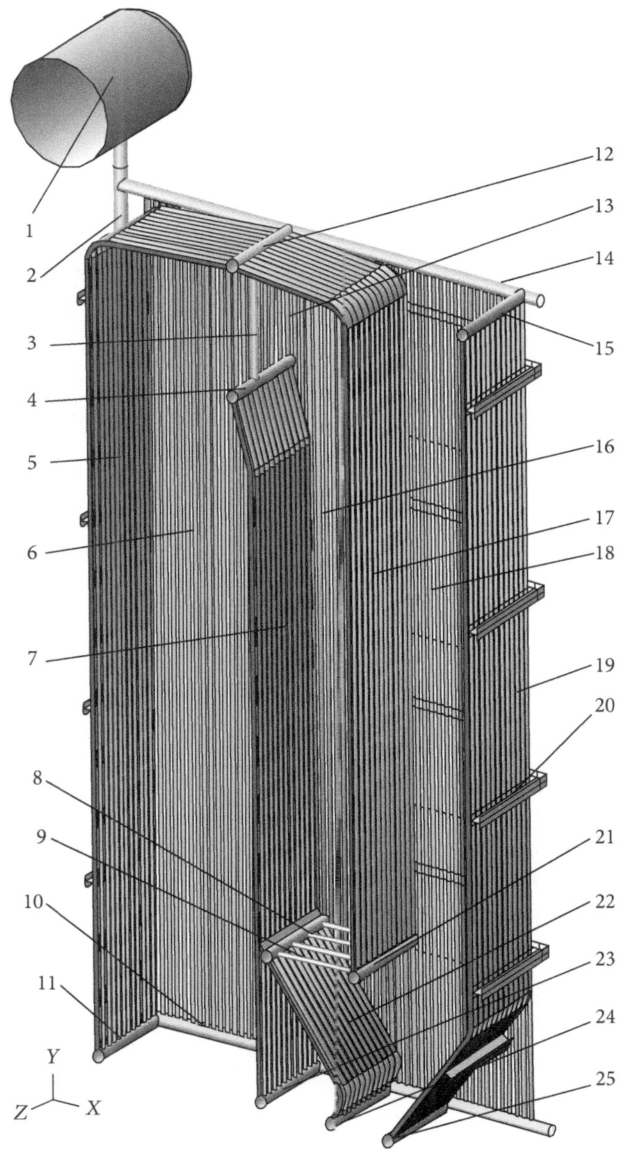

FIGURE 4: Real calculation geometry model.

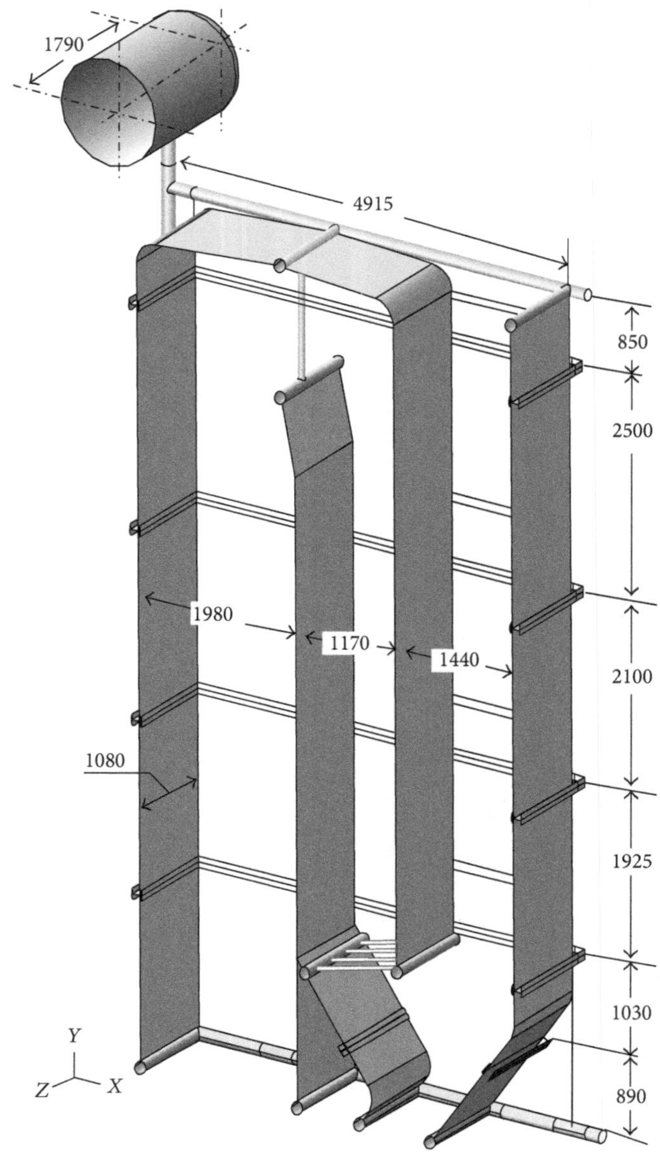

FIGURE 5: Calculation geometry model approximated as the equivalent orthotropic plate (dimensions are in mm).

The elasticity constants of equivalent orthotropic plate, for the membrane stiffness, can be calculated using the analytical expressions suggested in the paper [31, 32]. These expressions are obtained by equalizing the deformations of membrane wall and equivalent plate (thickness h), with the equivalent membrane load

$$E_{m,1} = \frac{E}{k \cdot h}\left[\frac{\pi}{4} \cdot \left(D^2 - (D - 2 \cdot \delta)^2\right) + (k - D) \cdot t\right],$$

$$E_{m,2} = \frac{k \cdot E}{h \cdot \left[3 \cdot (D/\delta - 1)^3 \cdot (\pi/8 - 1/\pi) + b/t\right]},$$

$$G_{m,12} \approx \frac{\sqrt{E_{m,1} \cdot E_{m,2}}}{2 \cdot \left(1 + \sqrt{\nu_{m,12} \cdot \nu_{m,21}}\right)},$$

$$\nu_{m,21} = \nu_{m,12} \cdot \frac{E_{m,2}}{E_{m,1}}, \tag{6}$$

where E is modulus of elasticity, and $\nu_{m,12} = \nu$ is Poisson's factor for the isotropic material of membrane wall. The expressions for the elasticity constants of equivalent orthotropic plate and for the bending stiffness of membrane wall are proposed in the paper [31, 32]. Same as for the membrane stiffness, these expressions are also obtained by equalizing

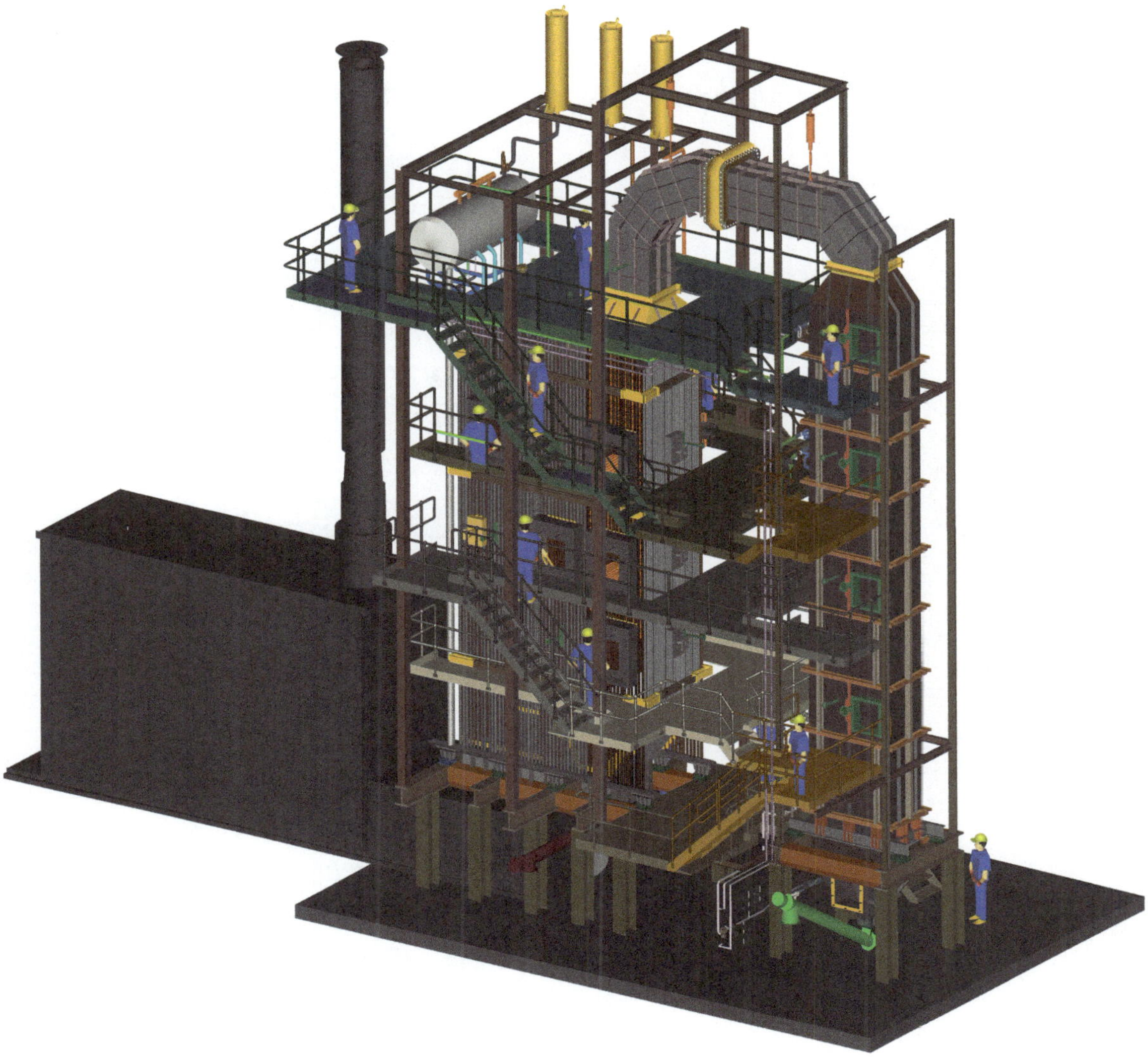

FIGURE 6: 3D-disposition of biomass incineration power plant "Milano."

the deformations of membrane wall and equivalent plate (thickness h), with the equivalent bending load

$$E_{b,1} = \frac{E}{k \cdot h^3} \cdot \left[\frac{3 \cdot \pi}{16} \cdot \left(D^4 - (D - 2 \cdot \delta)^4\right) + (k - D) \cdot t^3\right],$$

$$E_{b,2} = E \cdot \left(\frac{t}{h}\right)^3 \cdot \frac{(n \cdot k)^4}{\sum_{i=1}^{n} (i \cdot k)^4 - (i \cdot k - b)^4},$$

$$G_{b,12} \approx \frac{\sqrt{E_{b,1} \cdot E_{b,2}}}{2 \cdot \left(1 + \sqrt{\nu_{b,12} \cdot \nu_{b,21}}\right)},$$

$$\nu_{b,21} = \nu_{b,12} \cdot \frac{E_{b,2}}{E_{b,1}}, \tag{7}$$

where n is number of pipes in the membrane wall and $\nu_{b,12} = \nu$ is Poisson's factor for the isotropic material of membrane wall. After the homogenization of steam boiler membrane wall, the stiffness matrix of shell element can be calculated according to the expression

$$\mathbf{k} = \int_A \mathbf{B}^T \mathbf{D} \mathbf{B} \, dA, \tag{8}$$

where A is surface of shell element and $\mathbf{B}$ is operator of boundary magnitude.

3. Calculation of Reactions in the Steam Boiler Supports for a Real Model of Membrane Wall Geometry, for the Membrane Walls Approximated by the Equivalent Orthotropic Plate, and for the Ideal Rigid Model of Boiler Geometry

The finite element method using Abaqus 6.9-3 will be applied for the calculation of reactions in the supports. The real geometry of membrane walls (Figure 4) and the membrane walls approximated as the equivalent orthotropic plate (Figure 5)

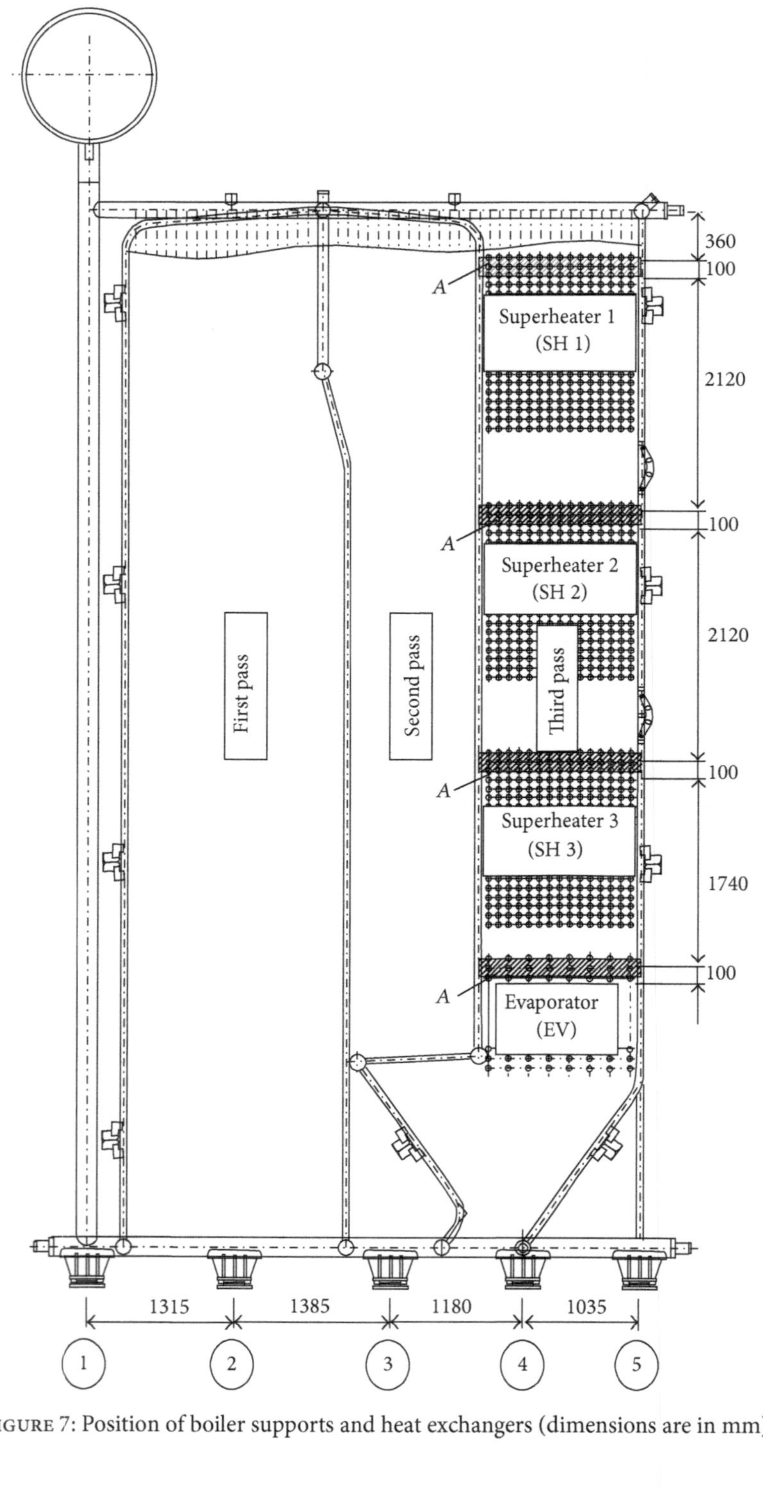

FIGURE 7: Position of boiler supports and heat exchangers (dimensions are in mm).

will be discretized by a four-node doubly curved thin shell element using the reduced integration with hourglass control. The finite element has the codename S4R5 and it is applied for small deformations and has five degrees of freedom per node.

The load of the mass of boiler structure will be taken into consideration for the calculation of reactions in the supports of "Milano" steam boiler (Figure 6). Other loads, such as the mass of working fluid, the fouling mass that is deposited on the heating surfaces during the combustion process, the mass of refractory, and the mass of insulation, will not be considered here. Pipe material of membrane walls is P235GH, material of membrane tape and buckstays is S235JRG2, and material of chambers is 16Mo3. All three steels belong to the group of low-carbon steels with $\leq$ 0.3% carbon. Since the boiler membrane walls are evaporator heating surfaces, the calculation temperature is $\vartheta = 275^{\circ}$C. Assuming that the mechanical properties of boiler structure are isotropic, [36] the modulus of elasticity for all three materials is

Table 1: List of positions according to Figure 4.

Pos.	Type	Dimensions	Pos.	Type	Dimensions
1	Shell and end of drum	$\phi 1200 \times 35$ mm	14	Collector header of side wall	$\phi 139.7 \times 12.5$ mm
2	Downcomers	$\phi 168.3 \times 16$ mm	15	Collector header of rear wall of third pass	$\phi 139.7 \times 12.5$ mm
3	Connecting pipe	$\phi 88.9 \times 8.8$ mm	16	Side wall of second pass	Pipe: $\phi 57 \times 4$ mm Tape: $t = 6$ mm Step: $k = 90$ mm Number of pipes: $n = 12$
4	Collector header of rear wall of first pass	$\phi 139.7 \times 12.5$ mm	17	Rear wall of second pass	Pipe: $\phi 57 \times 4$ mm Tape: $t = 6$ mm Step: $k = 90$ mm Number of pipes: $n = 23$
5	Front wall of first pass	Pipe: $\phi 57 \times 4$ mm Tape: $t = 6$ mm Step: $k = 90$ mm Number of pipes: $n = 23$	18	Side wall of third pass	Pipe: $\phi 48.3 \times 4.5$ mm Tape: $t = 6$ mm Step: $k = 90$ mm Number of pipes: $n = 15$
6	Side wall of first pass	Pipe: $\phi 57 \times 4$ mm Tape: $t = 6$ mm Step: $k = 90$ mm Number of pipes: $n = 21$	19	Rear wall of third pass	Pipe: $\phi 57 \times 4$ mm Tape: $t = 6$ mm Step: $k = 90$ mm Number of pipes: $n = 23$
7	Rear wall of first pass	Pipe: $\phi 57 \times 4$ mm Tape: $t = 6$ mm Step: $k = 90$ mm Number of pipes: $n = 23$	20	Buckstays	HE-B120
8	Collector header of wall of hopper	$\phi 139.7 \times 12.5$ mm	21	Distributor header of rear wall of second pass	$\phi 139.7 \times 12.5$ mm
9	Connecting pipes	$\phi 48.3 \times 4$ mm	22	Wall of hopper	Pipe: $\phi 57 \times 4$ mm Tape: $t = 6$ mm Step: $k = 90$ mm Number of pipes: $n = 23$
10	Distributor header of side wall	$\phi 168.3 \times 20$ mm	23	Distributor header of rear wall of first pass	$\phi 139.7 \times 12.5$ mm
11	Distributor header of front wall	$\phi 139.7 \times 12.5$ mm	24	Distributor header of wall of hopper	$\phi 139.7 \times 12.5$ mm
12	Collector header of front and rear wall of second pass	$\phi 139.7 \times 12.5$ mm	25	Distributor header of rear wall of third pass	$\phi 139.7 \times 12.5$ mm
13	Side wall in transition zone	Pipe: $\phi 57 \times 4$ mm Tape: $t = 6$ mm Step: $k = 90$ mm Number of pipes: $n = 34$			

approximately equal and amounts to $E = 187.12$ GPa. For the value of Poisson's factor, $\nu = 0.3$ is used.

Due to the single boiler symmetry, for the calculation using the finite element method, it is sufficient to model only one-half of the boiler geometry (Figures 4 and 5). The boiler is supported by ten supports, five on the distributor chamber of the left side wall and five on the distributor chamber of the right side wall (Figure 7). In the third pass of the boiler, the heat exchangers were installed, that is, one evaporator and three superheaters. The heat exchangers were supported on the side walls of the third pass. Due to the weight of heat exchangers, the load of the left and right side wall of the third

TABLE 2: Masses of constituent structural elements for half geometry of "Milano" boiler.

Pos.	Mass, kg	Pos.	Mass, kg	Pos.	Mass, kg	Pos.	Mass, kg	Pos.	Mass, kg
1	1967	7	639	13	315	19	786	25	43
2	595	8	43	14	197	20	668	SH1	1185
3	23	9	17	15	43	21	43	SH2	1185
4	43	10	402	16	702	22	157	SH3	527
5	893	11	43	17	721	23	43	EV	53
6	1184	12	43	18	1018	24	43		

TABLE 3: Elasticity constants and wall thickness of equivalent orthotropic plate for membrane walls of "Milano" boiler.

Membrane wall (position)	h, mm	$E_{m,1}$, GPa	$E_{m,2}$, GPa	$G_{m,12}$, GPa	$\nu_{m,21}$	$E_{b,1}$, GPa	$E_{b,2}$, GPa	$G_{b,12}$, GPa	$\nu_{b,21}$
5, 7, 17, 19, 22	23.48	76.492	1.367	4.915	0.005361	454.159	8.064	29.095	0.005327
6	23.48	76.492	1.367	4.915	0.005361	454.159	8.023	29.025	0.005300
13	23.48	76.492	1.367	4.915	0.005361	454.159	8.203	29.336	0.005419
16	23.48	76.492	1.367	4.915	0.005361	454.159	7.690	28.439	0.005080
18	20.40	88.592	3.880	8.722	0.013138	442.956	9.583	31.199	0.006490

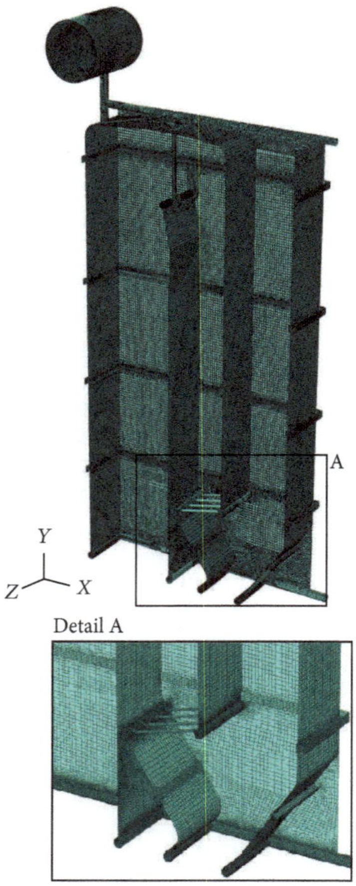

FIGURE 8: Finite element mesh-calculation model APPROX.

TABLE 4: Members of elasticity matrix of equivalent orthotropic plate for membrane walls of "Milano" boiler for membrane stiffness.

Membrane wall (position)	A_{11}, N/mm	A_{12}, N/mm	A_{22}, N/mm	A_{33}, N/mm
5, 6, 7, 13, 16, 17, 19, 22	1799282	9646	32152	115438
18	1814749	23843	79476	177966

TABLE 5: Members of elasticity matrix of equivalent orthotropic plate for membrane walls of "Milano" boiler for bending stiffness.

Membrane wall (position)	D_{11}, N·mm	D_{12}, N·mm	D_{22}, N·mm	D_{33}, N·mm
5, 7, 17, 19, 22	491000497	2615341	8717804	31404889
6	490996567	2602240	8674134	31329272
13	491014120	2660750	8869166	31665404
16	490964096	2494004	8313348	30696524
18	314156438	2038906	6796354	22084218

TABLE 6: Reactions in the boiler supports for the influence of total mass.

Reactions in boiler supports	F_{R1}, N	F_{R2}, N	F_{R3}, N	F_{R4}, N	F_{R5}, N
REAL	22748	27020	32657	38515	12676
APPROX	24183	25992	31015	40026	12499
RIGID	27234	22608	31088	36609	16077
Δ_A, %	6.31	−3.80	−5.03	3.92	−1.40
Δ_K, %	19.72	−16.33	−4.80	−4.95	26.83

pass is identical. For the calculation of reactions in the boiler supports, half of the weight of individual heat exchanger will be used as the equivalent tangential surface pressure on the side wall of the third pass. This pressure is applied on surface A of membrane wall on which the carriers of heat exchangers are welded (Figure 7). Bending of the side membrane walls of the third pass, due to the load of heat exchangers, was not taken into consideration.

TABLE 7: Reactions in the boiler supports for the influence of mass of boiler without the mass of heat exchangers.

Reactions in boiler supports without the mass of heat exchangers	F_{R1}, N	F_{R2}, N	F_{R3}, N	F_{R4}, N	F_{R5}, N
REAL	24329	29007	24633	21190	5521
APPROX	25790	27681	23752	21822	5574
RIGID	27397	21905	34552	10957	9869
Δ_A, %	6.01	−4.57	−3.58	2.98	0.96
Δ_K, %	12.61	−24.48	40.27	−48.29	78.75

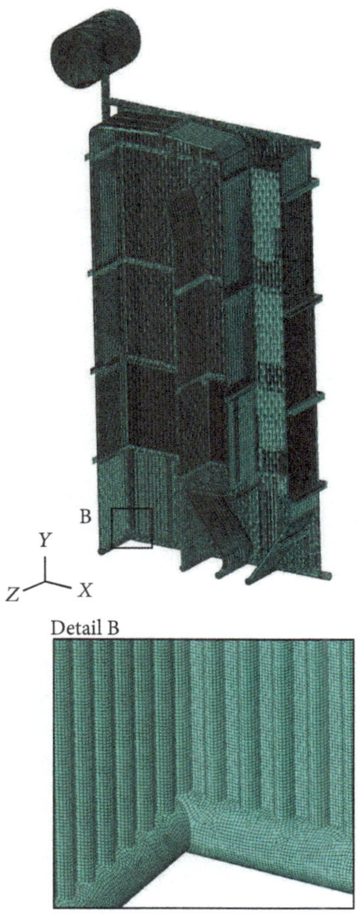

FIGURE 9: Finite element mesh-calculation model REAL.

The masses of individual structural elements of boiler, marked according to Figures 4 and 7 and Table 1, are given in Table 2. The inhomogeneity of material of boiler structure will not be taken into consideration here. It is assumed that the geometry of boiler is ideal, without the tolerances of production and assembly, as well as the tolerances of semifinished rolled products built into the boiler. It is assumed that the mechanical properties of material of boiler structure are isotropic, that is, that anisotropy of the welded joints and semifinished rolled products does not influence significantly the reactions in the boiler supports.

The mass of half of boiler without the heat exchangers is $m' = 10671$ kg, and with the heat exchangers it is $m = 13621$ kg. The membrane walls of "Milano" boiler are made with the step of 90 mm. Two dimensions of the cross-section of pipe ($\phi 57 \times 4$ mm and $\phi 48.3 \times 4.5$ mm) are applied, as well as the membrane tape (thickness, 6 mm). For the boiler membrane walls, it is possible to calculate the elasticity constants of equivalent orthotropic plate (Table 3) according to expressions (2) and (7). The members of elasticity matrix (Tables 4 and 5), which will be applied to the finite element of equivalent orthotropic plate, will be calculated according to expressions (3) and (4).

In this paper, the results of six different calculations of reactions in the supports of "Milano" boiler will be presented: the calculation of real geometry of boiler (REAL) discretized by the shell finite element, the calculation with the approximation of membrane walls as the plate/shell of equivalent stiffness (APPROX), and the calculation where the boiler is considered as a rigid body (RIGID). Each calculation is made with the load of the mass of boiler, with and without the mass of heat exchangers.

The calculation model for the real geometry of boiler consisted of 18943590 variables that include the degrees of freedom and Lagrange multipliers, and for the approximated geometry of boiler membrane walls as the plate/shell of equivalent stiffness, the calculation model consisted of 1619526 variables (degrees of freedom and Lagrange multipliers). In both cases, 4-node finite element S4R5 was predominantly used, and an extremely triangular element was created by the collapse of one node of 4-node element S4R5 (Figures 8 and 9).

The reaction forces in the supports for the influence of total mass of boiler are given in Table 6, and the reaction forces for the influence of mass of boiler without the mass of heat exchangers are given in Table 7. The reaction forces in the boiler supports are denoted according to Figure 7. In Tables 6 and 7, the deviations of reaction forces for calculation APPROX (compared to calculation REAL) were calculated, and the deviation was expressed as a percentage and marked as Δ_A. In the same way, the deviation of reaction forces for calculation RIGID (compared to calculation REAL) was calculated, and it was expressed as a percentage and marked as Δ_K. The deviations were calculated according to the following expressions:

$$\Delta_A = \left(\frac{F_{\mathrm{Ri}}\,(\mathrm{APPROX})}{F_{\mathrm{Ri}}\,(\mathrm{REAL})} - 1 \right) \cdot 100, \%,$$
$$\Delta_K = \left(\frac{F_{\mathrm{Ri}}\,(\mathrm{RIGID})}{F_{\mathrm{Ri}}\,(\mathrm{REAL})} - 1 \right) \cdot 100, \%. \qquad (9)$$

4. Conclusion

The investigation of the influence of approximation of steam boiler membrane walls, using the equivalent orthotropic plate for the calculation of reactions in the boiler supports, is presented in this paper. The expressions for the elasticity constants that have been published in previous papers [30–32] were applied.

(i) The reactions in the boiler supports were calculated with the assumption that the structure of boiler is ideally rigid.

(ii) The calculations were carried out for the load of the mass of boiler, with and without the mass of heat exchangers in the third pass of boiler.

(iii) The number of variables for the calculation model REAL was more than 11 times higher than for the calculation model APPROX. Therefore, the calculation of node values of model REAL lasted for 1 h, 23 min and 50 s, and for the model APPROX 1 min and 56 s. The calculation was carried out on a PC with CPU Intel(R) Core (TM) i7, 24 GB RAM, and 64-bit operating system.

(iv) The deviations of results for the reactions in the supports for the calculation APPROX, compared to the calculation REAL, are max. 6.31% for the load of the total mass of boiler and 6.01% for the load without the mass of heat exchangers.

(v) The deviations of results for the calculation RIGID, compared to the calculation REAL, are 78.75% for the load without the mass of heat exchangers.

(vi) If the results of calculation REAL are accepted as approximately accurate, then it can be concluded that the application of calculation model APPROX is acceptable for practical use. Due to the huge error, the calculation model RIGID is not recommended for practical use.

(vii) Using the method of equivalent stiffness of membrane wall, more accurate results of calculation could be acquired if, when calculating the modulus of elasticity $E_{b,2}$, the membrane pipe would be observed as a solid body, and not as an ideal rigid body [32].

(viii) Additional improvements could be achieved by taking into consideration the transverse shear stiffness.

Conflict of Interests

The authors declare that there is no conflict of interests regarding the publication of this paper.

Acknowledgment

The authors would like to thank PLANUM d.o.o. from Slavonski Brod in Croatia for usage of technical data which were needed for this research.

References

[1] V. Milošević-Mitić, B. Gaćeša, N. Anđelić, and T. Maneski, "Numerical calculation of the water-tube boiler using finite element of the orthotropic plate," *Structural Integrity and Life*, vol. 12, no. 3, pp. 185–190, 2012.

[2] M. T. Huber, "Die grundlagen einer rationellen berechnung der kreuzweise bewehrten eisenbetonplatten," *Zeitschrift des Österreichischen Ingenieur- und Architekten-Vereins*, vol. 66, pp. 557–564, 1914.

[3] M. T. Huber, "Die Theorie der kreuzweise bewehrten Eisenbeton-platten nebst Anwendungen auf mehrere bautechnisch wichtige Aufgaben über rechteckige Platten," *Bauingenieur*, vol. 4, pp. 354–360, 1923.

[4] M. T. Huber, *Probleme der Statik Technisch Wichtiger Orthotroper Platten*, Nakkadem Akadenji Nauk Technicznych, Warszawa, Poland, 1929.

[5] W. Flügge, "Die Stabilität der Kreiszylinderschale," *Ingenieur-Archiv*, vol. 3, no. 5, pp. 463–506, 1932.

[6] A. van der Neut, "The general instability of stiffened cylindrical shells under axial compression," Report S. 314, National Luchtvaartlaboratorium, 1947.

[7] F. A. Dale, R. C. T. Smith et al., "Grid sandwich panels in compression," Report ACA-16, Australian Council for Aeronautics, 1945.

[8] C. B. Smith, T. B. Heebink, and C. B. Norris, "The effective stiffness of a stiffener attached to a flat plywood plate," Report 1557, U. S. Department of Agriculture, Forest Products Laboratory, 1946.

[9] A. Pflüger, "Zum Beulproblem der anisotropen Rechteckplatte," *Ingenieur-Archiv*, vol. 16, no. 2, pp. 111–120, 1947.

[10] A. Gomza and P. Seide, "Minimum-weight design of simply supported transversely stiffened plates under compression," NACA TN 1710, 1948.

[11] C. Libove and R. E. Hubka, "Elastic constants for corrugated-core sandwich plates," NACA TN 2289, 1951.

[12] C. Libove and S. B. Batdorf, "A general small-deflection theory for flat sandwich plates," NACA Report 899, 1948.

[13] E. Reissner, "On small bending and stretching of sandwich-type shells," *International Journal of Solids and Structures*, vol. 13, no. 12, pp. 1293–1300, 1977.

[14] M. Stein and J. Mayers, "A small-deflection theory for curved sandwich plates," NACA Report 1008, 1951.

[15] N. J. Huffington Jr., "Theoretical determination of rigidity properties of orthogonally stiffened plates," *Journal of Applied Mechanics*, vol. 23, pp. 15–20, 1956.

[16] G. Gerard, "Minimum weight analysis of orthotropic plates under compressive loading," *Journal of the Aeronautical Sciences*, vol. 27, no. 1, pp. 21–26, 1960.

[17] M. Baruch and J. Singer, "Effect of eccentricity of stiffeners on the general instability of stiffened cylindrical shells under hydrostatic pressure," *Journal of Mechanical Engineering Science*, vol. 5, no. 1, pp. 23–27, 1963.

[18] R. J. Clifton, J. C. L. Chang, and T. Au, "Analysis of orthotropic plate bridges," *ASCE Journal of the Structural Division*, vol. 89, no. 5, pp. 133–171, 1963.

[19] W. J. Stroud, "Elastic constants for bending and twisting of corrugation-stiffened panels," NASA TR R 166, 1963.

[20] R. R. Meyer and R. J. Bellefante, "Fabrication and experimental evaluation of common domes having waffle-like stiffening—part I—program development," Report SM 47742, Douglas Missile & Space Systems Division, 1964.

[21] T.-C. Soong, "Buckling of cylindrical shells with eccentric spiral-type stiffeners," *AIAA Journal*, vol. 7, no. 10, pp. 65–72, 1969.

[22] R. R. Meyer, "Comments on: buckling of cylindrical shells with eccentric spiral-type stiffeners," *AIAA Journal*, vol. 7, no. 10, p. 2047, 1969.

[23] T.-C. Soong, "Reply by author to R. R. Meyer," *AIAA Journal*, vol. 7, no. 10, pp. 2047–2048, 1969.

[24] F. Nishino, R. P. Pama, and S.-L. Lee, "Orthotropic plates with eccentric stiffeners," *International Association of Bridge Structural Engineering*, vol. 34, no. 2, pp. 117–129, 1974.

[25] W. L. Ko, "Elastic constants for superplastically formed/diffusion-bonded sandwich structures," in *Proceedings of the AIAA/ASME/AHS 20th Structures, Structural Dynamics, and Materials Conference*, AIAA paper 79-0756, 1979.

[26] W. L. Ko, "Elastic constants for superplastically formed/diffusion-bonded sandwich structures," *AIAA Journal*, vol. 18, no. 8, pp. 986–987, 1980.

[27] W. L. Ko, "Elastic stability of superplastically formed/diffusion-bonded orthogonally corrugated core sandwich plates," in *Proceedings of the AIAA/ASME/AHS 21st Structures, Structural Dynamics, and Materials Conference*, AIAA paper 80-0683, 1980.

[28] N. Jaunky, N. F. Knight Jr., and D. R. Ambur, "Formulation of an improved smeared stiffener theory for buckling analysis of grid-stiffened composite panels," NASA TM 110162, 1995.

[29] N. Jaunky, F. Knight, and D. R. Ambur, "Formulation of an improved smeared stiffener theory for buckling analysis of grid-stiffened composite panels," *Composites B*, vol. 27, pp. 519–526, 1996.

[30] J. Sertić, D. Kozak, D. Damjanović, and P. Konjatić, "Homogenization of steam boiler membrane wall," in *Proceedings of 7th International Congress of Croatian Society of Mechanics*, Zadar, 2012.

[31] J. Sertić, I. Gelo, D. Kozak, D. Damjanović, and P. Konjetić, "Teoretical determination of elasticity constants for steam boiler membrane wall as the structurally ortotropic plate," in *Proceeding of 4th International Scientific and Expert Conference of the International TEAM Society*, Slavonski Brod, 2012.

[32] J. Sertić, I. Gelo, D. Kozak, D. Damjanović, and P. Konjatić, "Theoretical determination of elasticity constants for steam boiler membrane wall as the structurally orthotropic plate," *Technical Gazette*, vol. 20, no. 4, pp. 697–703, 2013.

[33] R. Solecki and R. J. Conant, *Advanced Mechanics of Materials*, Oxford University Press, Oxford, UK, 2003.

[34] E. Ventsel and T. Krauthammer, *Thin Plates and Shells. Theory, Analysis and Applications*, Marcel Dekker, Basel, Switzerland, 2001.

[35] V. V. Novozhilov, *Thin Shell Theory*, Noordhoff, Groningen, Germany, 1964.

[36] ASME B31.1, 1995.

15

Static Analysis of Large-Scale Multibody System Using Joint Coordinates and Spatial Algebra Operator

Mohamed A. Omar

Mechanical Engineering Department, Taibah University, Almadinah Almonawwarah 42353, Saudi Arabia

Correspondence should be addressed to Mohamed A. Omar; momar@taibahu.edu.sa

Academic Editor: Belal F. Yousif

Initial transient oscillations inhibited in the dynamic simulations responses of multibody systems can lead to inaccurate results, unrealistic load prediction, or simulation failure. These transients could result from incompatible initial conditions, initial constraints violation, and inadequate kinematic assembly. Performing static equilibrium analysis before the dynamic simulation can eliminate these transients and lead to stable simulation. Most exiting multibody formulations determine the static equilibrium position by minimizing the system potential energy. This paper presents a new general purpose approach for solving the static equilibrium in large-scale articulated multibody. The proposed approach introduces an energy drainage mechanism based on Baumgarte constraint stabilization approach to determine the static equilibrium position. The spatial algebra operator is used to express the kinematic and dynamic equations of the closed-loop multibody system. The proposed multibody system formulation utilizes the joint coordinates and modal elastic coordinates as the system generalized coordinates. The recursive nonlinear equations of motion are formulated using the Cartesian coordinates and the joint coordinates to form an augmented set of differential algebraic equations. Then system connectivity matrix is derived from the system topological relations and used to project the Cartesian quantities into the joint subspace leading to minimum set of differential equations.

1. Introduction

Virtual development procedures became the most economical venue in product design and optimization including earth-moving equipment and automotive systems. These systems are typically large and complex and composed of heterogeneous physical subsystems. Simulation of such systems may require multidomain modeling, for example, rigid and flexible bodies, nonlinear contact and friction force modules, terrain interaction, electrical and hydraulic subsystems, control systems, and nonholonomic constraints. Accurate multibody solver will be required in the heart of the analysis of such systems. As per Giampiero and Manfred [1], the computation of static equilibrium is often the first step in the analysis of multibody system models. The static equilibrium position defines the working point for the linearization of the system equation of motion and provides initial values for the dynamical simulation.

Large earth-moving equipment, agricultural machinery, tracked vehicles, tractors, bulldozers, motor graders, wheeled loaders, and forestry machines are some of the equipment that may work on slopped terrain. Stability assessment of such machinery on the slopes under different loading conditions and different articulation configurations requires accurate modeling of the machine and its interaction with the terrain. The behavior of existing terrain models, in most cases [2–5], is highly nonlinear and history dependent. Performing static equilibrium is very crucial step for accurate linear analysis and stability assessment of the multibody systems.

Another example would be when the multibody dynamics approach is used to model and simulate the startup and shut down of reciprocating engines and turbines which are supported by journal bearings. Journal bearings are typically nonlinear components and their dynamic characteristics (stiffness, damping, etc.) are history dependent [6, 7]. In such systems, the crankshaft and the rotor could be modeled as flexible (deformable bodies). The static equilibrium position of rotor center with respect to the bearing geometric center needs to be accurately determined in order to perform transient simulation or normal mode analysis. When the multibody system contains contacting bodies that may undergo slipping, static equilibrium cannot be obtained from force balance. In this case, the traditional static equilibrium solution will not converge to the true static equilibrium position.

In general, analysis and simulation scenarios of dynamic multibody systems could be summarized as follows:

(1) performing transient dynamic simulation of typical machine working cycle or operation;

(2) performing eigenvalue analysis to calculate the natural frequencies and normal modes and assess the system stability;

(3) determining the static equilibrium of an articulated system that may include rigid and flexible bodies connected with joints.

Three traditional approaches have been used to formulate the multibody system equations of motion: the Cartesian body coordinates (absolute coordinates), the joint coordinates, and velocity transformation method. The Cartesian body coordinates formulations are very popular and reported to be a simpler method to construct the equations of motion leading to a large set of differential algebraic equations [8, 9]. The configuration of a rigid body is described by a set of translational and rotational coordinates. Algebraic constraints are introduced to represent kinematic joints connecting bodies and then the Lagrange multiplier technique is used to describe joint reaction forces. The system of differential algebraic equations has to be solved simultaneously. Newton-Raphson iterations or similar techniques could be used to satisfy the applied constraints. Kinematic assembly of the system is necessary for starting a successful simulation.

Featherstone [10, 11] used spatial vectors to study the dynamics of articulated bodies. Featherstone and Orin [12] and Featherstone [13] presented an efficient approach for utilizing spatial algebra to model multibody systems and efficiently factor the inertia matrix for rigid body systems. Wehage and Haug [14] used a similar approach to develop set of automated procedures for robust and efficient solution of overconstrained multibody dynamics. Wehage and Belczynski [15] proposed structuring the kinematic and dynamic equations into block matrix structure and developed procedures to enable real-time simulation of multibody system. Rodriguez et al. [16, 17] presented a spatial operator based on the spatial algebra for developing multibody dynamic equations of motion. Determining the static equilibrium position using these formulations was not explicitly explained.

Traditional approach of solving for the static equilibrium is based on minimizing the potential energy of the system [1, 18]. This approach requires explicit expression for potential energy of all the bodies and the force elements in the system. This could be practically impossible for the abovementioned heterogeneous system simulations. Other approaches [19, 20] assume the system will be under static equilibrium when the first and second derivatives of the system generalized coordinates are set to zero. The differential equations of motion of the system then become a set of algebraic equations that could be solved along with the constraint equations. The resulting solution yields a set of initial values for the system states. Linear modes analysis could then be performed by carrying out Taylor expansion while only the linear terms are retained [1, 19]. In the abovementioned approaches, false equilibrium position could be reported. Also, violation of the constraints in the velocity level and acceleration level may exist.

Recently, Yang et al. [19] presented a general approach for solving the eigenvalue problem after calculating static equilibrium. The approach is similar to many existing solution approaches and could be summarized by [19] as follows. The system equations of motion could be derived from

$$\frac{d}{dt}\left(\frac{\partial L}{\partial \dot{\mathbf{x}}}\right)+\left(\frac{\partial L}{\partial \mathbf{x}}\right)=\left(\frac{\partial \mathbf{\Phi}}{\partial \mathbf{x}}\right)^{T}\boldsymbol{\lambda}+\mathbf{f},$$
$$\mathbf{\Phi}(\mathbf{x})=\mathbf{0}, \tag{1}$$

where $\mathbf{x}\in\mathbb{R}^{n}$ is the vector of general coordinates, $\boldsymbol{\lambda}\in\mathbb{R}^{m}$ is the vector of Lagrange multipliers, L is the Lagrange function associated with $(\mathbf{x},\dot{\mathbf{x}})$, $\mathbf{f}$ is the vector of nonconservative forces which are function of $(\mathbf{x},\dot{\mathbf{x}},t)$, $\mathbf{\Phi}$ is the vector of constraints associated with $(\mathbf{x})$, and $(\partial\mathbf{\Phi}/\partial\mathbf{x})^{T}\boldsymbol{\lambda}$ is the vector of generalized constraint forces.

The general approach to solve for the static equilibrium position starts by writing the objective function as follows:

$$\mathbf{F}(\ddot{\mathbf{x}},\dot{\mathbf{x}},\mathbf{x},t)=\left(-\frac{d}{dt}\left(\frac{\partial L}{\partial \dot{\mathbf{x}}}\right)+\frac{\partial L}{\partial \mathbf{x}}+\mathbf{f}\right)(\ddot{\mathbf{x}},\dot{\mathbf{x}},\mathbf{x},t),$$
$$\mathbf{G}(\mathbf{x})=\left(\frac{\partial \mathbf{\Phi}(\mathbf{x})}{\partial \mathbf{x}}\right)^{T}. \tag{2}$$

Then, the system equations could be written as

$$\mathbf{F}(\ddot{\mathbf{x}},\dot{\mathbf{x}},\mathbf{x},t)+\mathbf{G}(\mathbf{x})\boldsymbol{\lambda}=0,$$
$$\mathbf{\Phi}(\mathbf{x})=0. \tag{3}$$

The equilibrium state $(\mathbf{x}_0,\boldsymbol{\lambda}_0,t_0)$ satisfies the following conditions:

$$\mathbf{F}(0,0,\mathbf{x}_0,t_0)+\mathbf{G}(\mathbf{x}_0)\boldsymbol{\lambda}_0=0,$$
$$\mathbf{\Phi}(\mathbf{x}_0)=0. \tag{4}$$

With a small vibration about the equilibrium position $(\delta\ddot{\mathbf{x}},\delta\dot{\mathbf{x}},\mathbf{x}_0+\delta\mathbf{x},\boldsymbol{\lambda}_0+\delta\boldsymbol{\lambda},t)$, then

$$\mathbf{F}(\delta\ddot{\mathbf{x}},\delta\dot{\mathbf{x}},\mathbf{x}_0+\delta\mathbf{x},t)+\mathbf{G}(\mathbf{x}_0+\delta\mathbf{x})(\boldsymbol{\lambda}_0+\delta\boldsymbol{\lambda})=0,$$
$$\mathbf{\Phi}(\mathbf{x}_0+\delta\mathbf{x})=0. \tag{5}$$

Using Taylor expansion and reserving the linear terms, we get

$$\widehat{\mathbf{M}}(\mathbf{x}_0)\delta\ddot{\mathbf{y}}+\widehat{\mathbf{C}}(\mathbf{x}_0)\delta\dot{\mathbf{y}}+\widehat{\mathbf{K}}(\mathbf{x}_0,\boldsymbol{\lambda}_0)\delta\mathbf{y}=0, \tag{6}$$

where $\widehat{\mathbf{M}}(\mathbf{x}_0)=\begin{bmatrix}\mathbf{M}(\mathbf{x}_0) & 0\\ 0 & 0\end{bmatrix}$, $\widehat{\mathbf{C}}(\mathbf{x}_0)=\begin{bmatrix}\mathbf{C}(\mathbf{x}_0) & 0\\ 0 & 0\end{bmatrix}$, $\widehat{\mathbf{K}}(\mathbf{x}_0,\boldsymbol{\lambda}_0)=\begin{bmatrix}\mathbf{K}(\mathbf{x}_0,\boldsymbol{\lambda}_0) & \mathbf{G}(\mathbf{x}_0)\\ \mathbf{G}^{T}(\mathbf{x}_0) & 0\end{bmatrix}$, $\mathbf{M}(\mathbf{x}_0)=(\partial\mathbf{F}/\partial\ddot{\mathbf{x}})(\mathbf{x}_0)$, $\mathbf{C}(\mathbf{x}_0)=(\partial\mathbf{F}/\partial\dot{\mathbf{x}})(\mathbf{x}_0)$, $\mathbf{K}(\mathbf{x}_0,\boldsymbol{\lambda}_0)=(\partial(\mathbf{F}+\mathbf{G}\boldsymbol{\lambda})/\partial\mathbf{x})(\mathbf{x}_0,\boldsymbol{\lambda}_0)$, and $\delta\mathbf{y}=[\delta\mathbf{x}\ \ \delta\boldsymbol{\lambda}]^{T}$.

Writing the solution as $\delta \mathbf{y} = e^{rt}\mathbf{V}$, where r is one of the eigenvalues, $\mathbf{V}$ is the corresponding eigenvector and the eigenvalue problem becomes

$$\left(r^2\widehat{\mathbf{M}} + r\widehat{\mathbf{C}} + \widehat{\mathbf{K}}\right)\mathbf{V} = 0. \tag{7}$$

This equation represents a standard eigenvalue problem that could be solved using standard solver. The frequency shift could be used to insure that $\mathbf{K}$ is nonsingular.

Avilés et al. [21] presented a procedure for the solution of the nonlinear static equilibrium problem in complex multibody mechanical systems, including rigid and elastic elements. The error function was simple based on the potential function and set of nonlinear constraints. Lagrange multipliers along with various versions of the augmented Lagrange multipliers were used to form the system dynamic equations. A Newton-Raphson second-order method is used for seeking function minima for equilibrium positions. Wang et al. [22] used the bond graph to improve the efficiency for the kintostatic analysis of complex multibody systems. They proposed an effective decoupling method to simplify the resulting complexity of the equation of motion. As per Giampiero and Manfred [1], the linearized equation of motion will lead to nonlinear equations that can be solved using Newton's method.

Similarly, commercial multibody dynamic simulation solvers like SIMPACK use Newton-Raphson based solver for nonlinear equations and find the system equilibrium positions and/or calculate the preload in the springs and bushing elements. This solution approach has been used for trains, automotive, and airplane systems. Another approach to determine the static equilibrium is the quasistatic approach. In multibody systems, this approach may not be very efficient if there is significant difference in the inertia of the bodies in the system or if there are dynamic oscillators in the system.

This paper describes a general purpose formulation and implementation for modeling rigid and flexible body in multibody system based on the joint coordinates formulation. A new approach for solving static equilibrium will be presented in this paper. The presented static solver utilizes a recursive transient solver to evaluate the system equations of motion. An energy dissipative damping is introduced to eliminate the system oscillations around the equilibrium position. The system states are dampened using weighted damping factor that is function of the state velocity and the state error. The attributes associated with this method are the ability to handle large heterogeneous systems and ability to linearize the system in terms of arbitrary user-defined coordinates in a straightforward implementation. The spatial algebra operators are used to formulate the kinematic and dynamic equations of motion.

This paper is organized as follows. In the following section, the structure of the equation of motion of the multibody system using joint coordinate formulation is introduced for rigid body system based on the spatial algebra operator. In Section 3, the flexible body kinematic and dynamic equations of motion are presented. In Section 4, the equations of motion of the constrained multibody systems with closed loops will be presented. Section 5 outlines the constraint enforcement technique based on Baumgarte stabilization approach. Section 6 presents the proposed static equilibrium technique based on the energy drainage. Section 7 outlines the recursive algorithm implemented to solve the proposed multibody system equations of motion. Section 8 presents sample results of the proposed static solution approach for multibody systems. Finally, this paper is summarized and some conclusions are drawn in Section 9.

2. Multibody Formulation and System Equation of Motion

Figure 1 shows an illustration of the typical structure of articulated earth-moving equipment, a wheel loader. The major dynamic components of the machine, as shown in the figure, include the rear frame which contains the engine and drive train, the right and left rear wheels which are connected to the rear frame, the front frame which articulates with respect to the rear frame to steer the machine, the boom which is connected to the front frame and is driven by hydraulic cylinders, the bucket which is connected to the boom and is controlled by other hydraulic cylinders, and the left and right front wheels which are connected to the front frames. This machine model will be used to relate the theoretical multibody equations to physical model for clarification.

The proposed formulation in this investigation utilizes a joint coordinates based multibody approach. The kinematic tree or connectivity graph is developed to describe the system topology based on the connectivity between the different bodies in the system. The ground body (the inertial body) is considered the root (base) of the kinematic tree. Any dynamic body in the system is connected or referenced to one parent body through an arc joint that allows one or more DoF. While each body can have only one parent (ancestor body), it could have one or more child bodies (descendant bodies). In the kinematic tree the root body is numbered as 0 while the descendant bodies are numbered consecutively from 1 to n_b. The body and the joint connecting it to its parent are given the same number [10, 23]. It should be mentioned that the kinematic tree is not unique. Using the parent-child relation, a parent-child list could be developed and stored to be used later in the recursive calculations [13, 24, 25]. Any arbitrary body in the system could be modeled as rigid or flexible body. Figure 2 shows the kinematic tree of the wheel loader under investigation. The dynamic bodies are represented by the circles and the joints are represented by solid arrows. The rear frame is considered the base of our kinematic tree and is referenced to the ground with free joint (joint with 6-DoF). The rear wheels and the front frame are considered descendants of the rear frame and connected to their parent by revolute joints. The remaining bodies in the physical model are shown in the kinematic tree. It should be mentioned that, although the bucket and the boom are driven by two hydraulic cylinders each, only one cylinder is shown in the kinematic tree for simplicity,

$$\text{parent} = \text{list}(k) = \{0, 1, 1, 1, 2, 2, 2, 2, 2, 7, 8, 9\}. \tag{8}$$

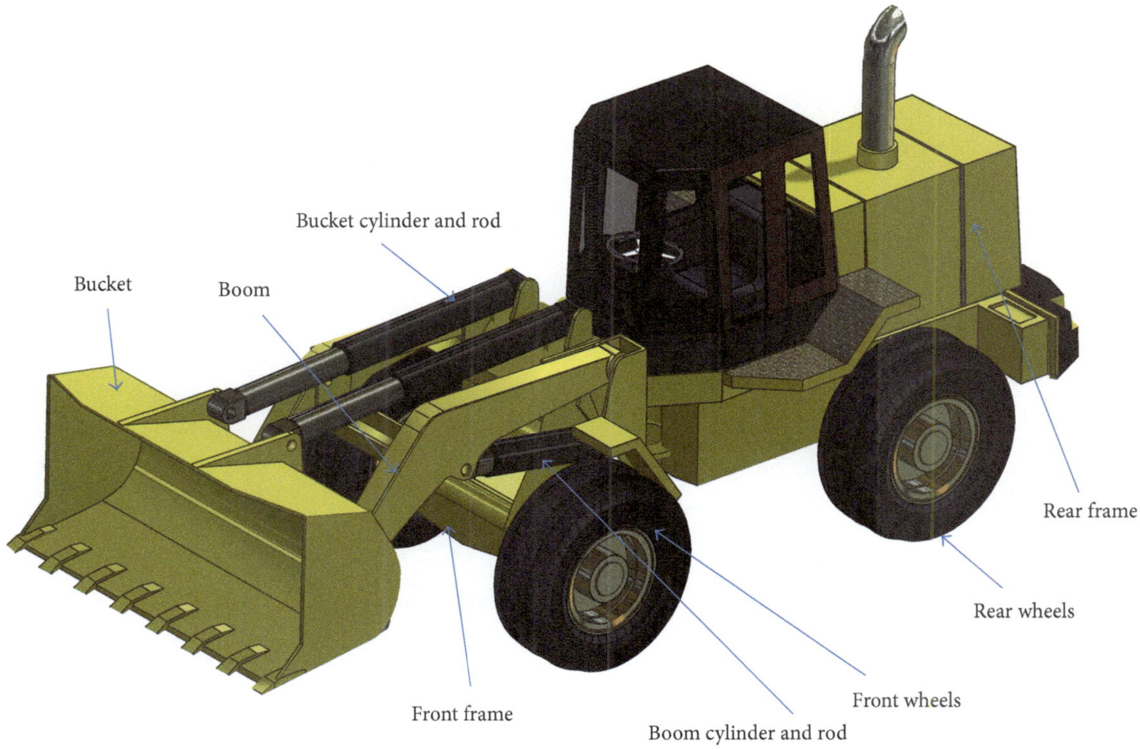

FIGURE 1: Typical structure of a wheel loader.

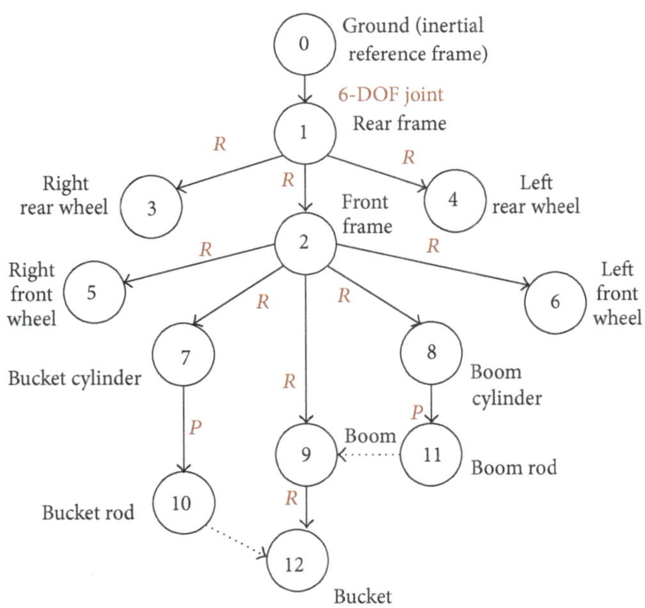

FIGURE 2: Kinematic tree of the wheel loader model.

In the proposed formulation the child body is joined to its parent through two markers; as shown in Figure 3, one marker is attached to the parent side called connector while the other marker is on the child side called the output marker. The output marker is the reference frame of the child body. The body kinematic quantities and dynamic matrices are expressed with respect to the body reference frame. The position and orientation of the connectors are measured relative to the parent reference marker through the spatial transformation matrix, as shown in Figure 3. The joint degrees of freedom, representing the joint variables, are defined as the allowed relative motion between the two connecting markers. Relative motion between the markers could be translation or rotation along one of the connector marker axes. Massless markers could be inserted between the two bodies to represent joints with more than one DoF. The joint variables displacement, velocities, and accelerations are used as the body states. The Cartesian displacements, velocities, acceleration vectors, and the joint reaction forces are considered as augmented algebraic variables.

The position vector of body i is defined in the global coordinate system by position vector of the origin of the body reference marker $\underline{r}^0_{0i,i}$, as shown in Figure 3, while the orientation of the body could be described by the 3×3 rotation matrix $\underline{R}^{0i}$ which is a function of set of spatial rotational angles. The spatial velocity vector of body i, $\mathbf{v}^0_{0i,i} = \left[\underline{\omega}^0_{0i} \;\; \dot{\underline{r}}^0_{0i,i}\right]^T$, could be obtained by differentiating the position vector and the body orientation parameters. As the recursive formulation is based on relative motion, the relative spatial velocity vector of the reference frame of a child body j with respect to its parent rigid body i observed at the origin of the local coordinate system of body j can be written as

$$\mathbf{v}^j_{ij,j} = \left[\underline{\omega}^j_{ij} \;\; \dot{\underline{r}}^j_{ij,j}\right]^T, \tag{9}$$

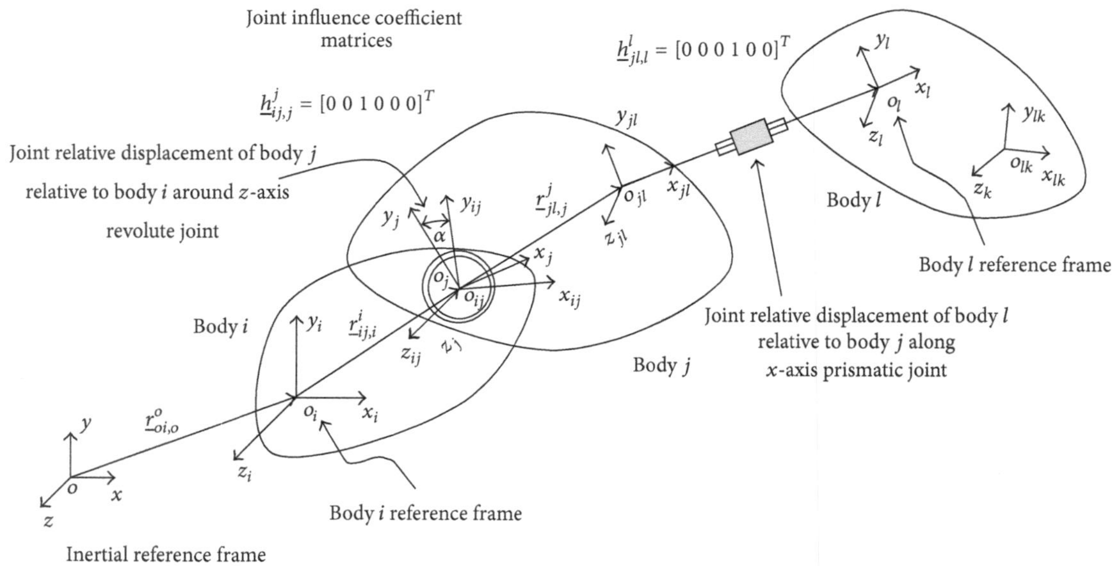

Figure 3: Body kinematics and joint connections.

where $\underline{\omega}^j_{ij}$ is the time derivative of the orientation parameters of body j with respect to body i and $\dot{\underline{r}}^j_{ij,j}$ is the derivative of the position vector of body j with respect to body i. The relative velocity across the joint between body i and body j can be written as follows:

$$\mathbf{v}^j_{ij,j} = \mathbf{h}^j \dot{\mathbf{q}}_j, \tag{10}$$

where $\mathbf{h}^j$ is a 6 by n_d joint influence coefficient matrix or partial velocity matrix corresponding to particular columns of the identity matrix depending on which n_d primitive degrees of freedom are represented by $\mathbf{q}_j$, as shown in Figure 3, and $\mathbf{q}_j$ and $\dot{\mathbf{q}}_j$ are the vector of joint variables and its time derivatives.

The spatial velocity vector can be transformed from the CS at j into the CS at i using spatial transformation matrix as follows [9, 10, 24]:

$$\mathbf{v}^i = X^{ij}\mathbf{v}^j, \tag{11}$$

where X^{ij} is the spatial transformation matrix and i and j indicate the coordinate systems. The general transformation matrix X^{ij} is the result of spatial rotation followed by a spatial translation as follows:

$$X^{ij} = L^{ij}R^{ij} = \begin{bmatrix} I & 0 \\ \tilde{\underline{r}}^i_{ij} & I \end{bmatrix} \begin{bmatrix} \underline{R}^{ij} & 0 \\ 0 & \underline{R}^{ij} \end{bmatrix} = \begin{bmatrix} \underline{R}^{ij} & 0 \\ \underline{R}^{ij}\tilde{\underline{r}}^i_{ij} & \underline{R}^{ij} \end{bmatrix}, \tag{12}$$

where L^{ij} is the 6×6 spatial translation matrix, $\underline{r}^i_{ij}$ is the 3×1 translational displacement vector of CS j relative to CS i defined in the CS i, $\tilde{\underline{r}}^i_{ij}$ is 3×3 skew symmetric matrix representing the cross-product operation of $\underline{r}^i_{ij}$, R^{ij} is the 6×6 spatial rotation matrix, and $\underline{R}^{ij}$ is the 3×3 transformation matrix relating the coordinate frames i and j.

The spatial velocity vector of body j defined in the global coordinate system could be computed as follows:

$$\mathbf{v}^0_{0j,0} = X^{0j}\mathbf{v}^j_{0j,j}, \quad \text{where } \mathbf{v}^j_{0j,j} = X^{ji}\mathbf{v}^i_{0i,i} + \mathbf{v}^j_{ij,j} = \mathbf{v}^j_{0i,j} + \mathbf{h}^j_{j,j}\dot{\mathbf{q}}_j. \tag{13}$$

The spatial velocity vector of a descendant body could be calculated recursively using its parent spatial velocity as follows:

$$\mathbf{v}^0_{0j,0} = \mathbf{v}^0_{0i,0} + \mathbf{v}^0_{ij,0} = \mathbf{v}^0_{0i,0} + X^{0j}\mathbf{h}^j\dot{\mathbf{q}}_j. \tag{14}$$

Assuming the system shown in Figure 3 is modeled using rigid bodies, the velocity of body l could be written in a similar way as follows:

$$\mathbf{v}^l_{0l,l} = X^{lj}\mathbf{v}^j_{0j,j} + \mathbf{h}^l\dot{\mathbf{q}}_l. \tag{15}$$

Rearranging the terms in the velocity equations, the recursive form of the system velocity could be expressed as follows [13]:

$$\begin{bmatrix} I & 0 & 0 \\ -X^{ji} & I & 0 \\ 0 & -X^{lj} & I \end{bmatrix} \begin{bmatrix} \mathbf{v}^i_{oi,i} \\ \mathbf{v}^j_{oj,j} \\ \mathbf{v}^l_{ol,l} \end{bmatrix} = \begin{bmatrix} \mathbf{h}^i & 0 & 0 \\ 0 & \mathbf{h}^j & 0 \\ 0 & 0 & \mathbf{h}^l \end{bmatrix} \begin{bmatrix} \dot{\mathbf{q}}_i \\ \dot{\mathbf{q}}_j \\ \dot{\mathbf{q}}_l \end{bmatrix}. \tag{16}$$

The velocity recursive equations in (16) could be written in a compact form as follows:

$$T^{\ell\ell}_a \mathbf{v}^\ell_a = \mathbf{H}^\ell_a \dot{\mathbf{q}}_a = \mathbf{v}^\ell_r, \tag{17}$$

where $T^{\ell\ell}_a$ represents the assembled system topology or connectivity matrix which could be constructed from the connectivity graph, the superscript ℓ refers to local matrices and the superscript a refers to the system assembly matrix, and $\mathbf{H}^\ell_a$ is the assembly influence coefficient matrices grouped in blocks corresponding to the joints' DoF.

It should be mentioned that $T_a^{\ell\ell}$ is a lower triangular matrix and thus has simple inverse that maintains the topological structure of the original matrix and becomes an upper triangular matrix as follows [26]:

$$\left(T_a^{\ell\ell}\right)^{-1} = \begin{bmatrix} I & -\left(X^{ji}\right)^T & 0 \\ 0 & I & -\left(X^{lj}\right)^T \\ 0 & 0 & I \end{bmatrix}. \quad (18)$$

Similarly the spatial acceleration vector of body j could be obtained by differentiating the velocity vector defined in (14) as follows:

$$\mathbf{a}_{0j,0}^0 = \mathbf{a}_{0i,0}^0 + \mathbf{a}_{ij,0}^0,$$
$$\text{where, } \mathbf{a}_{ij,0}^0 = \mathbf{h}_{j,0}^0\ddot{\mathbf{q}}_j + \tilde{\mathbf{v}}_{0j,0}^0\mathbf{h}_{j,0}^0\dot{\mathbf{q}}_j. \quad (19)$$

Assuming that the quadratic velocity term could be expressed as $\gamma_{0j,0}^0 = \tilde{\mathbf{v}}_{0j,0}^0\mathbf{h}_{j,0}^0\dot{\mathbf{q}}_j = \tilde{\mathbf{v}}_{0j,0}^0\mathbf{v}_{0j,0}^0$, the global acceleration vector could written as follows:

$$\mathbf{a}_{0j,0}^0 = \mathbf{a}_{0i,0}^0 + \mathbf{a}_{ij,0}^0 = \mathbf{a}_{0i,0}^0 + \mathbf{h}_{j,0}^0\ddot{\mathbf{q}}_j + \gamma_{0j,0}^0. \quad (20)$$

The recursive form to calculate the system local accelerations could be written as follows:

$$T_a^{\ell\ell}\mathbf{a}_a^\ell = \mathbf{H}_a^\ell\ddot{\mathbf{q}}_a + \dot{\mathbf{H}}_a^\ell\dot{\mathbf{q}}_a = \mathbf{H}_a^\ell\ddot{\mathbf{q}}_a + \tilde{\mathbf{v}}_r^\ell\mathbf{v}_r^\ell = \mathbf{H}_a^\ell\ddot{\mathbf{q}}_a + \gamma_a^\ell. \quad (21)$$

Equations (17) and (21) could be used to recursively calculate the velocity and acceleration of the system bodies marching outward from the root to the descendant bodies. Similarly, the system topological matrix could be used to project the joint forces in the joint subspace as follows:

$$Q_a = \left(H_a^l\right)^T\mathbf{F}_a^l. \quad (22)$$

The kinematic equations derived earlier could be used to express the rigid body equations of motion from the momentum equations. The spatial momentum vector of the rigid body j can be written as follows:

$$\mathbf{P}_{j,G}^j = M_{j,GG}^{jj}\mathbf{v}_{0j,G}^j, \quad (23)$$

where $\mathbf{P}_{j,G}^j$ is the spatial momentum of body j, $M_{j,GG}^{jj}$ is the spatial mass matrix defined at the body center of mass marker G, and $\mathbf{v}_{0j,G}^j$ is the global velocity of the marker G located at the body center of mass. The centroidal spatial mass matrix is defined as follows:

$$M_{j,GG}^{jj} = \begin{bmatrix} \underline{J}_{j,GG}^{jj} & 0 \\ 0 & m_j\underline{I} \end{bmatrix}, \quad (24)$$

where m_j is the mass body j, $\underline{I}$ is 3×3 identity matrix, and $\underline{J}_{j,GG}^{jj}$ is the 3×3 matrix representing the body j moment of inertia tensor defined at a marker located at the body center of mass, G. This rigid body momentum can be transformed into the global coordinate system and differentiated with respect to time to obtain the body equation of motion as follows:

$$M_{j,00}^{00}\mathbf{a}_{0j,0}^0 - \tilde{\mathbf{v}}_{0j,0}^{0T}M_{j,00}^{00}\mathbf{v}_{0j,0}^0 = \mathbf{G}_{j,0}^0, \quad (25)$$

where the symbol 0 refers to global coordinate system, $\mathbf{G}_{j,0}^0 = \left[\underline{\tau}_{j,0}^0 \ \underline{g}_j^0\right]^T$ is the spatial force vector including the gravitational forces, $\underline{g}_j^0$ is the sum of all external forces acting on the body j, and $\underline{\tau}_{j,0}^0$ contains the sum of all torques and the moments of all forces about the origin of frame 0.

The equation of motion of any two connected bodies i and j could be driven from the free body diagram as follows. For body j, the equation of motion in its local coordinate system could be written as

$$M_{j,jj}^{jj}\mathbf{a}_{0j,j}^j - \mathbf{F}_{j,j}^j = \mathbf{G}_{j,j}^j + \tilde{\mathbf{v}}_{0j,j}^{jT}\mathbf{P}_{j,j}^j, \quad (26)$$

where $\mathbf{F}_{j,j}^j$ is the vector of reaction forces of the joint connecting body j to its parent. While for body i, the equation of motion in its local coordinate system could be written as

$$M_{i,ii}^{ii}\mathbf{a}_{0i,i}^i - \mathbf{F}_{i,i}^i + \left(X^{ji}\right)^T\mathbf{F}_{j,j}^j = \mathbf{G}_{i,i}^i + \left(\tilde{\mathbf{v}}_{0i,i}^i\right)^T\mathbf{P}_{i,i}^i. \quad (27)$$

The equations could be rearranged in matrix form as follows:

$$\begin{bmatrix} M_{i,ii}^{ii} & 0 \\ 0 & M_{j,jj}^{jj} \end{bmatrix}\begin{bmatrix} \mathbf{a}_{0i,i}^i \\ \mathbf{a}_{0j,j}^j \end{bmatrix} + \begin{bmatrix} I & -\left(X^{ji}\right)^T \\ 0 & I \end{bmatrix}\begin{bmatrix} -\mathbf{F}_{i,i}^i \\ -\mathbf{F}_{j,j}^j \end{bmatrix} = \begin{bmatrix} \mathbf{G}_{i,i}^i + \left(\tilde{\mathbf{v}}_{0i,i}^i\right)^T\mathbf{P}_{i,i}^i \\ \mathbf{G}_{j,j}^j + \left(\tilde{\mathbf{v}}_{0j,j}^j\right)^T\mathbf{P}_{j,j}^j \end{bmatrix}. \quad (28)$$

Similar to the velocity and acceleration equations (14), (15), (16), and (17), it could be shown that the connectivity matrix could be used to simplify the expression of the system equations of motion. The system equation of motion could be written for the whole system assembly in a more compact form as follows:

$$M^{\ell\ell}\mathbf{a}_a^\ell - \left(T_a^{\ell\ell}\right)^T\mathbf{F}_a^\ell = \mathbf{G}_a^\ell + \left(\tilde{\mathbf{v}}_a^\ell\right)^T\mathbf{P}. \quad (29)$$

The system of equations in (29) contains the unknown Cartesian acceleration of the bodies as well as the joint reaction forces. In order to be able to solve the equation of motion, the kinematic relations from (21), the constraint equations from (22), and the system dynamic equation (29) are rearranged in matrix form as follows:

$$\begin{bmatrix} M^{\ell\ell} & 0 & \left(T_a^{\ell\ell}\right)^T \\ 0 & 0 & -\mathbf{H}_a^\ell \\ T_a^{\ell\ell} & -\left(\mathbf{H}_a^\ell\right)^T & 0 \end{bmatrix}\begin{bmatrix} \mathbf{a}_a^\ell \\ \ddot{\mathbf{q}}_a \\ -\mathbf{F}_a^\ell \end{bmatrix} = \begin{bmatrix} \mathbf{G}_a^\ell + \left(\tilde{\mathbf{v}}_a^\ell\right)^T\mathbf{P}_a^\ell \\ Q_a \\ \gamma_a^\ell \end{bmatrix}, \quad (30)$$

where $M^{\ell\ell}$ is a block diagonal inertia matrix expressed in the Cartesian form, $T_a^{\ell\ell}$ is the body connectivity matrix, H_a^ℓ is a block diagonal joint influence coefficient matrix, a_a^ℓ is the body Cartesian accelerations, $\mathbf{F}_a^\ell$ is a vector of the Lagrange multipliers, $\ddot{\mathbf{q}}_a^\ell$ is the second time derivative of joint variables, $\mathbf{G}_a^\ell$ is a vector of the externally applied loads, $\gamma_a^{\ell\ell}$ is the quadratic constraint derivatives, and Q_a^ℓ is the generalized forces. Since the inertia matrix $M^{\ell\ell}$ is invertible, (30) could be solved to get the unknown joint accelerations $\ddot{\mathbf{q}}_a$ which can be used to calculate the joint Cartesian accelerations, $\mathbf{a}_a^{\ell}$, and the joint reaction forces $\mathbf{F}_a^{\ell}$ as follows:

$$\left(\left(\mathbf{H}_a^{\ell}\right)^T \left(T_a^{\ell\ell}\right)^{-T} \mathbf{M}^{\ell\ell} \left(T_a^{\ell\ell}\right)^{-1} \mathbf{H}_a^{\ell}\right) \ddot{\mathbf{q}}_a = Q_a + \left(\mathbf{H}_a^{\ell}\right)^T \left(T_a^{\ell\ell}\right)^{-T} \left(\mathbf{G}_a^{\ell} + \left(\tilde{\mathbf{v}}_a^{\ell}\right)^T \mathbf{P}_a^{\ell}\right). \tag{31}$$

Equation (31) could be written in a compact form as follows:

$$\mathbf{M}_{qq} \ddot{\mathbf{q}}_a = \mathbf{Q}, \tag{32}$$

where $\mathbf{M}_{qq}$ is the generalized mass matrix projected in the joint subspace and can be calculated as follows:

$$\mathbf{M}_{qq} = \left(\mathbf{H}_a^{\ell}\right)^T \left(T_a^{\ell\ell}\right)^{-T} \mathbf{M}^{\ell\ell} \left(T_a^{\ell\ell}\right)^{-1} \mathbf{H}_a^{\ell} \tag{33}$$

and $\mathbf{Q}$ is the generalized force vector projected in the joint subspace and can be calculated as follows:

$$\mathbf{Q} = Q_a + \left(\mathbf{H}_a^{\ell}\right)^T \left(T_a^{\ell\ell}\right)^{-T} \left(\mathbf{G}_a^{\ell} + \left(\tilde{\mathbf{v}}_a^{\ell}\right)^T \mathbf{P}_a^{\ell}\right). \tag{34}$$

This equation of motion (32) represents minimum set of differential equations that are function of the system generalized coordinates. However, this form is limited to open-loop unconstrained mechanisms and therefore is suitable only for robotic type mechanisms.

3. Equation of Motion of Rigid and Flexible Body System

The flexible body dynamics can be modeled in the multibody system as a reduced form of the finite element model. The finite element can be reduced using nodal approach or modal approach. Modal formulation uses a set of kinematically admissible modes to represent the deformation of the flexible body [27, 28]. Component mode synthesis approach is widely used in the multibody dynamics codes to accurately simulate the flexible body dynamics in the multibody system. The main advantage of the modal formulation is that fewer elastic modes can be used to accurately capture the flexible body dynamics at reasonable computational cost [29, 30]. Craig-Bampton approach was introduced to account for the effect of boundary conditions and the attachment joints of the flexible body [31, 32]. The structure of the reduced flexible body equation-of-motion can be written as follows [23]:

$$\begin{bmatrix} M_{rb}^{rr} & M^{rf} \\ M^{fr} & M^{ff} \end{bmatrix} \begin{bmatrix} a^r \\ \ddot{q}^f \end{bmatrix} = \begin{bmatrix} G^r \\ Q^f \end{bmatrix}, \tag{35}$$

where r refers to the flexible body's local reference frame, f refers to flexible or elastic coordinates, M_{rb}^{rr} represents the 6×6 inertia matrix associated with the reference motion, M^{rf}, M^{fr} are the inertia coupling terms between the reference motion and the elastic coordinates, M^{ff} is the inertia matrix associated with the elastic coordinates, a^r is the reference frame acceleration in the Cartesian space, q^f represents the elastic displacements measured relative to that frame, G^r is the vector of external forces including centrifugal and Coriolis forces [1, 23, 27], and Q^f is the vector of elastic forces projected into the modal space. It should be mentioned that the matrices M_{rb}^{rr}, M^{rf}, and M^{fr} depend on the elastic coordinates and should be updated every time step. To improve the computation efficiency of the flexible body simulation, a set of inertia shape invariants (ISI) are identified and calculated before the simulation starts. The ISI are used during the dynamic simulation to update the inertia terms. The ISI depend on the kinematic expression of the system equations.

A general form of the equation of motion of multibody system with flexible bodies could be written in a compact form as follows [23]:

$$\left[\begin{array}{cc|c} \mathbf{M}^{rr} & \mathbf{T}^T & \mathbf{M}^{rf} \\ \mathbf{T} & 0 & -\mathbf{H} \\ \hline \left(\mathbf{M}^{rf}\right)^T & -\mathbf{H}^T & \mathbf{M}^{ff} \end{array}\right] \left[\begin{array}{c} \mathbf{a} \\ -\mathbf{F} \\ \hline \ddot{\mathbf{q}} \end{array}\right] = \left[\begin{array}{c} \mathbf{G} \\ \boldsymbol{\gamma} \\ \hline \mathbf{Q} \end{array}\right], \tag{36}$$

where the block diagonal matrix $\mathbf{M}^{rr}$ is composed of six by six inertia matrices that represent the Cartesian inertia matrices associated with the reference frame for the rigid and flexible bodies in the system, $\mathbf{M}^{rf}$ is the inertia coupling terms between the modal elastic coordinates and Cartesian reference accelerations, $\mathbf{M}^{ff}$ is a block diagonal matrix containing the flexible body inertias, $\mathbf{T}$ is a lower triangular topology matrix, $\mathbf{H}$ is a block matrix of the joint influence coefficient matrices, $\mathbf{a}$ is the vector of Cartesian accelerations, $\mathbf{F}$ is the vector of joint reaction forces, $\ddot{\mathbf{q}}$ is the vector of joint accelerations with the appended elastic coordinates, $\mathbf{G}$ is the vector of external forces, $\boldsymbol{\gamma}$ is a column matrix of 6×1 spatial quadratic acceleration vectors, and $\mathbf{Q}$ is the vector of joint and elastic forces.

The system of equations given by (36) can be easily solved for the joint accelerations and the modal coordinates accelerations. Then the Cartesian accelerations and the reaction forces could be easily computed. Since the inverse of system topology matrix $\mathbf{T}$ is just its transpose, manipulating the terms in (36), we can get the joint and modal accelerations as follows:

$$\left(\mathbf{M}^{ff} + \mathbf{H}^T \left(\mathbf{T}^T \mathbf{M}^{rr} \mathbf{T}\right) \mathbf{H} + \mathbf{H}^T \mathbf{T}^T \mathbf{M}^{rf} + \mathbf{M}^{fr} \mathbf{T} \mathbf{H}\right) \ddot{\mathbf{q}} = \mathbf{Q} + \mathbf{H}^T \mathbf{T}^T \mathbf{G} - \mathbf{H}^T \left(\mathbf{T}^T \mathbf{M}^{rr} \mathbf{T}\right) \boldsymbol{\gamma} - \mathbf{M}^{fr} \mathbf{T} \boldsymbol{\gamma}. \tag{37}$$

This equation could be written in a compact form as follows:

$$\mathbf{M}_{qq} \ddot{\mathbf{q}}_a = \mathbf{Q}. \tag{38}$$

The Cartesian accelerations could be calculated as follows:

$$\mathbf{a} = \mathbf{T}\mathbf{H}\ddot{\mathbf{q}} + \mathbf{T}\boldsymbol{\gamma}, \tag{39}$$

and the Cartesian joint reaction forces could be obtained from

$$\mathbf{F} = \mathbf{T}^T \mathbf{M}^{rf} + \left(\mathbf{T}^T \mathbf{M}^{aa} \mathbf{T}\right) \mathbf{H}\ddot{\mathbf{q}} - \mathbf{T}^T \mathbf{G} + \left(\mathbf{T}^T \mathbf{M}^{aa} \mathbf{T}\right) \boldsymbol{\gamma}. \tag{40}$$

The sequence of evaluating the terms in (37) to (40) could be optimized in order to minimize the computational efforts and to avoid repeated calculations. Both (32) and (38) represent the minimum set of differential equations that can completely describe the system dynamics.

4. Equation of Motion of Constrained System

The abovementioned development could be efficiently used to model open-loop systems like robots and human body systems. In the open-loop systems, each body is connected with one joint only. This open-loop systems approach will not be adequate to model the wheel loader system shown in Figure 1. In the kinematic tree, shown in Figure 2, the solid arrows represent a joint where the dotted lines represent a joint that will form a closed kinematic loop. According to this description, the kinematic tree of the wheel loader contains two closed loops formed of the following list of bodies: $\Gamma^1 = \{2, 8, 11, 9\}$ and $\Gamma^2 = \{2, 9, 12, 10, 7\}$. Each of those two loops could be represented by a set of constraint equations that are function of the joint variable of all the bodies forming the loop as follows: $\Phi^1(\mathbf{q}_2, \mathbf{q}_8, \mathbf{q}_{11}, \mathbf{q}_9) = 0$ and $\Phi^2(\mathbf{q}_2, \mathbf{q}_9, \mathbf{q}_{12}, \mathbf{q}_{10}, \mathbf{q}_7) = 0$. The first and second derivatives of the constraint equations could be easily evaluated from the kinematic equations. The multibody system's equations of motion with appended constraints could be formed as follows:

$$\begin{bmatrix} \mathbf{M}_{qq} & \mathbf{J}^T \\ \mathbf{J} & 0 \end{bmatrix} \begin{bmatrix} \ddot{\mathbf{q}}_a \\ \boldsymbol{\lambda}_q \end{bmatrix} = \begin{bmatrix} \mathbf{Q} \\ \gamma_q \end{bmatrix}, \tag{41}$$

where $\mathbf{M}_{qq}$ is the generalized mass matrix from (32) and (38), $\mathbf{J}$ is the Jacobian matrix, and the other column matrices are similar to those described earlier.

The coefficient matrix in the left hand side of (41) is often singular, and general-purpose sparse matrix solver algorithms are routinely used to analyze, factor, and solve the equations. An alternative approach based on generalized coordinate partitioning [14] may be employed at this stage. The factored forms of (33) and (34) used to obtain (32) play a significant role in the solver efficiency. The generalized coordinate partitioning method [14] first analyzes the model topology and decomposes it into a number of kinematically uncoupled superelements, each with a small Jacobian matrix [2]. The uncoupled acceleration constraint equations could be arranged as follows:

$$\begin{bmatrix} \Phi_q^1 & \cdots & 0 \\ \vdots & \vdots & \vdots \\ 0 & \cdots & \Phi_q^n \end{bmatrix} \begin{bmatrix} \ddot{q}^1 \\ \vdots \\ \ddot{q}^n \end{bmatrix} = \begin{bmatrix} \gamma^1 \\ \vdots \\ \gamma^n \end{bmatrix}. \tag{42}$$

The Jacobian matrices in (42) are factored into LU form using complete pivoting to partition the variables into dependent and independent sets. The results are then combined into a revised set of constraint equations [33]

$$\ddot{q}^d = B\ddot{q}^i + \gamma^d, \tag{43}$$

where $\ddot{q}^d$ is the vector of dependent variables and $\ddot{q}^i$ is the vector of independent variables. Appending (43) with a set of Lagrange multipliers to the partitioned and permuted equations of motion gives

$$\left[\begin{array}{cc|c} M_q^{dd} & I & M_q^{di} \\ I & 0 & -B \\ \hline M_q^{id} & -B^T & M_q^{ii} \end{array}\right] \begin{bmatrix} \ddot{q}^d \\ \lambda^d \\ \hline \ddot{q}^i \end{bmatrix} = \begin{bmatrix} Q^d \\ \gamma^d \\ Q^i \end{bmatrix}. \tag{44}$$

The coefficient matrix in the upper-left part of (44) is structurally nonsingular, so it can be reduced to

$$\overline{M}_q^{ii}\ddot{q}^i = \overline{Q}^i, \tag{45}$$

where

$$\begin{aligned} \overline{M}_q^{ii} &= M_q^{ii} + M_q^{id}B + B^T M_q^{di} + B^T M_q^{dd}B, \\ \overline{Q}^i &= \left(Q^i - M_q^{id}\gamma^d\right) + B^T\left(Q^d - M_q^{dd}\gamma^d\right). \end{aligned} \tag{46}$$

Equation (45) could be solved by Cholesky factorization.

5. Constraint Enforcement

The equations of motion in (41) and (44) are index-2 differential algebraic equations (DAE) [1]. This set of DAE are traditionally linearized and solved by implicit numerical integrators [34–36]. The generalized coordinate partitioning approach [14] is used to avoid the complexity and cost of implicit solvers, but this approach does not guarantee continued satisfaction of the system constraints. Baumgarte [37] introduced a constraint stabilization method based on the control theory. In this approach, the original acceleration level constraints are replaced with

$$\ddot{\boldsymbol{\Phi}} + \alpha\dot{\boldsymbol{\Phi}} + \beta\boldsymbol{\Phi} = 0, \tag{47}$$

where α and β are often chosen to appropriately achieve critical damping. The Baumgarte stabilization is implemented in the following form:

$$\ddot{\boldsymbol{\Phi}} + 2\omega_n\dot{\boldsymbol{\Phi}} + \omega_n^2\boldsymbol{\Phi} = 0, \tag{48}$$

where ω_n is a user-specified gain (natural frequency). Unfortunately, the value of the parameter ω_n is problem dependent, and no general procedure exists for its determination. If $\boldsymbol{\Phi} = \dot{\boldsymbol{\Phi}} = \ddot{\boldsymbol{\Phi}} = 0$ represents the constraint equations associated with (41), then substituting (48) into (41) leads to

$$\begin{bmatrix} \mathbf{M}_q & \mathbf{J}^T \\ \mathbf{J} & 0 \end{bmatrix} \begin{bmatrix} \ddot{\mathbf{q}} \\ \boldsymbol{\lambda}_q \end{bmatrix} = \begin{bmatrix} \mathbf{Q} \\ \gamma_q - 2\omega_n\dot{\boldsymbol{\Phi}} - \omega_n^2\boldsymbol{\Phi} \end{bmatrix}. \tag{49}$$

A similar modification would be made to the right hand side of (44). The value set for ω_n is critical and affects the integrator performance. If ω_n is too large, constraint violations will be very small, but the equation of motion will tend to be stiff and the integrator will be slow. If ω_n is too small, the stabilization algorithm will not efficiently reduce the constraint violations. From experience, the maximum value of best ω_n is probably near the highest system frequencies in the system. To this end, the developed set of dynamic equations of motion is sufficient to perform the dynamic simulation of open- and closed-loop systems. It is also sufficient for modeling rigid and flexible bodies in the multibody system.

6. Static Equilibrium Analysis

The mechanical system can be considered at static equilibrium if the kinetic energy is zero or constant target value and the potential energy is minimum. Many methods have been proposed to solve static equilibrium in nonlinear multibody system models. Most of those proposed techniques require minimizing potential functions and/or linearizing the equations of motion [36] as mentioned earlier. Writing potential functions for arbitrary nonlinear force modules or linearizing them is a daunting task. The problem becomes more challenging when models contain friction and other history-based force models because unique iterative static solutions may not exist. The proposed approach uses the dynamic model itself to solve the static equilibrium position. This approach may have some drawbacks because convergence can be slow in lightly damped systems. In order to improve the solver efficiency, an energy-drainage mechanism is used.

In the energy-drainage approach, a technique similar to the Baumgarte equation was implemented into the solver. After solving (45) to obtain the accelerations $\ddot{\mathbf{q}}$, the solver imposes the additional requirement on the second derivatives

$$\ddot{q}_d = -2\omega_d \dot{q}_i - \omega_d^2 \left(q_i - q_{i-1}\right), \tag{50}$$

where ω_d is a natural damping frequency and q_i and q_{i-1} represent the system states at times t_i and t_{i-1}. Choosing ω_d is tricky as it was in the constraint stabilization. If ω_d is too small, the damping will be weak and the model will oscillate too long. And if ω_d is too large, damping will be strong and the model will drift slowly toward equilibrium. With this damping method, equilibrium is approached as $\|q_i - q_{i-1}\|$ gets acceptably small, not necessarily when the kinetic energy gets small.

Another feature of energy-drainage damping is its ability to inhibit the effects of bad initial conditions that often cause models to fly apart and the numerical integrator to fail. Extremely high forces from the offending contact models can cause unrealistic accelerations, which are then integrated to give unrealistic velocities and displacements. When energy-drainage damping is active, the resulting large system velocities are immediately fed back through the above damping equation, and this generates large opposing acceleration terms, canceling those from the model. With energy-drainage damping, the contacting bodies will tend to drift slowly apart until the forces and accelerations become reasonable.

7. Recursive Algorithm

At the beginning of the dynamic simulation, the structure of the equation of motion is determined based on a preliminary analysis of the system topology. The dependent and independent variable sets are determined using the generalized coordinate portioning approach. The solver integrates both dependent and independent variable sets and the kinematic constraints are enforced using Baumgarte stabilization, as explained in the previous section. The input states to the integrator represent the first and second time derivative of the joint variables and the output states are the joint displacements and the joint velocities. The recursive algorithm of the multibody dynamic simulation can be summarized as follows.

(1) Using the joint variables which returned from the integrator, the solver calculates the Cartesian displacements, velocities, and accelerations. Forward evaluation scheme is utilized (starting from the root body to the descendant/branch bodies).

(2) Update and factor the Jacobian matrix.

(3) Apply generalized coordinate portioning technique to determine the quality of the independent and dependent set of variables.

(4) Calculate the constraint violations and apply the Baumgarte stabilization.

(5) Calculate the internal and external forces and apply them to the different bodies (interaction forces between bodies, driver forces, soil/terrain forces, etc.).

(6) Transform the inertia matrices into global coordinate system.

(7) Calculate the inertia forces and centrifugal and Coriolis forces in Cartesian space.

(8) Propagate the external and inertia forces from the descendant bodies into their parents.

(9) Project the Cartesian forces into the body joint space.

(10) Factor the mass matrix based on the current independent and dependent variables selection.

(11) Calculate second derivatives of the joint variables.

(12) Apply the energy drainage operator according to (50).

(13) Send the states to the integrator.

The solver utilizes a predictor-corrector integrator with variable-order interpolation polynomials and variable time step. This explicit integrator insures the stability of the solution and the ability to capture the high-speed impacts between the different machine parts.

8. Examples

This section presents some examples of the simulation results multibody system ranging from simple pendulum to a full wheeled machine. The first example represents a uniform beam connected to the ground with a revolute joint to form a single pendulum, as shown in Figure 4. The link has mass of 1.7595 Kg. The beam center of mass is located in the middle. The pendulum was originally horizontal and is allowed to fall under the effect of gravity. The static equilibrium position of this pendulum is known to be vertical.

The proposed multibody formulation and solver were used to model the pendulum and compute the static equilibrium position. The transient simulation is run for a few iterations to allow the system to pick up some kinetic energy. Once the pendulum starts to move the energy draining mechanism is applied. A stopping criterion is set to be a target

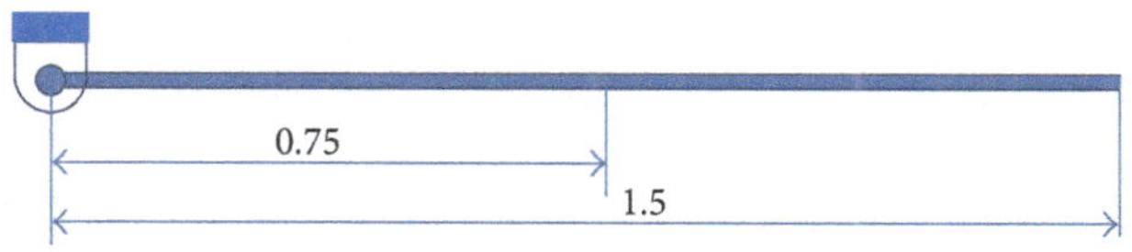

FIGURE 4: Single pendulum beam.

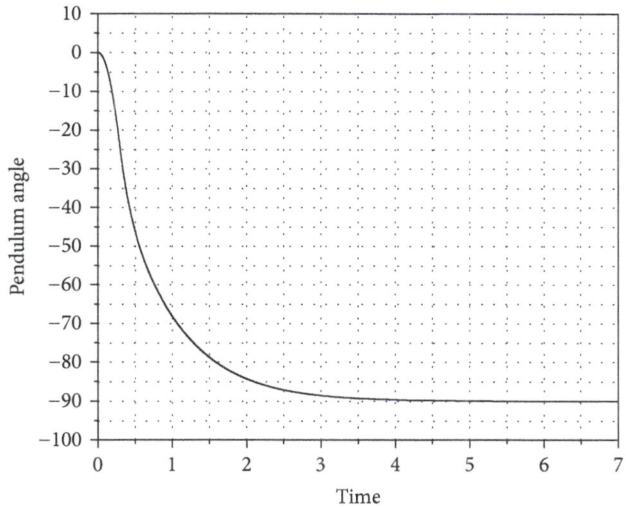

FIGURE 5: The pendulum angle as it falls under the effect of gravity.

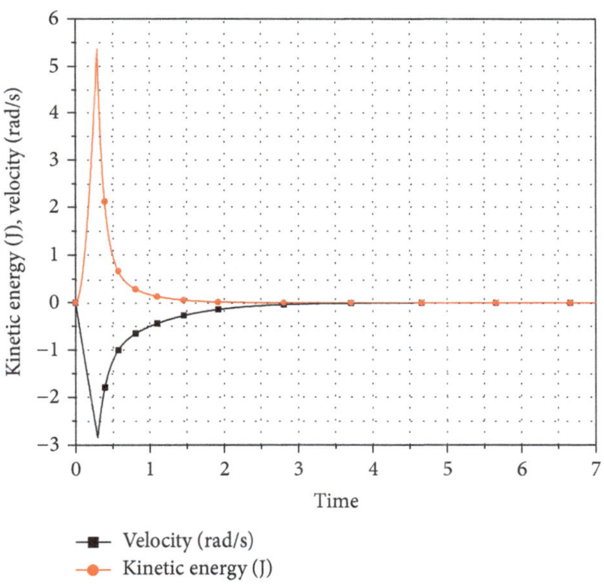

FIGURE 6: The velocity and kinetic energy of the pendulum.

value of the kinetic energy. When the pendulum reaches this target value, it would have reached the static equilibrium.

Figure 5 shows the value of the pendulum angle as it falls under the effect of gravity till the system reaches equilibrium position. The kinetic energy is drained out from the system as shown in Figure 6.

The change in the pendulum potential energy from the reference is shown in Figure 7. As the pendulum starts moving, the potential energy of the pendulum decreases

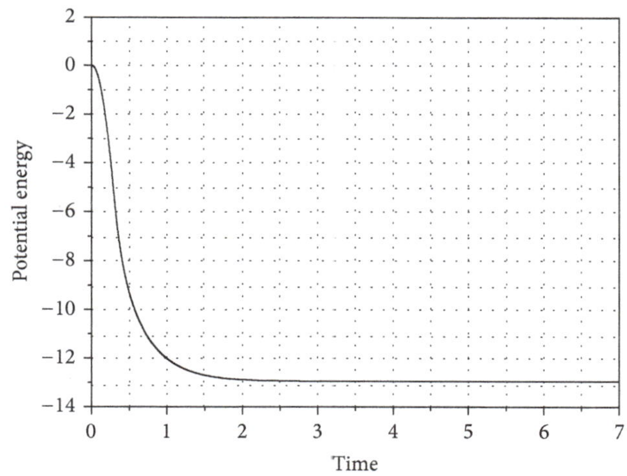

FIGURE 7: The potential energy of the pendulum.

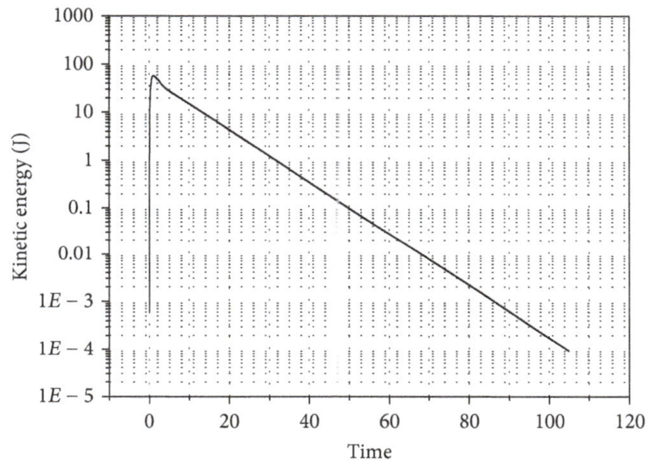

FIGURE 8: Kinetic energy of the 50-link chain.

and the minimum value is reached at the static equilibrium position. The final value of the potential energy is −12.9455J which matches the closed form solution for the pendulum.

It should be mentioned that time in the plots is only a representation of the number of iteration to achieve solution convergence. Also, it should be mentioned that this example represents an extreme case where the pendulum initial position was extremely far from the equilibrium position. This example shows that the solver is stable and can achieve realistic solutions even in extreme cases.

The second example represents a system that might have minor assembly errors or constraint violations. The system represents a chain of 50 links connected with revolute joints. The chain links had same properties of the link in first example. In this example, the first link is attached to the ground and it has initial angle of 5° from the vertical position. Every following link has an initial deviation of 1° relative to its parent. The static equilibrium analysis was performed on the system. The kinetic energy of the system is shown in Figure 8. The potential energy of the system is shown in Figure 9 while the angles of all the links are shown in Figure 10.

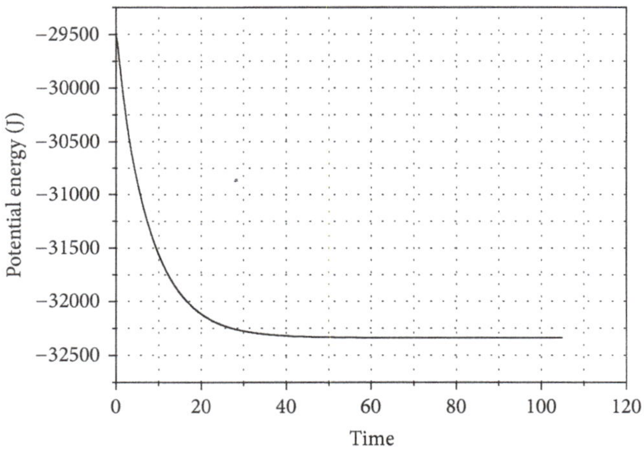

FIGURE 9: Potential energy of the 50-link chain.

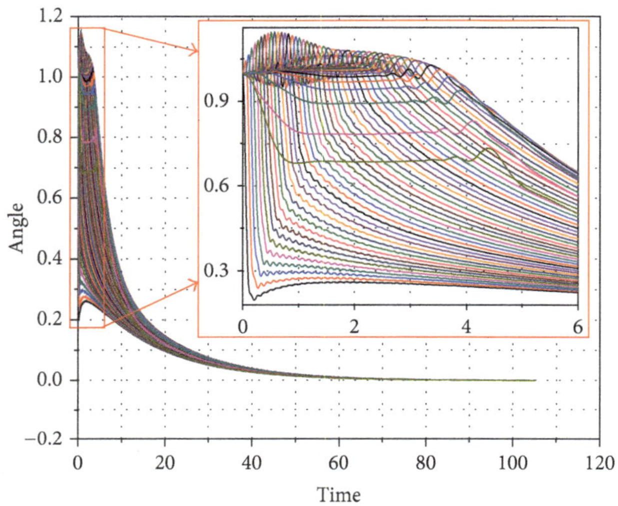

FIGURE 10: Angles of the links in the 50-link chain.

To demonstrate the completeness of the proposed approach a full machine with closed kinematic loops will be presented in the following example. Forestry equipment is typically required to work on different terrains ranging from swampy, compactable, and hard soil. It may be required to work on slopes dragging wood logs and tree. During operation and articulations, the location of the center of mass of such machine may change and causes the machine to slip, flip over, or fall. Design of such machines requires accurate prediction of the machine stability under the different loading conditions and at different articulation angles. Also, power requirements analysis of such machine requires the simulation to start from a stable static equilibrium position. A grapple skidder machine, shown in Figure 11, will be used as a demonstration example.

The kinematic tree of the machine could be developed, as shown in Figure 12, in order to establish the parent child list. It could be seen that the kinematic tree of grapple skidder is very similar to that of the wheel loader shown in Figure 2. The solver algorithm analyzes the kinematic tree to identify the closed loops of the mechanism and define the constraint equations associated with the closed-loop joints. The constraint equations of the two kinematic closed loops could be written as $\Phi^1(\mathbf{q}_2, \mathbf{q}_8, \mathbf{q}_{11}, \mathbf{q}_9) = 0$ and $\Phi^2(\mathbf{q}_2, \mathbf{q}_9, \mathbf{q}_{12}, \mathbf{q}_{10}, \mathbf{q}_7) = 0$.

The next step for the solver is to analyze the closed loops and merge any two loops that share any joint variable into one larger superelement. This operation could be iterative and may lead to very large superelements. In our example, since the two closed loops are sharing the variables of body 9, they have to be merged into a large superelement that contains all the bodies of the two loops. The Jacobian matrix of this superelement could be calculated as follows:

$$\mathbf{J} = \begin{bmatrix} \dfrac{\partial \Phi^1}{\partial \mathbf{q}_2} & \dfrac{\partial \Phi^1}{\partial \mathbf{q}_8} & \dfrac{\partial \Phi^1}{\partial \mathbf{q}_{11}} & \dfrac{\partial \Phi^1}{\partial \mathbf{q}_9} & 0 & 0 & 0 \\ \dfrac{\partial \Phi^2}{\partial \mathbf{q}_2} & 0 & 0 & \dfrac{\partial \Phi^2}{\partial \mathbf{q}_9} & \dfrac{\partial \Phi^2}{\partial \mathbf{q}_{12}} & \dfrac{\partial \Phi^2}{\partial \mathbf{q}_{10}} & \dfrac{\partial \Phi^2}{\partial \mathbf{q}_7} \end{bmatrix}. \quad (51)$$

The first raw block of the Jacobian matrix represents the partial derivatives of the first kinematic loop constraint equations with respect to the joint variables of the bodies in the first loop while the second row block represents the partial derivatives of the second loop constraint equations. The system topology matrix will be generated by the solver and should be as follows:

$$T_a^{\ell\ell} = \begin{bmatrix}
I \\
-X^{2,1} & I \\
-X^{3,1} & 0 & I \\
-X^{4,1} & 0 & 0 & I \\
-X^{2,1} & -X^{5,2} & 0 & 0 & I \\
-X^{2,1} & -X^{6,2} & 0 & 0 & 0 & I \\
-X^{2,1} & -X^{7,2} & 0 & 0 & 0 & 0 & I \\
-X^{2,1} & -X^{8,2} & 0 & 0 & 0 & 0 & 0 & I \\
-X^{2,1} & -X^{9,2} & 0 & 0 & 0 & 0 & 0 & 0 & I \\
-X^{2,1} & -X^{7,2} & 0 & 0 & 0 & 0 & -X^{10,7} & 0 & 0 & I \\
-X^{2,1} & -X^{8,2} & 0 & 0 & 0 & 0 & 0 & -X^{11,8} & 0 & 0 & I \\
-X^{2,1} & -X^{9,2} & 0 & 0 & 0 & 0 & 0 & 0 & -X^{12,9} & 0 & 0 & I \\
-X^{2,1} & -X^{9,2} & 0 & 0 & 0 & 0 & 0 & 0 & -X^{12,9} & 0 & 0 & -X^{13,12} & I
\end{bmatrix}. \quad (52)$$

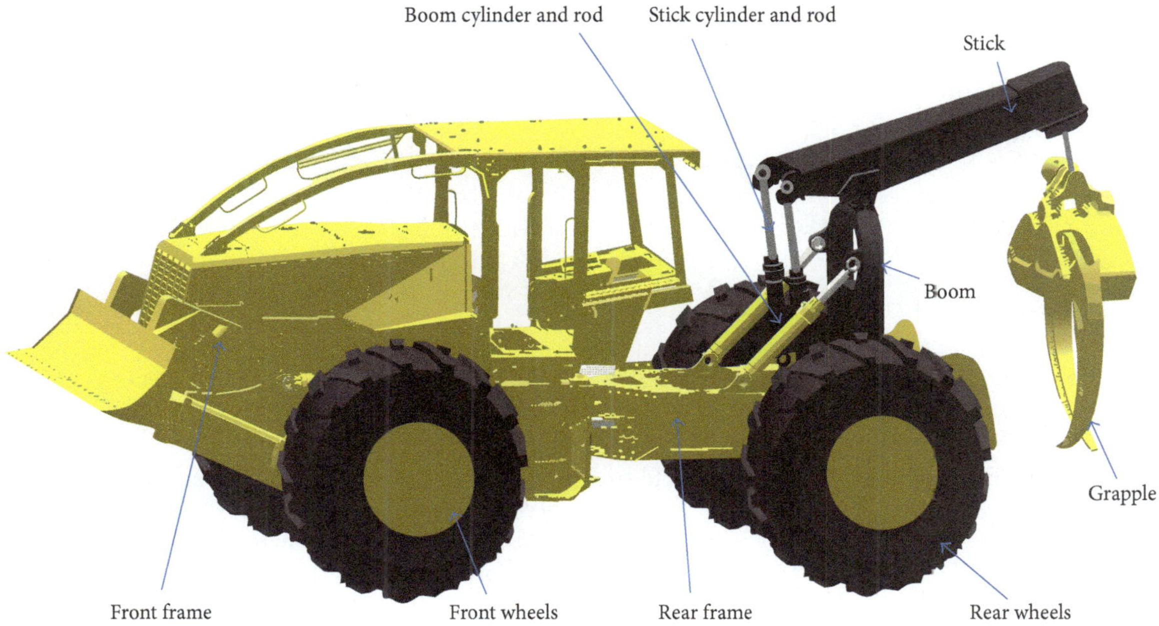

FIGURE 11: Grapple skidder forestry machine.

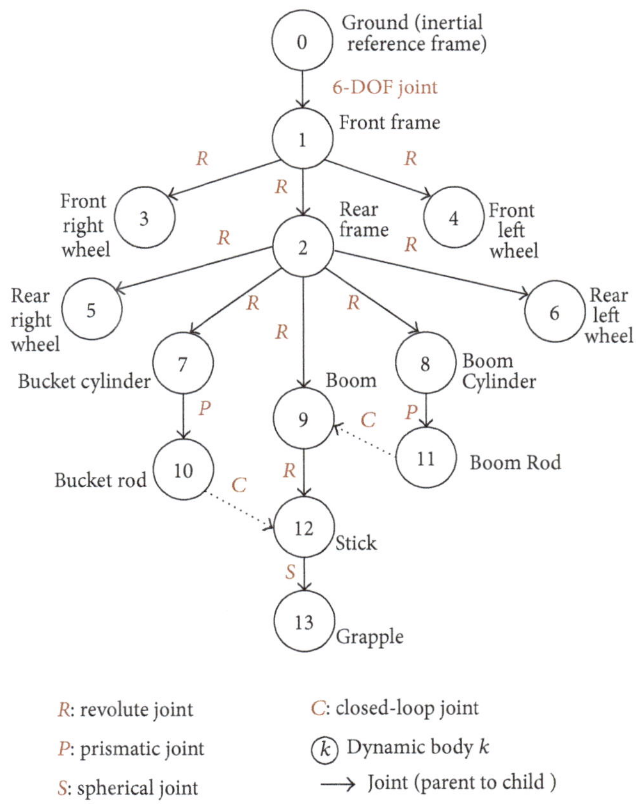

FIGURE 12: Kinematic tree of log dragging forestry equipment.

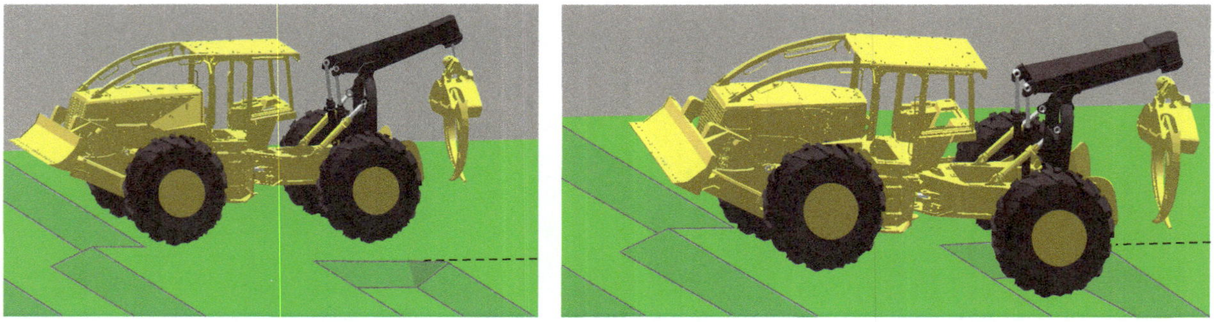

FIGURE 13: Static equilibrium of wheeled machine (grapple skidder) on a washboard surface.

The system assembly influence coefficient matrix could be written as follows:

$$\mathbf{H}_a^\ell = \begin{bmatrix} \mathbf{h}^1 & 0 & 0 & 0 & 0 & 0 & 0 & 0 & 0 & 0 & 0 & 0 & 0 \\ 0 & \mathbf{h}^2 & 0 & 0 & 0 & 0 & 0 & 0 & 0 & 0 & 0 & 0 & 0 \\ 0 & 0 & \mathbf{h}^3 & 0 & 0 & 0 & 0 & 0 & 0 & 0 & 0 & 0 & 0 \\ 0 & 0 & 0 & \mathbf{h}^4 & 0 & 0 & 0 & 0 & 0 & 0 & 0 & 0 & 0 \\ 0 & 0 & 0 & 0 & \mathbf{h}^5 & 0 & 0 & 0 & 0 & 0 & 0 & 0 & 0 \\ 0 & 0 & 0 & 0 & 0 & \mathbf{h}^6 & 0 & 0 & 0 & 0 & 0 & 0 & 0 \\ 0 & 0 & 0 & 0 & 0 & 0 & \mathbf{h}^7 & 0 & 0 & 0 & 0 & 0 & 0 \\ 0 & 0 & 0 & 0 & 0 & 0 & 0 & \mathbf{h}^8 & 0 & 0 & 0 & 0 & 0 \\ 0 & 0 & 0 & 0 & 0 & 0 & 0 & 0 & \mathbf{h}^9 & 0 & 0 & 0 & 0 \\ 0 & 0 & 0 & 0 & 0 & 0 & 0 & 0 & 0 & \mathbf{h}^{10} & 0 & 0 & 0 \\ 0 & 0 & 0 & 0 & 0 & 0 & 0 & 0 & 0 & 0 & \mathbf{h}^{11} & 0 & 0 \\ 0 & 0 & 0 & 0 & 0 & 0 & 0 & 0 & 0 & 0 & 0 & \mathbf{h}^{12} & 0 \\ 0 & 0 & 0 & 0 & 0 & 0 & 0 & 0 & 0 & 0 & 0 & 0 & \mathbf{h}^{13} \end{bmatrix}. \quad (53)$$

When the solver starts, the Cartesian velocities and accelerations are calculated using forward substitution from the root body to the descendant bodies. The Jacobian matrices are updated based on the new values of the spatial displacements and velocities. The external forces acting on the system, the contact forces between the different bodies, and the interaction forces between the system and terrain are calculated and applied to each body's force vector. The system topology matrix is used to project the Cartesian quantities into the joint subspace and define the minimum set of the system variables as shown in (32) to (34). The system Jacobian matrix is then appended to the inertia matrix to form the augmented constrained multibody equation of motion. The solver then utilizes the generalized coordinate portioning approach to identify the dependent and independent variables by factoring the Jacobian matrix into LU form. The independent variable is identified, as shown in (45). The constraint violation penalty is applied to the system equation of motion using Baumgarte equation, as shown in (48). The entries to the mass matrix are permuted based on the order in the LU factorization order.

To perform the static analysis, at this point, the system accelerations are modified using the energy drainage function as shown in (50), and both the dependent and independent variables are sent to the integrator. The integrated states are used to calculate the Cartesian position, velocities, and accelerations of all bodies. The Cartesian velocities are used to calculate the kinetic energy of the system and the global position of the bodies is used to calculate the potential energies. If the value of the kinetic energy of the system reached the target, the solver stops and reports the current state as the body equilibrium position.

Figure 13 shows a successful simulation to achieve the static equilibrium of wheeled machine. The machine is dropping from a small height over a washboard surface. As the machine drops, the rear tires fall on the inclined plane of the surface causing the machine to roll down and backward. As the machine contacts the opposite surface of the washboard ditch, the machine comes to static equilibrium.

9. Conclusions

This paper presented an approach for evaluating the static equilibrium position of multibody systems. The multibody system can include rigid and flexible bodies. The proposed approach is implemented in a joint-based multibody dynamics formulation that uses the spatial algebra to derive the equation of motion of the multibody system. The kinematic equation of the closed-loop systems was optimized to employ Baumgarte constraint stabilization approach to eliminate the constraint violation while at the same time avoiding using iterative constraint enforcement schemes. In order to determine the system static equilibrium position, an energy drainage mechanism was introduced to modify the system states before integrating it. The static equilibrium is achieved by running a transient simulation of the dynamics system while the integrated states were dampened with the energy drainage operator.

Conflict of Interests

The author declares that there is no conflict of interests regarding the publication of this paper.

References

[1] M. Giampiero and P. Manfred, *Simulation Algorithms in Vehicle System Dynamics*, CRC Press, 2004.

[2] M. A. Omar, "An applied approach for large-scale multibody dynamics simulation and machine-terrain interaction," *SAE International Journal of Passenger Cars—Electronic and Electrical Systems*, vol. 1, no. 1, pp. 820–828, 2009.

[3] J. Y. Wong, "Dynamics of tracked vehicles," *Vehicle System Dynamics*, vol. 28, no. 2-3, pp. 197–219, 1997.

[4] H. C. Lee, J. H. Choi, and A. A. Shabana, "Spatial dynamics of multibody tracked vehicles part II: contact forces and simulation results," *Vehicle System Dynamics*, vol. 29, no. 2, pp. 113–137, 1998.

[5] Z.-D. Ma and N. C. Perkins, "A track-wheel-terrain interaction model for dynamic simulation of tracked vehicles," *Vehicle System Dynamics*, vol. 37, no. 6, pp. 401–421, 2002.

[6] Z.-D. Ma and N. C. Perkins, "An efficient multibody dynamics model for internal combustion engine systems," *Multibody System Dynamics*, vol. 10, no. 4, pp. 363–391, 2003.

[7] P. Roland, D. Daniel, S. Emilio, and Á. N. Miguel, "Validation of a multibody model for an X-by-wire vehicle prototype through field testing," in *Proceedings of the ECCOMAS Thematic Conference on Multibody Dynamics*, J. C. Samin and P. Fisette, Eds., Brussels, Belgium, July 2011.

[8] A. A. Shabana, *Dynamics of Multibody Systems*, Cambridge University Press, Cambridge, UK, 3rd edition, 2005.

[9] P. E. Nikravesh, *An Overview of Several Formulations for Multibody Dynamics*, Product Engineering, 2005.

[10] R. Featherstone, *Rigid Body Dynamics Algorithms*, Springer, New York, NY, USA, 2008.

[11] R. Featherstone, "The acceleration vector of a rigid body," *International Journal of Robotics Research*, vol. 20, no. 11, pp. 841–846, 2001.

[12] R. Featherstone and D. Orin, "Robot dynamics: equations and algorithms," in *Proceedings of the IEEE International Conference on Robotics and Automation (ICRA '00)*, pp. 826–834, April 2000.

[13] R. Featherstone, "Efficient factorization of the joint-space inertia matrix for branched kinematic trees," *International Journal of Robotics Research*, vol. 24, no. 6, pp. 487–500, 2005.

[14] R. A. Wehage and E. J. Haug, "Generalized coordinate partitioning for dimension reduction in analysis of constrained dynamic systems," *Journal of Mechanical Design*, vol. 134, pp. 247–255, 1982.

[15] R. A. Wehage and M. J. Belczynski, "High resolution vehicle simulations using precomputer coefficients," in *Proceedings of the Winter Annual Meeting of the American Society of Mechanical Engineers*, vol. 44, pp. 311–325, November 1992.

[16] G. Rodriguez, A. Jain, and K. Kreutz-Delgado, "Spatial operator algebra for multibody system dynamics," *Journal of the Astronautical Sciences*, vol. 40, no. 1, pp. 27–50, 1992.

[17] G. Rodriguez, A. Jain, and K. Kreutz-Delgado, "Spatial operator algebra for manipulator modeling and control," *International Journal of Robotics Research*, vol. 10, no. 4, pp. 371–381, 1991.

[18] E. Eich-Soellner and C. Führer, *Numerical Methods in Multibody Dynamics*, Teubner, Stuttgart, Germany, 1998.

[19] C. Yang, D. Cao, Z. Zhao, Z. Zhang, and G. Ren, "A direct eigenanalysis of multibody system in equilibrium," *Journal of Applied Mathematics*, vol. 2012, Article ID 638546, 12 pages, 2012.

[20] J. García de Jalón and E. Bayo, *Kinematic and Dynamic Simulation of Multibody Systems, the Real-Time Challenge*, Springer, New-York, NY, USA, 1994.

[21] R. Avilés, G. Ajuria, V. Gómez-Garay, and S. Navalpotro, "Comparison among nonlinear optimization methods for the static equilibrium analysis of multibody systems with rigid and elastic elements," *Mechanism and Machine Theory*, vol. 35, no. 8, pp. 1151–1168, 2000.

[22] Z. S. Wang, Y. Y. Tao, and Q. Y. Wen, "A vector bond graph method of kineto-static analysis for spatial multibody systems," *Applied Mechanics and Materials*, vol. 321–324, pp. 1725–1729, 2013.

[23] M. A. Omar, "Modeling flexible bodies in multibody systems in joint-coordinates formulation using spatial algebra," *Advances in Mechanical Engineering*, vol. 2014, Article ID 468986, 18 pages, 2014.

[24] R. W. Wehage, "Automated procedures for robust and efficient solution of over-constrained multibody dynamics," in *Proceeding of ASME International Mechanical Engineering Congress and Exposition (IMECE '12)*, vol. 85259, Huston, Tex, USA, November 2012.

[25] P. E. Nikravesh, "Construction of the equations of motion for multibody dynamics using point and joint coordinates," in *Computer-Aided Analysis of Rigid and Flexible Mechanical Systems*, vol. 268 of *NATO ASI Series E: Applied Sciences*, pp. 31–60, Kluwer Academic Publishers, 1994.

[26] A. Jain, "Multibody graph transformations and analysis—part I: tree topology systems," *Nonlinear Dynamics*, vol. 67, no. 4, pp. 2779–2797, 2012.

[27] M. A. Omar, *Finite Element Modeling of Leaf Springs in Multibody Systems*, VDM Verlag Dr. Müller Aktiengesellschaft & Co. KG, Saarbrücken, Germany, 2010.

[28] M. A. Omar, A. A. Shabana, A. Mikkola, W.-Y. I. Loh, and R. Basch, "Multibody system modeling of leaf springs," *Journal of Vibration and Control*, vol. 10, no. 11, pp. 1601–1638, 2004.

[29] K. J. Bathe, *Finite Element Procedures*, Prentice-Hall, Englewood Cliffs, NJ, USA, 1996.

[30] T. J. R. Hughes, *The Finite Element Method: Linear Static and Dynamic Finite Element Analysis*, Dover, New York, NY, USA, 2000.

[31] R. R. Craig and M. C. Bampton, "Coupling of substructures for dynamic analyses," *AIAA Journal*, vol. 6, no. 7, pp. 1313–1319, 1968.

[32] R. R. Craig and C. J. Chang, "On the use of attachment modes in substructure coupling for dynamic analysis," in *Proceedings of the 18th Conference on Structures, Structural Dynamics and Materials*, San Diego, Calif, USA, 1977.

[33] A. A. Shabana, *Computational Dynamics*, John Wiley & Sons, 3rd edition, 2010.

[34] U. M. Ascher and L. R. Petzold, *Computer Methods for Ordinary Differential Equations and Differential-Algebraic Equations*, Society for Industrial and Applied Mathematics, Philadelphia, Pa, USA, 1998.

[35] K. E. Brenan, S. L. Campbell, and L. R. Petzold, *Numerical Solutions of Initial-Value Problems in Differential-Algebraic Equations*, Society for Industrial and Applied Mathematics, New York, NY, USA, 1996.

[36] J. García de Jalón and E. Bayo, *Kinematic and Dynamic Simulation of Multibody Systems, the Real-Time Challenge*, Springer, NewYork, NY, USA, 1994.

[37] J. Baumgarte, "Stabilization of constraints and integrals of motion in dynamical systems," *Computer Methods in Applied Mechanics and Engineering*, vol. 1, no. 1, pp. 1–16, 1972.

16

Reduction of Free Edge Peeling Stress of Laminated Composites Using Active Piezoelectric Layers

Bin Huang and Heung Soo Kim

Department of Mechanical, Robotics and Energy Engineering, Dongguk University-Seoul, 30 Pildong-ro 1-gil, Jung-gu, Seoul 100-715, Republic of Korea

Correspondence should be addressed to Heung Soo Kim; heungsoo@dgu.edu

Academic Editor: Pizhong Qiao

An analytical approach is proposed in the reduction of free edge peeling stresses of laminated composites using active piezoelectric layers. The approach is the extended Kantorovich method which is an iterative method. Multiterms of trial function are employed and governing equations are derived by taking the principle of complementary virtual work. The solutions are obtained by solving a generalized eigenvalue problem. By this approach, the stresses automatically satisfy not only the traction-free boundary conditions, but also the free edge boundary conditions. Through the iteration processes, the free edge stresses converge very quickly. It is found that the peeling stresses generated by mechanical loadings are significantly reduced by applying a proper electric field to the piezoelectric actuators.

1. Introduction

Laminated composite materials [1] have been increasingly attracted in the engineering industries, due to their high stiffness-to-weight ratio, high strength-to-weight ratio, high corrosion resistance, high chemical stability, and so on. They have been popularized in the aircraft manufacturing, vehicle industries, medical fields, and even sport equipment industries. With the improvement of manufacturing process and productivity, they have potential in replacing the conventional metal alloys and nonmetal materials massively in the future. However, it is well known that due to the mismatch of elastic properties between the adjacent layers, there exist large undesirable interlaminar peeling stresses. Because of these highly concentrated peeling stresses, failures such as delamination or crack may occur near the free edges and further lead to the damage of structures.

The concentrations of peeling stress often occur within a short distance from the free edges, known as the free edge boundary effect [2, 3]. The stresses decrease and vanish in the interior region of the laminates. Also, even an in-plane load applied on the laminate will cause 3D state of stress distributions near the free edge. Thus, some theories are not capable of predicting the free edge peeling stress distributions which assume the plane stress condition, such as classical laminate theory (CLT) [4, 5].

The concentrated free edge peeling stresses are critical to the fatigues of laminated structures. Some passive control methods have been developed either by adjusting the stacking sequence of laminates [6] or by changing the edge shape of fibers [7] to prevent the onset of delamination. Nevertheless, due to their passive nature, they cannot work effectively and other desired goals cannot be achieved. Thus, in this study, we will investigate the peeling stress reduction by using piezoelectric actuators. Owing to their electromechanical coupling nature, piezoelectric actuators have potential to generate stresses which have opposite signs of those peeling stresses generated by mechanical loadings and, in return, to reduce the magnitudes of those peeling stresses.

Since the last decades, a large volume of literature sources has introduced the modeling of smart composite laminated plates and beams [8–12]. Chen and Qiao investigated the electromechanical behavior and fracture of piezoelectric bilayer beams [13–15]. They also investigated the interlaminar stresses of plated beam with adhesive layers [16]. They found that the adhesive layers can affect the local deformations and in return alleviate the stress concentrations. While predicting the interlaminar stresses of laminated structures, most approaches neglect the adhesive layer effect. Numerous approximate approaches, without considering the adhesive layer effect, have been developed since there is no method of obtaining exact 3D elastic solutions. The developed theories vary from the displacement field based approaches [17] to stress field based approaches [18, 19]. All of these theories have their own characteristics and specific applications.

For the displacement field based equivalent single layer theories, such as classical laminate theory, first order shear deformation theory [20, 21], and higher order shear deformation theory [22–26], these theories are efficient in predicting the global response of laminates in bending, buckling, and vibration analysis, while they are not capable of predicting the localized free edge stress concentrations. The stress components are usually calculated from the constitutive relation or the equilibrium equations. Thus, the free edge stress concentration at the layer interface cannot be revealed accurately. For the stress function based equivalent single layer theories [27, 28], the localized free edge stress concentrations can be accurately predicted, while, on the other hand, displacement information cannot be achieved. Moreover, prescribed boundary conditions are mandatory in those equivalent single layer theories either for the preassumed displacement trial functions or for the stress trial functions. The layerwise theories [29] are much more accurate in analyzing the stress concentration at the ply level. For the displacement based layerwise theories, they allow the in-plane displacement to vary continuously through the laminate thickness, while their derivatives with respect to the thickness coordinate may be discontinuous through the thickness, thereby allowing for the possibility of continuous transverse stresses at the ply interfaces. For the stress based layerwise theories [30, 31], the stress fields are assumed layerwisely and transverse stress continuity conditions have to be enforced at the ply interfaces. Both the displacement and stress based theories can accurately predict the stress concentration at the free edges. However, the layerwise theories are much more computationally expensive than the equivalent single layer theories.

Among the analytical methods, another approximate method named as the extended Kantorovich method [32, 33] can be referred to. The extended Kantorovich method is an iterative method for solving partial differential equations. Although the initial assumption of admissible functions influences the convergence time a lot, the solutions can be improved gradually by repeating solving the governing equations and applying boundary and continuity conditions during the iteration processes. The assumed admissible functions with prescribed boundary conditions are of priority and any types of boundary condition problems can be solved by using this approach. The interlaminar stresses at the straight free edges of laminates under extension loads were investigated by Cho and Yoon [34]. And later Cho and Kim [35] analyzed the free edge interlaminar stresses of composite laminates under extension, bending, twisting, and thermal loads by using the extended Kantorovich method. In their study, they preferred using the stress trial functions, and traction-free boundary conditions and free edge boundary conditions were satisfied automatically during the iterative processes. Ungbhakorn and Singhatanadgid [36] proposed an extended Kantorovich method to investigate the buckling problem of rectangular laminates with various boundary conditions. The initial trial functions were used arbitrarily and the principle of minimum total potential energy was used to derive governing equations. Andakhshideh and Tahani [37] investigated the interlaminar stresses near free edges of finite length general composite laminates subjected to axial and shearing loads using three-dimensional extended Kantorovich method. They provided convergent 3D elasticity solutions with four terms of trial functions. Due to their arbitrary choice of initial trial functions and various boundary conditions, the extended Kantorovich method could be a good methodology in the interlaminar stress analysis of laminated composite structures.

In the present study, a stress function based extended Kantorovich method is proposed to investigate the reduction of free edge peeling stresses using piezoelectric actuators. Symmetrically layered laminates will be taken into consideration as symmetric laminates are most commonly used to avoid coupling of bending and twisting modes. Most studies start from the initial assumption that the strain or stress is independent of the longitudinal coordinate, and this study also adopts this point. The three-dimensional stress solutions are obtained by iterative processes involving multiterms of initially assumed trial functions. By the extended Kantorovich method, the influence of initial assumptions can be well eliminated and the accuracy of stress distributions can be improved a lot. Finally, convergent results are given as numerical examples and compared with those obtained by finite element method. It is expected to obtain significant peeling stress reduction by applying proper electric fields to piezoelectric actuators.

2. Mathematical Modeling

2.1. Derivation of Governing Equation. For the given geometry of laminated composite plate, as shown in Figure 1, two piezoelectric actuators are symmetrically surface bonded on the top and bottom surfaces of the laminates without considering the adhesive layer affect. This kind of structure is usually considered to be smart composite laminates [38]. Two piezoelectric layers are used to control the free edge peeling stresses that are generated by mechanical loadings. The longitudinal dimension of the laminates is considered long enough so that the stress fields are independent of the x-axis. Therefore, generalized plane strain states are assumed to be in the x direction.

The extended Kantorovich method requires iterative calculation methodology. By preassumption of trial functions and solving a set of ordinary differential equations, the solution of the first process can be obtained. Due to the rough initial assumption, it cannot guarantee the accuracy of solution. Thus, to overcome this weak point, iterative processes can be conducted to improve the accuracy of solution. The governing equations can be derived by using

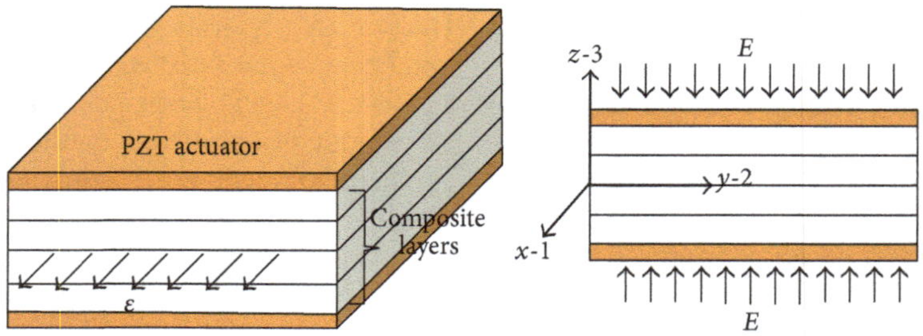

FIGURE 1: Geometry of laminated composite plate with surface bonded piezoelectric actuators under combined electromechanical loadings.

either Galerkin's method or the principle of virtual work. In this work, the principle of complementary virtual work is employed. The complementary strain energy of the laminated composite plate is calculated by the following equation:

$$\delta U = \iint_s \varepsilon_i \delta\sigma_i dA$$
$$= \iint_s \varepsilon_2\delta\sigma_2 + \varepsilon_3\delta\sigma_3 + \varepsilon_4\delta\sigma_4 + \varepsilon_5\delta\sigma_5 + \varepsilon_6\delta\sigma_6 dA = 0. \quad (1)$$

Before substituting strain and stress components into the complementary strain energy equation, an initially assumed stress field should be introduced. Instead of using displacement assumption, stress field assumption is preferred. A good choice for stress field is Lekhnitskii stress functions [39] which can express the three-dimensional stress state of piezo-bonded composite laminates. Lekhnitskii stress functions can automatically satisfy the pointwise equilibrium equations as follows:

$$\sigma_2 = \frac{\partial^2 F}{\partial \eta^2}, \qquad \sigma_3 = \frac{\partial^2 F}{\partial \xi^2}, \qquad \sigma_4 = -\frac{\partial^2 F}{\partial \xi \partial \eta}$$
$$\sigma_5 = -\frac{\partial \psi}{\partial \xi}, \qquad \sigma_6 = \frac{\partial \psi}{\partial \eta}, \quad (2)$$

where $\eta = z/H$ and $\xi = y/H$ are nondimensionalized coordinates and H is the total thickness of the laminates. The solutions F and ψ are series combinations and assumed to be a combination of in-plane stress functions ($f_i(\xi)$ and $p_i(\xi)$) and out-of-plane stress functions ($g_i(\eta)$) using separation of variables as follows:

$$F = \sum_{i=1}^{n} f_i(\xi)\, g_i(\eta), \qquad \psi = \sum_{i=1}^{n} p_i(\xi)\, g_i^{\mathrm{I}}(\eta). \quad (3)$$

With the initial assumption of stress field, the strain tensor $\{\varepsilon_i\}$ can be found from the linear elastic constitutive relation. This relation is assumed for each individual layer with piezo induced strain term:

$$\{\varepsilon_i\}_{(k)} = [\overline{S}]_{(k)}\{\sigma_i\}_{(k)} + [d]_{(k)}\{E\}_{(k)}, \quad (i = 1, 2, \ldots, 6), \quad (4)$$

where $[\overline{S}]$ is the generalized compliance matrix for orthotropic materials. $[d]$ is the piezoelectric constant matrix and $[E]$ is the applied electric field.

As the mechanical extension load is applied in the x direction, all other strain components can be expressed using ε_1 and applied piezoelectric fields. Consider

$$\varepsilon_i = \widehat{S}_{ij}\sigma_j + \frac{S_{i1}}{S_{11}}\varepsilon_1 + \widehat{d}_{i3}E_3, \quad (i = 2, 3, \ldots, 6,\ j = 2, 3, \ldots, 6), \quad (5)$$

where

$$\widehat{S}_{ij} = S_{ij} - \frac{S_{i1}S_{1j}}{S_{11}}, \qquad \widehat{d}_{i3} = d_{i3} - \frac{S_{i1}}{S_{11}}d_{13}. \quad (6)$$

Substituting (2) and (5) into (1), the complementary strain energy becomes

$$\delta U = \iint_s \varepsilon_i \delta\sigma_i dA$$
$$= \iint_s \left[\left(\widehat{S}_{ij}\sigma_j + \frac{S_{i1}}{S_{11}}\varepsilon_1 + \widehat{d}_{3i}E_3\right)\delta\sigma_i\right] d\xi\, d\eta = 0, \quad (7)$$
$$(i, j = 2, 3, \ldots, 6).$$

To simplify the above equation, integration by parts and free edge boundary conditions ($\sigma_2 = \sigma_4 = \sigma_6 = 0$, at $\xi = 0$) are applied. Then, the double integration is simplified to the following equation and the governing equations can be obtained by using the variational principle. Consider

$$\int \left[a_{ij}^{(4)} f_j^{\mathrm{IV}} + a_{ij}^{(2)} f_j^{\mathrm{II}} + a_{ij}^{(0)} f_j + b_{ij}^{(2)} p_j^{\mathrm{II}} + b_{ij}^{(0)} p_j + r_i\right]\delta f_i d\xi$$
$$+ \int \left[d_{ij}^{(2)} f_j^{\mathrm{II}} + d_{ij}^{(0)} f_j + c_{ij}^{(2)} p_j^{\mathrm{II}} + c_{ij}^{(0)} p_j + s_i\right]\delta p_i d\xi = 0,$$
$$(i, j = 1, 2, \ldots, n), \quad (8)$$

where

$$a_{ij}^{(4)} = \int \widehat{S}_{33} g_i g_j d\eta,$$

$$a_{ij}^{(2)} = \int \widehat{S}_{23} \left(g_i^{\mathrm{II}} g_j + g_i g_j^{\mathrm{II}}\right) d\eta - \int \widehat{S}_{44} g_i^{\mathrm{I}} g_j^{\mathrm{I}} d\eta,$$

$$a_{ij}^{(0)} = \int \widehat{S}_{22} g_i^{\mathrm{II}} g_j^{\mathrm{II}} d\eta,$$

$$b_{ij}^{(2)} = \int \widehat{S}_{36} g_i g_j^{\mathrm{II}} d\eta - \int \widehat{S}_{45} g_i^{\mathrm{I}} g_j^{\mathrm{I}} d\eta, \quad b_{ij}^{(0)} = \int \widehat{S}_{26} g_i^{\mathrm{II}} g_j^{\mathrm{II}} d\eta,$$

$$c_{ij}^{(2)} = -\int \widehat{S}_{55} g_i^{\mathrm{I}} g_j^{\mathrm{I}} d\eta, \quad c_{ij}^{(0)} = \int \widehat{S}_{66} g_i^{\mathrm{II}} g_j^{\mathrm{II}} d\eta,$$

$$d_{ij}^{(2)} = \int \widehat{S}_{36} g_i^{\mathrm{II}} g_j d\eta - \int \widehat{S}_{45} g_i^{\mathrm{I}} g_j^{\mathrm{I}} d\eta, \quad d_{ij}^{(0)} = \int \widehat{S}_{26} g_i^{\mathrm{II}} g_j^{\mathrm{II}} d\eta,$$

$$r_i = \int \left(\frac{S_{21}}{S_{11}} \varepsilon_1 + \widehat{d}_{23} E_3\right) g_i^{\mathrm{II}} d\eta,$$

$$s_i = \int \left(\frac{S_{61}}{S_{11}} \varepsilon_1 + \widehat{d}_{63} E_3\right) g_i^{\mathrm{II}} d\eta. \tag{9}$$

2.2. Solutions of Governing Equations and Generalized Eigenvalue Problem. To solve the fourth-order and second-order coupled ordinary differential equations, the trial functions $g_i(\eta)$ are specified first in (10). Although prescribed boundary conditions are not necessarily satisfied for trial functions when using the extended Kantorovich method, the mode shapes of the clamped-clamped beam are used as the initial out-of-plane trail functions to satisfy the traction-free boundary conditions automatically ($\sigma_3 = \sigma_4 = \sigma_5 = 0$ at $\eta = \pm 1/2$). Consider

$$g_i(\eta) = \cos(\beta_i \eta) + \frac{\cos(\beta_i/2)}{\cosh(\beta_i/2)} \cosh(\beta_i \eta), \tag{10}$$

where β_i are the roots of the following characteristic equation. Consider

$$\tan\left(\frac{\beta}{2}\right) + \tanh\left(\frac{\beta}{2}\right) = 0. \tag{11}$$

Thus, all coefficients in the governing equations can be obtained and the general solutions of governing equations can be assumed as the sum of homogeneous solutions ($f_i^{(H)}$ and $p_i^{(H)}$) and particular solutions ($f_i^{(P)}$ and $p_i^{(P)}$). The homogeneous solutions of $f_i^{(H)}$ and $p_i^{(H)}$ are assumed to be the following form:

$$f_i^{(H)} = v_i^f e^{\lambda \xi}, \qquad p_i^{(H)} = v_i^p e^{\lambda \xi}. \tag{12}$$

Substituting (12) into the governing equations and defining $v_j^{f^{\mathrm{II}}} = \lambda^2 v_j^f$, the governing equations can be rewritten in the matrix form and a standard eigenvalue problem is formed. Consider

$$\begin{bmatrix} 0 & -a_{ij}^{(4)} & -b_{ij}^{(2)} \\ 0 & 0 & c_{ij}^{(0)} \\ I & 0 & 0 \end{bmatrix}^{-1} \begin{bmatrix} a_{ij}^{(0)} & a_{ij}^{(2)} & b_{ij}^{(0)} \\ d_{ij}^{(0)} & d_{ij}^{(2)} & b_{ij}^{(0)} \\ 0 & I & 0 \end{bmatrix} \begin{bmatrix} v_j^f \\ v_j^{f^{\mathrm{II}}} \\ v_j^p \end{bmatrix}$$
$$= \lambda^2 \begin{bmatrix} v_j^f \\ v_j^{f^{\mathrm{II}}} \\ v_j^p \end{bmatrix}, \quad (i, j = 1, 2, \ldots, n), \tag{13}$$

where λ^2 are eigenvalues of the above equation. Since the interlaminar stresses decay in the interior region of piezo-bonded composite laminates, only negative roots of λ^2 are selected. Equation (13) results in $3n$ eigenvalues and eigenvectors, so that the homogeneous solutions consist of $3n$ terms linear combination as follows:

$$f_i^{(H)} = t_j v_{ij}^f e^{-\lambda_j \xi},$$

$$p_i^{(H)} = t_j v_{ij}^p e^{-\lambda_j \xi}, \quad (i = 1, 2, \ldots, n), \; (j = 1, 2, \ldots, 3n), \tag{14}$$

where t_j are constants that will be determined by the prescribed boundary conditions.

The particular solutions are obtained by considering $f_i^{(P)}$ and $p_i^{(P)}$ as constants; thus all their derivatives are zeros in the governing equations and the remaining parts can be rewritten in the following equation:

$$\begin{bmatrix} f_j^{(P)} \\ p_j^{(P)} \end{bmatrix} = -\begin{bmatrix} a_{ij}^{(0)} & b_{ij}^{(0)} \\ b_{ij}^{(0)} & c_{ij}^{(0)} \end{bmatrix}^{-1} \begin{bmatrix} r_i \\ s_i \end{bmatrix} \quad (i, j = 1, 2, \ldots, n). \tag{15}$$

The final step of solving the governing equations is to obtain the constants t_j in the homogeneous solutions. They can be calculated by enforcing the free edge boundary conditions ($\sigma_2 = \sigma_4 = \sigma_6 = 0$ at $\xi = 0$), and this will lead f_i, f_i^{I}, and p_i to be zeros at the free edges. Consider

$$t_j v_{ij}^f e^{-\lambda_j \xi}\Big|_{\xi=0} + f_i^{(P)} = 0,$$

$$-t_j \lambda_j v_{ij}^f e^{-\lambda_j \xi}\Big|_{\xi=0} = 0, \tag{16}$$

$$t_j v_{ij}^p e^{-\lambda_j \xi}\Big|_{\xi=0} + p_i^{(P)} = 0.$$

After solving the governing equations, all stress components can be obtained by substituting the solutions into the Lekhnitskii stress functions and they have the following expressions:

$$\sigma_2 = \sum_{i=1}^{n}\sum_{j=1}^{3n}\left(t_j v_{ij}^{f} e^{-\lambda_j \xi} + f_i^{(P)}\right) g_i^{\mathrm{II}},$$

$$\sigma_3 = \sum_{i=1}^{n}\sum_{j=1}^{3n}\lambda_i^2 t_j v_{ij}^{f} e^{-\lambda_j \xi} g_i,$$

$$\sigma_4 = \sum_{i=1}^{n}\sum_{j=1}^{3n}\lambda_j t_j v_{ij}^{f} e^{-\lambda_j \xi} g_i^{\mathrm{I}}, \quad (17)$$

$$\sigma_5 = \sum_{i=1}^{n}\sum_{j=1}^{3n}\lambda_j t_j v_{ij}^{p} e^{-\lambda_j \xi} g_i^{\mathrm{I}},$$

$$\sigma_6 = \sum_{i=1}^{n}\sum_{j=1}^{3n}\left(t_j v_{ij}^{p} e^{-\lambda_j \xi} + p_i^{(P)}\right) g_i^{\mathrm{II}}.$$

2.3. Iterative Procedures. The accuracy of the first step largely depends on the initial assumption of trail functions and the number of series terms involved in the trail functions. The results may also be influenced by the choice of trail functions which may cause singularity problem, harmonic distributions of stresses, and so forth. Thus, it is important to overcome the weaknesses appeared in the first process. We prefer using the extended Kantorovich method which is an iterative process. Through the iterative processes, the influence of the initial assumption can be eliminated and the convergent results can be obtained.

In the second step, the solutions of the first step ($f_i(\xi)$ and $p_i(\xi)$) are considered to be the known functions and the principle of complementary virtual work is used again to derive governing equations. Here, the unknown out-of-plane stress functions ($g_i(\eta)$) are assumed to be layer dependent and the continuity conditions are enforced at the layer interface. Thus, from the principle of complementary virtual work in (7) and integration by parts, the governing equation for the out-of-plane stress functions can be obtained as follows:

$$\int\left[m_{ij}^{(4)(k)} g_j^{(k)(\mathrm{IV})} + m_{ij}^{(2)(k)} g_j^{(k)(\mathrm{II})} + m_{ij}^{(0)(k)} g_j^{(k)} + x_i^{(k)}\right] \times \delta g_j^{(k)} d\eta + \Lambda = 0 \quad (i,j = 1,2,\ldots,n), \quad (k = 1,2,\ldots,N), \quad (18)$$

where Λ represents the boundary terms induced by integration by parts and the superscript (k) refers to the kth ply. The layerwise coefficients are listed as follows:

$$m_{ij}^{(4)(k)} = \int\left[\widehat{S}_{22}^{(k)} f_i f_j + \widehat{S}_{26}^{(k)}\left(f_i p_j + p_i f_j\right) + \widehat{S}_{66}^{(k)} p_i p_j\right] d\xi,$$

$$m_{ij}^{(2)(k)} = \int\left[\widehat{S}_{23}^{(k)}\left(f_i f_j^{\mathrm{II}} + f_i^{\mathrm{II}} f_j\right) + \widehat{S}_{36}^{(k)}\left(f_i^{\mathrm{II}} p_j + p_i f_j^{\mathrm{II}}\right) - \widehat{S}_{44}^{(k)} f_i^{\mathrm{I}} f_j^{\mathrm{I}} - \widehat{S}_{45}^{(k)}\left(f_i^{\mathrm{I}} p_j^{\mathrm{I}} + p_i^{\mathrm{I}} f_j^{\mathrm{I}}\right) - \widehat{S}_{55}^{(k)} p_i^{\mathrm{I}} p_j^{\mathrm{I}}\right] d\xi,$$

$$m_{ij}^{(0)(k)} = \int \widehat{S}_{33}^{(k)} f_i^{\mathrm{II}} f_j^{\mathrm{II}} d\xi,$$

$$x_i^{(k)} = \int\left(\frac{S_{13}^{(k)}}{S_{11}^{(k)}}\varepsilon_1 + \widehat{d}_{33} E_3\right) f_i^{\mathrm{II}} d\xi. \quad (19)$$

The solution procedure is the same as in the first step. A generalized eigenvalue problem is solved and the general solution is calculated by adding the homogeneous solutions $g_i^{(k)(H)}$ and the particular solutions $g_i^{(k)(P)}$:

$$g_i^{(k)} = b_j^{(k)} v_i^{g(k)} e^{\mu^{(k)}\eta} + g_i^{(k)(P)}, \quad (k = 1,2,\ldots,N), \quad (i = 1,2,\ldots,n), \quad (20)$$

where $\mu^{(k)}$ are the roots of eigenvalues and $v_i^{g(k)}$ are the eigenvectors.

The constants $b_j^{(k)}$ in the homogeneous solutions can be determined from the traction-free boundary conditions on the top and bottom surfaces ($\sigma_i = 0$ at $\eta = \pm 1/2$, $i = 3,4,5$), interlaminar stress continuity conditions at the ply interfaces ($\sigma_i^+(k) = \sigma_i^-(k+1)$, $i = 3,4,5$), and boundary terms Λ set to be zero. These conditions lead to the following equations:

$$g_i^{(k)} = b_j^{(k)} v_i^{g(k)} e^{\mu^{(k)}\eta}\Big|_{\eta=\pm 1/2} + g_i^{(k)(P)} = 0, \quad (k = 1, N),$$

$$g_i^{(k)^{\mathrm{I}}} = b_j^{(k)} \mu^{(k)} v_i^{g(k)} e^{\mu^{(k)}\eta}\Big|_{\eta=\pm 1/2} = 0, \quad (k = 1, N),$$

$$b_j^{(k)} v_i^{g(k)} e^{\mu^{(k)}\eta}\Big|_{\eta=\eta_k^+} + g_i^{(k)(P)} = b_j^{(k+1)} v_i^{g(k+1)} e^{\mu^{(k+1)}\eta}\Big|_{\eta=\eta_{k+1}^-} + g_i^{(k+1)(P)}, \quad (k=1,2,\ldots,N-1),$$

$$b_j^{(k)} \mu^{(k)} v_i^{g(k)} e^{\mu^{(k)}\eta}\Big|_{\eta=\eta_k^+} = b_j^{(k+1)} \mu^{(k+1)} v_i^{g(k+1)} e^{\mu^{(k+1)}\eta}\Big|_{\eta=\eta_{k+1}^-}, \quad (k = 1,2,\ldots,N-1). \quad (21)$$

Thus, after obtaining the solutions of governing equation, all stress components can be obtained and are listed as follows:

$$\sigma_2 = \sum_{i=1}^{n}\sum_{j=1}^{4n} f_i(\xi)\left[\mu_j^{(k)^2} v_{ij}^{g(k)} b_j^{(k)} e^{\mu_j^{(k)}\eta}\right],$$

$$\sigma_3 = \sum_{i=1}^{n}\sum_{j=1}^{4n} f_i^{\mathrm{II}}(\xi)\left[v_{ij}^{g(k)} b_j^{(k)} e^{\mu_j^{(k)}\eta} + g_i^{(k)(P)}\right],$$

$$\sigma_4 = \sum_{i=1}^{n}\sum_{j=1}^{4n} - f_i^{\mathrm{I}}(\xi)\left[\mu_j^{(k)} v_{ij}^{g(k)} b_j^{(k)} e^{\mu_j^{(k)}\eta}\right], \quad (22)$$

$$\sigma_5 = \sum_{i=1}^{n}\sum_{j=1}^{4n} - p_i^{\mathrm{I}}(\xi)\left[\mu_j^{(k)} v_{ij}^{g(k)} b_j^{(k)} e^{\mu_j^{(k)}\eta}\right],$$

$$\sigma_6 = \sum_{i=1}^{n}\sum_{j=1}^{4n} p_i(\xi)\left[\mu_j^{(k)^2} v_{ij}^{g(k)} b_j^{(k)} e^{\mu_j^{(k)}\eta}\right].$$

After conducting the second step, if the results are still not convergent, more processes can be conducted. The iterative algorithm can be concluded as follows.

(1) Assume an initial set of out-of-plane stress functions $(g_i(\eta))$ and then take the principle of complementary virtual work to solve the in-plane stress functions $(f_i(\xi)$ and $p_i(\xi))$.

(2) Use the in-plane stress functions $(f_i(\xi)$ and $p_i(\xi))$ obtained from the first step as the known functions and repeat taking the principle of complementary virtual work to solve the out-of-plane stress functions $(g_i(\eta))$.

(3) Use the out-of-plane stress functions $(g_i(\eta))$ obtained from the second step as the known functions and repeat taking the principle of complementary virtual work to solve the in-plane stress functions $(f_i(\xi)$ and $p_i(\xi))$.

Three steps are narrated in the above algorithm and more processes can be conducted, if required. However, one essential difference should be emphasized. In the first step, the initially assumed out-of-plane stress functions represent the global state of stress distributions which implies that the first step is a stress based equivalent single layer theory, while from the second step, the out-of-plane stress functions are assumed layerwisely and continuity conditions should be enforced at the layer interfaces. All stress components satisfy the boundary conditions automatically when using the extended Kantorovich method.

3. Numerical Results

Based on the theoretical formulations discussed in the previous section, numerical examples are investigated. The laminated composite structures are subjected to uniaxial extension and electric load. The material properties of graphite/epoxy composite laminates [40] and piezoelectric actuator (PZT-5H) are given in Table 1. The lamina thickness is 0.125 mm, which is almost one-fourth of the PZT thickness (0.5 mm). For convenience, 4 laminae, which have the same fiber orientations, are considered one lamina, so that each lamina thickness is then the same as the PZT thickness. The width of the laminate b is equal to four times the total thickness H.

3.1. Convergence Study. The laminates are subjected to uniaxial tension only in this part. To investigate the convergence of the proposed method, different terms of initially assumed stress functions are used and several iterative processes are conducted. To validate the proposed modeling, 3D finite element analysis is also conducted using commercial software ANSYS. The Solid 64 element is used for composite laminates and Solid 5 is used for PZTs. To obtain accurate interlaminar stress distributions, fine mesh is required and it should be refined at the layer interface. Thus, we used 20 elements along the thickness direction for a single layer. The total degree of freedom is 640,281 that is much larger than our proposed method. The three-dimensional solid laminate has a relatively long length, compared with the width and height, to express the generalized plane strain state effectively.

Table 1: Material properties of composite lamina and PZT-5H actuator.

Composite	Actuator
$E_1 = 138\,\text{GPa}$	$E_1 = E_2 = 62\,\text{GPa}$
$E_2 = E_3 = 14.5\,\text{GPa}$	$E_3 = 48\,\text{GPa}$
$G_{12} = G_{13} = G_{23} = 5.9\,\text{GPa}$	$G_{12} = 23.5\,\text{GPa}$, $G_{13} = G_{23} = 23.5\,\text{GPa}$
$\nu_{12} = \nu_{23} = \nu_{21} = 0.21$	$\nu_{12} = \nu_{23} = \nu_{21} = 0.29$
	$d_{31} = d_{32} = 274 \times 10^{-12}\,\text{m/V}$
	$d_{33} = -593 \times 10^{-12}\,\text{m/V}$, $d_{15} = -741 \times 10^{-12}\,\text{m/V}$

Figure 2 shows interlaminar normal stress (σ_3) distribution through the thickness direction at the free edge of $[\text{PZT}/0_4/90_4]\text{s}$ laminate under 0.1% uniaxial strain obtained from the first process. The result is compared with the FEM result, and the influence of term number used in the first process is investigated. It is found that the accuracy of the stress distribution largely depends on the number of terms in the first process. With 8 terms of initially assumed stress functions, the result cannot converge well and undesired oscillation appears due to the initial assumption of harmonic functions. Moreover, the stress concentration at the layer interfaces also cannot be predicted accurately. As we expected the peak values should be exactly located at the PZT/0 layer interface $(z/H = 1/3)$ and 0/90 layer interface $(z/H = 1/6)$, while it cannot be recovered by using 8 terms. Only by increasing the number of terms, the locations can be predicted more precisely. Until using 40 terms of initially assumed stress functions, the stress concentrations can be found exactly located at the layer interfaces, but undesired oscillations still exist. The proposed result can be verified by the FEM result. Both of them show the same tendency and magnitude, whereas the result obtained by finite element method shows difficulties in predicting the stress concentrations at the layer interfaces even using fine mesh. A proper choice of the term number not only guarantees the accuracy of results, but also improves the computational ability. It is to be noted that the computational time using 40 terms for solving the first process only costs seconds, while FEM takes 10–20 minutes to run a single case.

The results of iterative process for $[\text{PZT}/0_4/90_4]\text{s}$ laminates under 0.1% uniaxial strain are given in Figures 3–5. These results are obtained by only 8 terms and different iterative processes. Figure 3 shows the interlaminar normal stress (σ_3) distribution through the thickness direction at the free edge. From the graph, it is found that the iterative processes can significantly improve the accuracy of the stress prediction. The locations of stress concentration can be exactly predicted at the layer interfaces and the oscillations appeared in the first iteration disappeared since the third iteration. Figures 4 and 5 show interlaminar normal stress distribution (σ_3) and interlaminar shear stress distribution

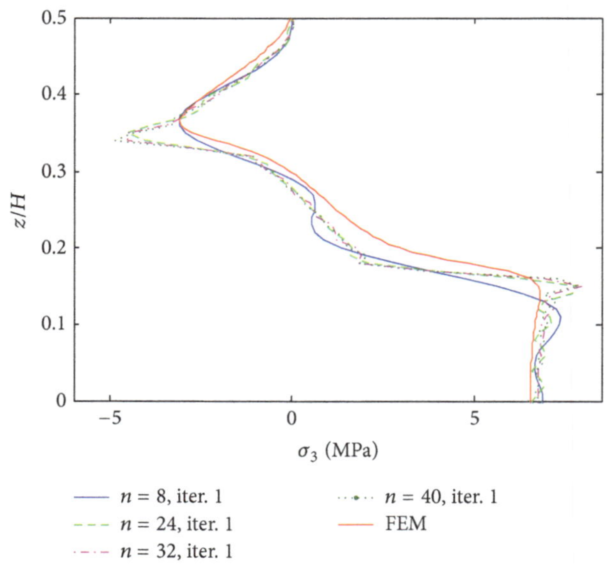

FIGURE 2: σ_3 distribution of [PZT/0_4/90_4]s laminate at the free edge, using different terms, under 0.1% uniaxial strain.

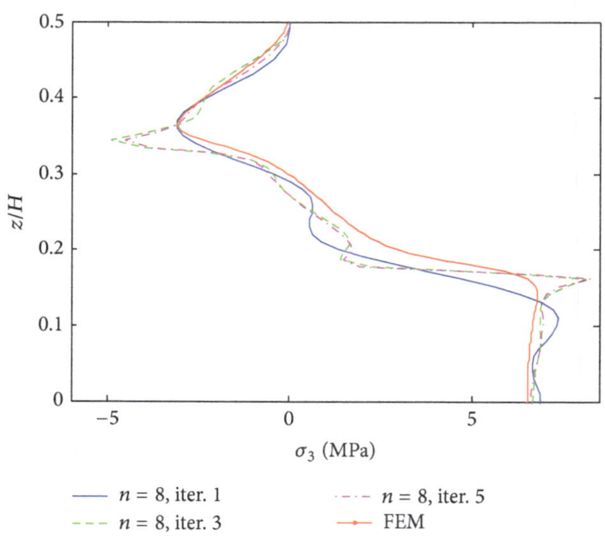

FIGURE 3: σ_3 distribution of [PZT/0_4/90_4]s laminate at the free edge from different iterations, under 0.1% uniaxial strain.

(σ_4) through the in-plane direction, respectively. First iteration with 8 terms of initially assumed stress functions cannot guarantee the convergence of the stresses, while the third and fifth iterations provide almost converged stresses. The stresses obtained by FEM show great coincidence with those obtained by the proposed method but finite element analysis underestimates stress concentrations at the free edge. Interlaminar shear stress (σ_4) also cannot satisfy the free edge boundary condition exactly in finite element analysis. It is concluded that the iterative processes can provide the converged stresses with small number of terms. Not only the influence of initial assumption can be eliminated, but also the stress concentration locations can be predicted accurately by the iterative processes. In the following work, the results of the fifth iteration with 8 terms of initially assumed that the stress functions are considered to investigate the reduction of peeling stresses by piezoelectric excitation.

3.2. Reduction of Peeling Stresses Using Piezoelectric Excitations. The existence of peeling stresses generated by the mechanical loadings can be reduced by applying proper electric fields to the actuators. Due to the electromechanical coupling phenomenon, the induced strain or stress generated by the electric loadings has the possibility of compensating the strain or stress generated by the mechanical loadings, in turn, to reduce the peeling stresses and to expand the service life of laminated structures. To investigate the feasibility of peeling stress reduction using piezoelectric actuators, two general lay-ups are considered and numerical examples are discussed in this part.

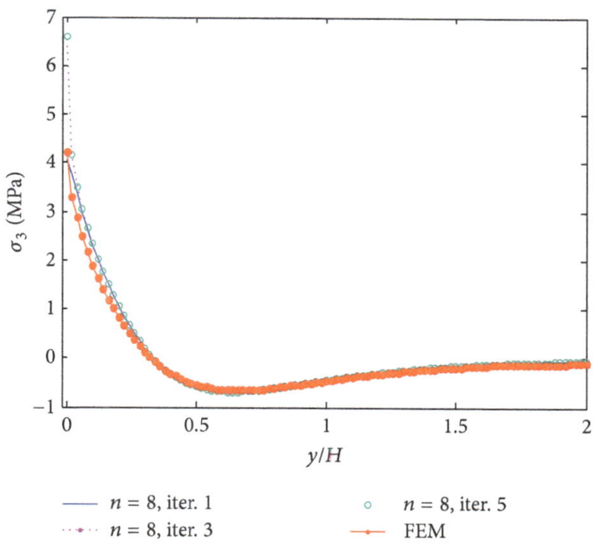

FIGURE 4: σ_3 distribution of $[PZT/0_4/90_4]s$ laminate at the interface of $z/H = 1/6$ from different iterations, under 0.1% uniaxial strain.

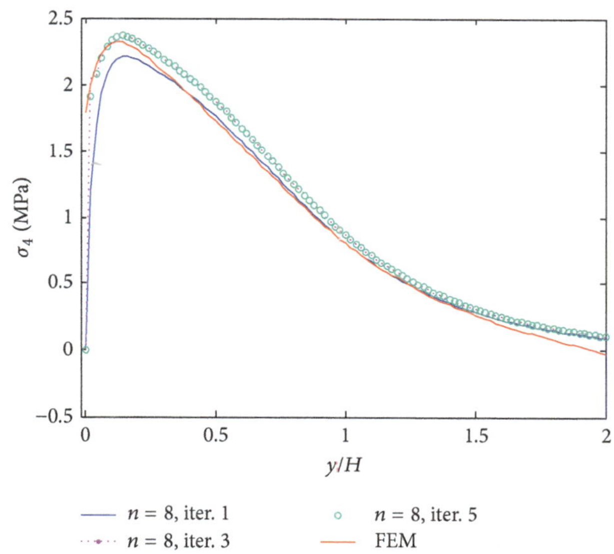

FIGURE 5: σ_4 distribution of $[PZT/0_4/90_4]s$ laminate at the interface of $z/H = 1/6$ from different iterations, under 0.1% uniaxial strain.

The first example is given by cross ply composite laminate ($[PZT/0_4/90_4]s$) which is subjected to 0.1% uniaxial strain. Moreover, different electric voltages are applied to the actuators that are expected to reduce the free edge peeling stresses. Figure 6 shows interlaminar normal stress (σ_3) distribution through the thickness direction at the free edge. It can be clearly found that the peeling magnitude of σ_3 is significantly reduced at the 0/90 layer interface by applying a positive electric voltage to the actuators. Without electric potential, the concentrated peeling stress is 8.3 MPa. It is reduced to 4.9 MPa, when 300 V electric potential is applied. This is of great meaningfulness to the reduction of positive peeling stress at the interface. Figures 7 and 8 show interlaminar normal stress (σ_3) and interlaminar shear stress (σ_4) distribution through the in-plane direction at the 0/90 interface, where $z/H = 1/6$. Our ambition is to reduce the peeling stress at the composite layer interface by using piezoelectric actuator. Thus, this interface is what we concerned. σ_3 and σ_4 show large positive values at the free edge and near the free edge under the mechanical loading. With the increasing of the electric potential, the magnitudes of peeling stresses are reduced dramatically. The peeling stress σ_3 can be reduced to almost 50.2 percent when 300 V electric potential is applied which is summarized in Table 2. For interlaminar shear stress (σ_4), 17.3 percent decrement is observed when 300 V electric potential is applied.

The second example is given by angle ply composite laminate ($[PZT/45_4/{-45_4}]s$) which is subjected to −0.1% uniaxial strain load. The laminate is subjected to compressive load while it sustains large peeling stresses at the 45/−45 layer

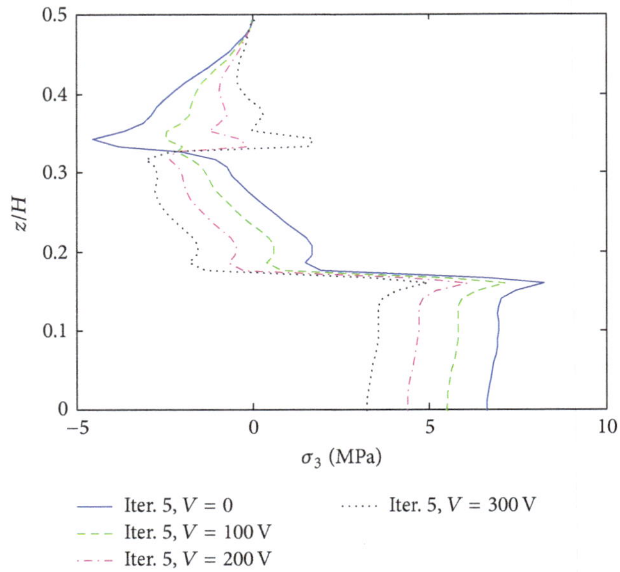

FIGURE 6: σ_3 distribution of [PZT/0_4/90_4]s laminate at the free edge, under 0.1% uniaxial strain.

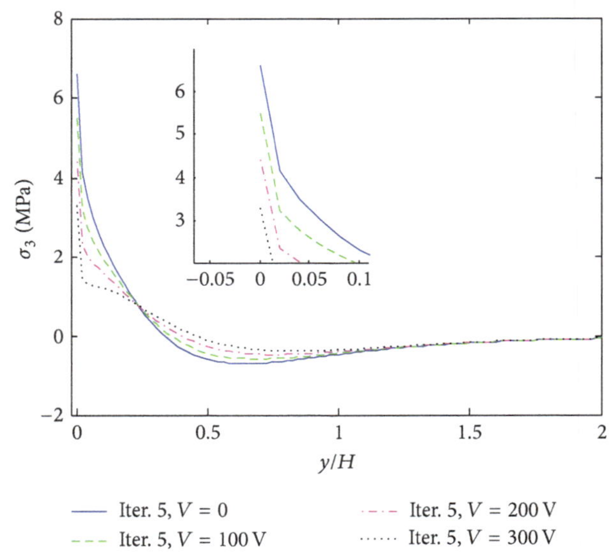

FIGURE 7: σ_3 distribution of [PZT/0_4/90_4]s laminate at the interface of $z/H = 1/6$, under 0.1% uniaxial strain.

TABLE 2: Comparison of peak stresses for cross ply composite laminate ([PZT/0_4/90_4]s) at the 0/90 layer interface, where $z/H = 1/6$, under 0.1% uniaxial strain, fifth iteration, $n = 8$.

Applied electric potential (V)	σ_3		σ_4	
	(MPa)	Reduction (%)	(MPa)	Reduction (%)
0	6.62	0	2.37	0
100	5.51	16.8%	2.19	7.6%
200	4.41	33.4%	2.06	13.1%
300	3.30	50.2%	1.96	17.3%

interfaces. Figure 9 shows interlaminar normal stress (σ_3) distribution through the thickness direction at the free edge. It is found that the peeling stress is severely concentrated at the 45/−45 layer interface which could lead to the onset of delamination of composite layers. However, this peeling stress can also be significantly reduced in angle ply laminate, when a positive electric potential is applied to the actuators. The peak value is reduced from 11.83 MPa to 8.52 MPa. The interlaminar normal stress (σ_3) and interlaminar shear stress (σ_4) distribution through the in-plane direction at the 45/−45 layer interface are given in Figures 10 and 11, respectively. The interlaminar normal stress (σ_3) and interlaminar shear stress (σ_4) well match with those obtained by the finite element analysis for the mechanical loading case. However, finite element analysis still underestimates the stress concentration

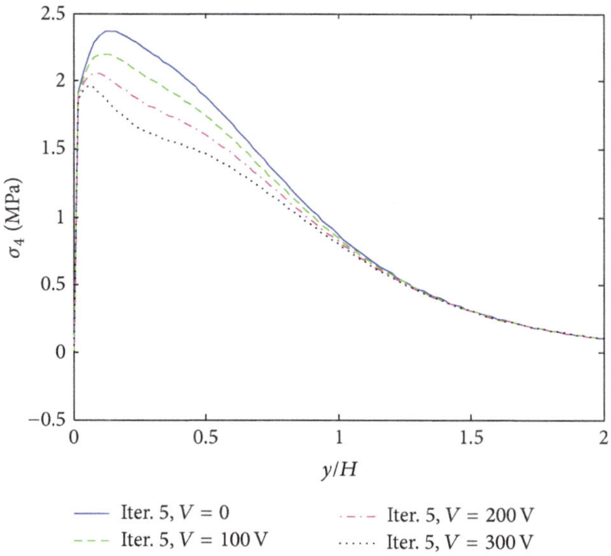

FIGURE 8: σ_4 distribution of $[PZT/0_4/90_4]s$ laminate at the interface of $z/H = 1/6$, under 0.1% uniaxial strain.

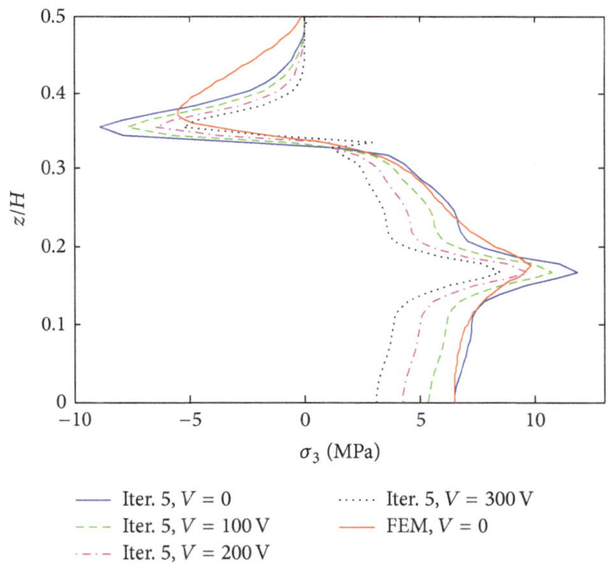

FIGURE 9: σ_3 distribution of $[PZT/45_4/-45_4]s$ laminate at the free edge, under −0.1% uniaxial strain.

at the free edge. The detailed reductions of peeling stresses can be found in Table 3. Since angle ply composite laminates sustain larger shear stresses than cross ply composite laminates, the reduction of σ_3 is smaller than that of cross ply, while the reduction of σ_4 is larger than that of the cross ply. The proposed reduction algorithm also works for the in-plane stress (σ_2) which is obtained at the interior of the laminate, shown in Figure 12. It is interesting to find that the magnitude of σ_2 also decreases when a positive electric potential is applied. This result can be verified by using classical laminate theory (CLT). The fifth result can accurately recover the CLT solution. Thus, the proposed reduction approach is well suited to the angle ply composite laminate.

4. Conclusions

In this study, the reduction of free edge peeling stresses of laminated composite structures using piezoelectric actuators was analyzed by using the extended Kantorovich method. Multiterms of the initially assumed stress functions were involved, and three-dimensional stress distributions were obtained. The accuracy and effectiveness of the present approach were demonstrated by comparing the stresses obtained by ANSYS. The provided results showed excellent agreements with those obtained by FEM, while it showed accuracy and great efficiency in computation. It was found that small terms of trail functions could lead to accurate

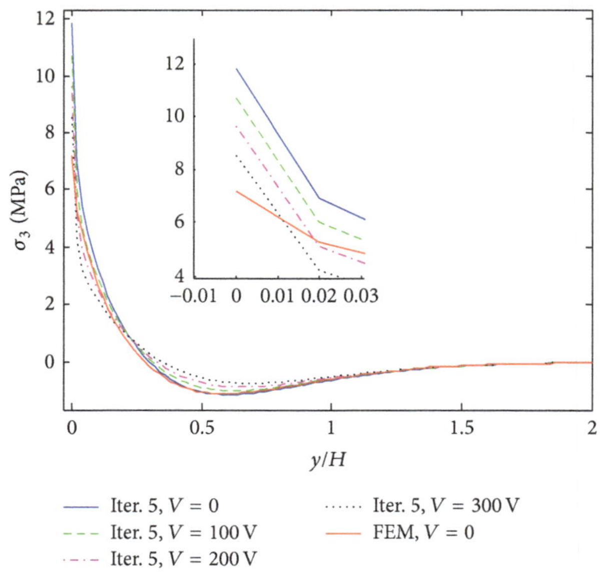

FIGURE 10: σ_3 distribution of [PZT/45_4/-45_4]s laminate at the interface of $z/H = 1/6$, under −0.1% uniaxial strain.

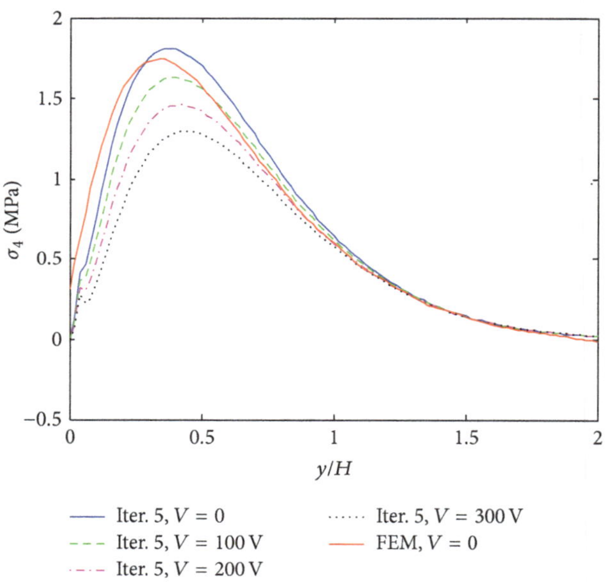

FIGURE 11: σ_4 distribution of [PZT/45_4/-45_4]s laminate at the interface of $z/H = 1/6$, under −0.1% uniaxial strain.

TABLE 3: Comparison of peak stresses for angle ply laminate ([PZT/45_4/-45_4]s) at the 45/−45 layer interface, where $z/H = 1/6$, under −0.1% uniaxial strain, fifth iteration, $n = 8$.

Applied electric potential (V)	σ_3		σ_4	
	(MPa)	Reduction (%)	(MPa)	Reduction (%)
0	11.28	0	1.81	0
100	10.73	9.3%	1.63	10.0%
200	9.63	18.6%	1.46	19.3%
300	8.52	24.5%	1.30	28.2%

solutions. The present method is general and applicable to various lay-up configurations. The peeling stresses generated by mechanical load at the free edges were significantly reduced by applying a proper electric potential. The present method could serve as an efficient tool for reducing the free edge peeling stresses and extending the service life of laminated composite structures.

Conflict of Interests

The authors declare that there is no conflict of interests regarding the publication of this paper.

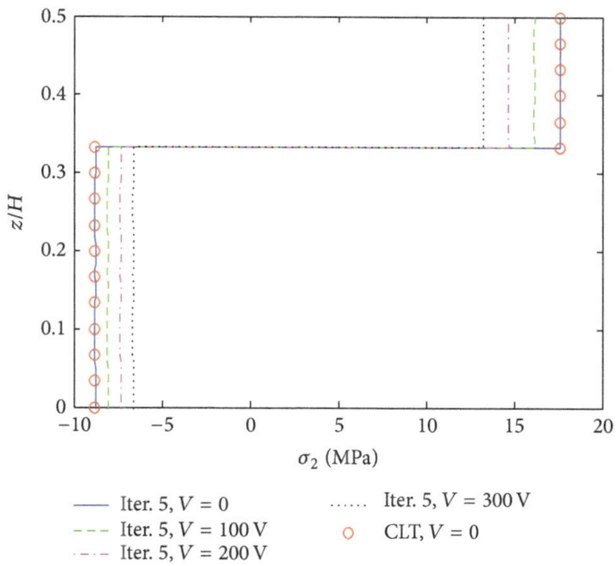

FIGURE 12: σ_2 distribution of $[\mathrm{PZT}/45_4/-45_4]\mathrm{s}$ laminate at the interior, under −0.1% uniaxial strain.

Acknowledgment

This research was supported by Basic Science Research Program through the National Research Foundation of Korea (NRF) funded by the Ministry of Education, Science and Technology (NRF-2011-0021720).

References

[1] C. T. Herakovich, "Mechanics of composites: a historical review," *Mechanics Research Communications*, vol. 41, pp. 1–20, 2012.

[2] R. L. Spilker and S. C. Chou, "Edge effects in symmetric composite laminates: importance of satisfying the traction-free-edge condition," *Journal of Composite Materials*, vol. 14, no. 1, pp. 2–20, 1980.

[3] S. S. Wang and I. Choi, "Boundary-layer effects in composite laminates—part 1: free-edge stress singularities," *Journal of Applied Mechanics*, vol. 49, no. 3, pp. 541–548, 1982.

[4] J. N. Reddy and A. A. Khdeir, "Buckling and vibration of laminated composite plates using various plate theories," *AIAA Journal*, vol. 27, no. 12, pp. 1808–1817, 1989.

[5] S. Konieczny and C. Woźniak, "Corrected 2D-theories for composite plates," *Acta Mechanica*, vol. 103, no. 1–4, pp. 145–155, 1994.

[6] N. J. Pagano and R. B. Pipes, "Some observations on the interlaminar strength of composite laminates," *International Journal of Mechanical Sciences*, vol. 15, no. 8, pp. 679–688, 1973.

[7] N. J. Pagano and L. M. Lackman, "Prevention of delamination of composite laminates," *AIAA Journal*, vol. 13, no. 3, pp. 399–401, 1975.

[8] Y. M. Ghugal and R. P. Shimpi, "A review of refined shear deformation theories of isotropic and anisotropic laminated plates," *Journal of Reinforced Plastics and Composites*, vol. 21, no. 9, pp. 775–813, 2002.

[9] S. Kapuria, P. Kumari, and J. K. Nath, "Efficient modeling of smart piezoelectric composite laminates: a review," *Acta Mechanica*, vol. 214, no. 1-2, pp. 31–48, 2010.

[10] S. Lee and B. D. Youn, "A design and experimental verification methodology for an energy harvester skin structure," *Smart Materials and Structures*, vol. 20, no. 5, Article ID 057001, 2011.

[11] S. Lee, B. D. Youn, and B. C. Jung, "Robust segment-type energy harvester and its application to a wireless sensor," *Smart Materials and Structures*, vol. 18, no. 9, Article ID 095021, 2009.

[12] S. Lee and B. D. Youn, "A new piezoelectric energy harvesting design concept: multimodal energy harvesting skin," *IEEE Transactions on Ultrasonics, Ferroelectrics, and Frequency Control*, vol. 58, no. 3, pp. 629–645, 2011.

[13] P. Qiao and F. Chen, "On the compliance and energy release rate of generically-unified beam-type fracture specimens," *Journal of Composite Materials*, vol. 45, no. 1, pp. 65–101, 2011.

[14] F. Chen and P. Qiao, "Electromechanical behavior of interface deformable piezoelectric bilayer beams," *Journal of Engineering Mechanics*, vol. 136, no. 4, Article ID 007004QEM, pp. 413–428, 2010.

[15] P. Qiao and F. Chen, "Interface crack between two interface deformable piezoelectric layers," *International Journal of Fracture*, vol. 156, no. 2, pp. 185–201, 2009.

[16] F. Chen and P. Qiao, "On the intralaminar and interlaminar stress analysis of adhesive joints in plated beams," *International Journal of Adhesion and Adhesives*, vol. 36, pp. 44–55, 2012.

[17] M. Izadi and M. Tahani, "Analysis of interlaminar stresses in general cross-ply laminates with distributed piezoelectric actuators," *Composite Structures*, vol. 92, no. 3, pp. 757–768, 2010.

[18] G. Flanagan, "An efficient stress function approximation for the free-edge stresses in laminates," *International Journal of Solids and Structures*, vol. 31, no. 7, pp. 941–952, 1994.

[19] H. S. Kim, M. Cho, J. Lee, A. Deheeger, M. Grédiac, and J. D. Mathias, "Three dimensional stress analysis of a composite patch using stress functions," *International Journal of Mechanical Sciences*, vol. 52, no. 12, pp. 1646–1659, 2010.

[20] S. Kapuria, G. P. Dube, and P. C. Dumir, "First-order shear deformation theory solution for a circular piezoelectric composite plate under axisymmetric load," *Smart Materials and Structures*, vol. 12, no. 3, pp. 417–423, 2003.

[21] H. T. Thai and D. H. Choi, "A simple first-order shear deformation theory for the bending and free vibration analysis of functionally graded plates," *Composite Structures*, vol. 101, pp. 332–340, 2013.

[22] J. N. Reddy and C. F. Liu, "A higher-order shear deformation theory of laminated elastic shells," *International Journal of Engineering Science*, vol. 23, no. 3, pp. 319–330, 1985.

[23] A. Chattopadhyay and C. E. Seeley, "A higher order theory for modeling composite laminates with induced strain actuators," *Composites B: Engineering*, vol. 28, no. 3, pp. 243–252, 1997.

[24] A. Ghoshal, H. S. Kim, A. Chattopadhyay, and W. H. Prosser, "Effect of delamination on transient history of smart composite plates," *Finite Elements in Analysis and Design*, vol. 41, no. 9-10, pp. 850–874, 2005.

[25] H. S. Kim, A. Chattopadhyay, and A. Ghoshal, "Dynamic analysis of composite laminates with multiple delamination using improved layerwise theory," *AIAA Journal*, vol. 41, no. 9, pp. 1771–1779, 2003.

[26] H. S. Kim, A. Chattopadhyay, and C. Nam, "Implementation of a coupled thermo-piezoelectric-mechanical model in the LQG controller design for smart composite shells," *Journal of Intelligent Material Systems and Structures*, vol. 13, no. 11, pp. 713–724, 2002.

[27] H. S. Kim, S. Y. Rhee, and M. Cho, "Simple and efficient interlaminar stress analysis of composite laminates with internal ply-drop," *Composite Structures*, vol. 84, no. 1, pp. 73–86, 2008.

[28] H. S. Kim, J. Lee, and M. Cho, "Free-edge interlaminar stress analysis of composite laminates using interface modeling," *Journal of Engineering Mechanics*, vol. 138, no. 8, pp. 973–983, 2012.

[29] D. H. Robbins Jr. and J. N. Reddy, "Modelling of thick composites using a layerwise laminate theory," *International Journal for Numerical Methods in Engineering*, vol. 36, no. 4, pp. 655–677, 1993.

[30] C. Kassapoglou, "Determination of interlaminar stresses in composite laminates under combined loads," *Journal of Reinforced Plastics and Composites*, vol. 9, no. 1, pp. 33–58, 1990.

[31] W. L. Yin, "Free-edge effects in anisotropic laminates under extension bending and twisting—part I: a stress-function-based variational approach," *Journal of Applied Mechanics*, vol. 61, no. 2, pp. 410–415, 1994.

[32] A. D. Kerr and H. Alexander, "An application of the extended Kantorovich method to the stress analysis of a clamped rectangular plate," *Acta Mechanica*, vol. 6, no. 2-3, pp. 180–196, 1968.

[33] A. D. Kerr, "An extended Kantorovich method for the solution of eigenvalue problems," *International Journal of Solids and Structures*, vol. 5, no. 6, pp. 559–572, 1969.

[34] M. Cho and J. Y. Yoon, "Free-edge interlaminar stress analysis of composite laminates by extended Kantorovich method," *AIAA Journal*, vol. 37, no. 5, pp. 656–660, 1999.

[35] M. Cho and H. S. Kim, "Iterative free-edge stress analysis of composite laminates under extension, bending, twisting and thermal loadings," *International Journal of Solids and Structures*, vol. 37, no. 3, pp. 435–459, 2000.

[36] V. Ungbhakorn and P. Singhatanadgid, "Buckling analysis of symmetrically laminated composite plates by the extended Kantorovich method," *Composite Structures*, vol. 73, no. 1, pp. 120–128, 2006.

[37] A. Andakhshideh and M. Tahani, "Interlaminar stresses in general thick rectangular laminated plates under in-plane loads," *Composites B: Engineering*, vol. 47, pp. 58–69, 2013.

[38] B. N. Singh, A. Umrao, K. K. Shukla, and N. Vyas, "Second-order statistics of natural frequencies of smart laminated composite plates with random material properties," *Smart Structures and Systems*, vol. 4, no. 1, pp. 19–34, 2008.

[39] S. G. Lekhnitskii, *Theory of Elasticity of an Anisotropic Body*, Holden-Day, San Francisco, Calif, USA, 1963.

[40] M. Tahani and A. Nosier, "Three-dimensional interlaminar stress analysis at free edges of general cross-ply composite laminates," *Materials and Design*, vol. 24, no. 2, pp. 121–130, 2003.

17

Numerical Manifold Method for the Forced Vibration of Thin Plates during Bending

Ding Jun,[1] Chen Song,[1] Wen Wei-Bin,[2] Luo Shao-Ming,[2] and Huang Xia[1]

[1] *College of Mechanical Engineering, Chongqing University of Technology, Chongqing 400054, China*
[2] *Zhongkai University of Agriculture and Engineering, Guangzhou 510225, China*

Correspondence should be addressed to Ding Jun; dingjunawen@126.com

Academic Editor: Yumin Cheng

A novel numerical manifold method was derived from the cubic B-spline basis function. The new interpolation function is characterized by high-order coordination at the boundary of a manifold element. The linear elastic-dynamic equation used to solve the bending vibration of thin plates was derived according to the principle of minimum instantaneous potential energy. The method for the initialization of the dynamic equation and its solution process were provided. Moreover, the analysis showed that the calculated stiffness matrix exhibited favorable performance. Numerical results showed that the generalized degrees of freedom were significantly fewer and that the calculation accuracy was higher for the manifold method than for the conventional finite element method.

1. Introduction

According to the concepts of topological manifold and differentiable manifold and based on the discontinuous dynamics of DDA block systems, the numerical manifold method (NMM) adopts the finite cover technique to set up a unified calculation form for the finite element method, the DDA, and the analytical method [1]. NMM combines the advantages of the finite element method and DDA and can deal with various types of complex geometric and material boundaries. Thus, NMM has an obvious advantage in the simulation of large geometric deformations and joint deformations, as well as dynamic and static crossings [2]. The application research on NMM mainly focuses on certain fields, such as geotechnical engineering [3, 4] and crack propagation [5–8]. However, NMM has recently been successfully applied to fatigue failure [9], fluid calculations [10], and seepage [11]. In theoretical research, most NMMs adopt shape functions in the finite element method as the weight functions for the manifold method and use the local cover functions of high-order polynomials to improve the continuity and interpolation accuracy of elements [12–14]. Notably, a polynomial form based on the global coordinate system will increase the calculated condition number of the stiffness matrix or even cause a solution failure. For this reason, Lin et al. [15] improved the form of local cover functions to enhance the computational efficiency and interpolation accuracy of elements. However, the number of generalized degrees of freedom remains large. The mesh-less numerical manifold method is more flexible in the selection of mathematical meshes and can avoid problems resulting from mesh generation [16]. However, plenty of matching points have to be used in the calculation, and matrix inversions have to be intensively performed. Thus, the number of calculations remains large, but the integral accuracy is low, which restricts the application [17, 18]. Recently, Wen and Luo [19] studied the numerical manifold method for polygonal elements, which has a stronger adaptability to physical meshes. The newly developed numerical manifold method based on quadratic B-spline interpolation can further improve the continuity of elements and prevent the linear correlation problem in the conventional numerical manifold method [20, 21].

The discrete finite element solution is the most widely adopted solution for structural dynamics. The interpolation function required by the calculation is based on the static node structure, which is inconsistent with actual problems.

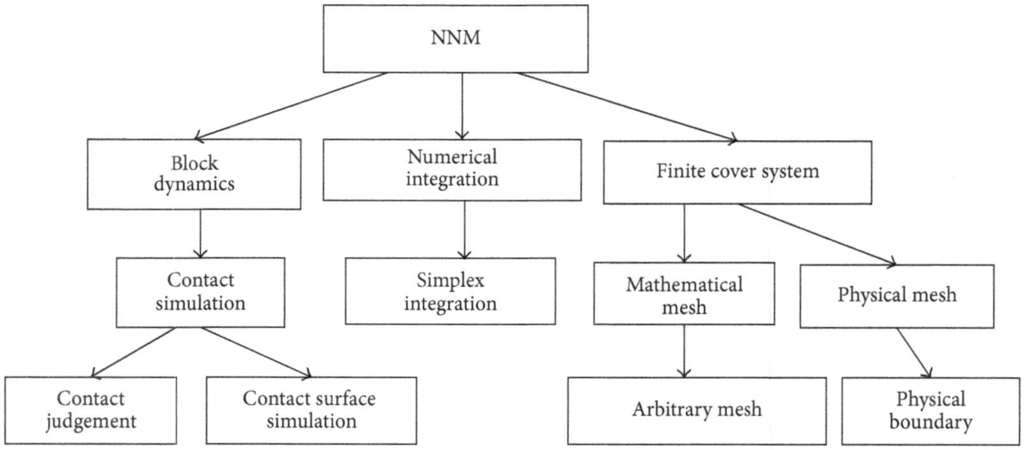

FIGURE 1: Basic structure of the NMM.

The interpolation shape function is related to the element shape and the coordinate system, such that the mesh requirements are relatively high, and the interpolation accuracy is obviously reduced as the structure changes significantly. For the finite elements with mesh structures, the shape functions usually have a low order, which indicates that they cannot meet the requirements of the continuity of displacement by structural deformation. NMM based on finite cover can solve the above problems effectively. The local cover function corresponding to the node displacement of the finite element is no longer relevant to the nodes. If the order of the local cover function is increased, the continuity and interpolation accuracy of the elements can easily be improved. In particular, the manifold method can adopt fixed mathematical meshes that are independent of the physical meshes (material boundary) in setting up the unified interpolation scheme. This method has the advantages of the Lagrangian and Eulerian methods. The manifold method can also avoid problems such as mesh distortion, moving boundary, and migration terms that are difficult to deal with in the finite element and conventional numerical manifold methods [22]. Based on the concept of fixed meshes and considering the deficiency of the existing manifold method in terms of interpolation accuracy, continuity, and solution efficiency, a manifold element with a new cover form and interpolation is constructed. To apply the new element to the solution of structural dynamics, the dynamic theory based on the numerical manifold method is investigated. The numerical manifold scheme for the bending vibration of thin plates is derived, and the final calculation results are comparatively analyzed.

2. Basic Principles of NMM

2.1. Basic Concept of NMM. NMM consists of block dynamics, simplex integration, and a finite cover system. Figure 1 shows the concrete structure [23]. NMM adopts two sets of independent meshes: a mathematical mesh (cover) and a physical mesh (cover). The mathematical mesh is used to define the interpolation accuracy of the solution based on user-defined forms, and the selected mathematical covers can overlap with one another. Consistency with the physical cover is unnecessary, but the mathematical cover has to include the whole material under study. The physical cover is used to describe the geometric properties of the material and the domain of integration, such as the material boundaries, cracks, blocks, and the interfaces of different materials.

In the finite cover system, the intersection of the mathematical cover and the physical mesh is defined as a manifold element; that is, the mathematical cover is subdivided by a physical mesh. The weight function in the manifold method is equivalent to a shape function in the finite element method. If the proper weight function is selected, the weighted summation of the local cover functions can generate the overall cover function of the whole solution domain.

2.2. Generation of Overall Cover Function. For simplicity, the manifold method adopts a mathematical cover with a regular structure or a finite element mesh to generate the desired weight function. The generated weight function should meet the property of element decomposition. Meanwhile, the local cover function corresponding to each mathematical cover is not directly relevant to the nodes. This function can be a constant, linear, and high-order polynomial or take other local series forms. The two-dimensional problem can be taken as an example. The local cover function of any mathematical cover U_i is defined as follows:

$$u_i(x,y);\quad v_i(x,y)\quad (x,y)\in U_i. \tag{1}$$

The local cover function (u_i, v_i) can be expressed as the linear combination of the basis function f_{ij} with a given m-order and the corresponding unknown coefficients $d_{i,2j-1}$ and $d_{i,2j}$. The matrix form is as follows:

$$\begin{Bmatrix} u_i(x,y) \\ v_i(x,y) \end{Bmatrix} = \sum_{j=1}^{m} \begin{bmatrix} f_{i,j}(x,y) & 0 \\ 0 & f_{i,j}(x,y) \end{bmatrix} \times \begin{Bmatrix} d_{i,2j-1} \\ d_{i,2j} \end{Bmatrix} = [\mathbf{f}_i]\,\mathbf{D}_i. \tag{2}$$

Any manifold element e is formed by the intersection of the q cover $U_{e(i)}$ $(i = 1,2,\ldots,q)$ subdivided by a physical cover. The weight function $N_{e(i)}(x, y)$ corresponding to each cover $U_{e(i)}$ is used to solve the overall cover function (u, v) in the domain, which can be represented as follows:

$$\begin{Bmatrix} u(x,y) \\ v(x,y) \end{Bmatrix} = \sum_{i=1}^{q} N_{e(i)}(x,y) \begin{Bmatrix} u_i(x,y) \\ v_i(x,y) \end{Bmatrix} = \sum_{i=1}^{q}\sum_{j=1}^{m} \begin{bmatrix} N_{e(i)}(x,y) f_{e(i),j}(x,y) & 0 \\ 0 & N_{e(i)}(x,y) f_{e(i),j}(x,y) \end{bmatrix} \times \begin{Bmatrix} d_{e(i),2j-1} \\ d_{e(i),2j} \end{Bmatrix} = \sum_{i=1}^{q} N_{e(i)}(x,y) [\mathbf{f}_i] (\mathbf{D}_{e(i)}) = [\mathbf{T}]_e \bullet \mathbf{D}_e. \quad (3)$$

3. Two-Dimensional Manifold Method for Cubic B-Spline Interpolation

3.1. Definition and Property of the Cubic B-Spline Basis Function. The B-spline basis function can be defined in many ways to facilitate computer programming. In this paper, the reverse derivation method is used to define it [24]. $\{x_1, x_2, \ldots, x_p\}$ is assumed to be a monotonically nondecreasing sequence; that is, $x_i \le x_{i+1}$ and $i = 0, 1, \ldots, p-1$, where x_i is the B-spline node. $B_{i,k}(x)$ is the ith k B-spline curve with the following definition:

$$B_{i,0}(x) = \begin{cases} 1 & x_i \le x < x_{i+1} \\ 0 & \text{others} \end{cases}$$

$$B_{i,k}(x) = \frac{x - x_i}{x_{i+k} - x_i} B_{i,k-1}(x) + \frac{x_{i+k+1} - x}{x_{i+k+1} - x_{i+1}} B_{i+1,k-1}(x), \quad (4)$$

where $[x_i, x_{i+1})$ is the node interval of ith B-spline. During the process of reverse derivation, $0/0 = 0$ is required.

B-spline has a local support feature. In other words, when $x \notin [x_i, x_{i+k+1})$, $B_{i,k}(x) = 0$, when $x \in (x_i, x_{i+1})$, $B_{i,k}(x) > 0$. All B-spline curves in the B-spline node interval $[x_i, x_{i+1})$ meet the property of the element decomposition [19]; that is,

$$\sum_i B_{i,3}(x) = 1. \quad (5)$$

According to the definition of the B-spline basis function, the mathematical expression of the cubic B-spline curve expressed in piecewise can be derived as

$$B_{i,3}(x) = \begin{cases} \dfrac{(x-x_i)^3}{(x_{i+3}-x_i)(x_{i+2}-x_i)(x_{i+1}-x_i)} & x \in [x_i, x_{i+1}) \\[2ex] \dfrac{(x-x_i)^3}{(x_{i+3}-x_i)(x_{i+2}-x_i)(x_{i+1}-x_i)} + \dfrac{(x-x_{i+1})^3 (x_{i+4}-x_i)}{(x_{i+1}-x_i)(x_{i+1}-x_{i+2})(x_{i+1}-x_{i+3})(x_{i+1}-x_{i+4})} & x \in [x_{i+1}, x_{i+2}) \\[2ex] \dfrac{(x_{i+3}-x)^3 (x_{i+4}-x_i)}{(x_{i+3}-x_i)(x_{i+3}-x_{i+1})(x_{i+3}-x_{i+2})(x_{i+3}-x_{i+4})} + \dfrac{(x_{i+4}-x)^3}{(x_{i+4}-x_{i+1})(x_{i+4}-x_{i+2})(x_{i+4}-x_{i+3})} & x \in [x_{i+2}, x_{i+3}) \\[2ex] \dfrac{(x_{i+4}-x)^3}{(x_{i+4}-x_{i+1})(x_{i+4}-x_{i+2})(x_{i+4}-x_{i+3})} & x \in [x_{i+3}, x_{i+4}] \\[2ex] 0 & x \notin [x_i, x_{i+4}]. \end{cases} \quad (6)$$

Figure 2 shows the cubic B-spline curves when the B-spline nodes are evenly distributed. Formula (6) can be used to solve all the B-spline curves in the figure. Figure 3 shows the B-spline curves within the B-spline interval $[x_i, x_{i+1})$ or within the i element. The mathematical expression is as follows:

$$B_{i-3,3}(x) = \frac{(x_{i+1}-x)^3}{(x_{i+1}-x_{i-2})(x_{i+1}-x_{i-1})(x_{i+1}-x_i)},$$

$$B_{i-2,3}(x) = \frac{(x_{i+2}-x)^3}{(x_{i+2}-x_{i-1})(x_{i+2}-x_i)(x_{i+2}-x_{i+1})} + \frac{(x_{i+1}-x)^3 (x_{i+2}-x_{i-2})}{(x_{i+1}-x_{i-2})(x_{i+1}-x_{i-1})(x_{i+1}-x_i)(x_{i+1}-x_{i+2})},$$

$$B_{i-1,3}(x) = \frac{(x-x_{i-1})^3}{(x_{i+2}-x_{i-1})(x_{i+1}-x_{i-1})(x_i-x_{i-1})} + \frac{(x-x_i)^3 (x_{i+3}-x_{i-1})}{(x_i-x_{i-1})(x_i-x_{i+1})(x_i-x_{i+2})(x_i-x_{i+3})},$$

$$B_{i,3}(x) = \frac{(x-x_i)^3}{(x_{i+3}-x_i)(x_{i+2}-x_i)(x_{i+1}-x_i)}. \quad (7)$$

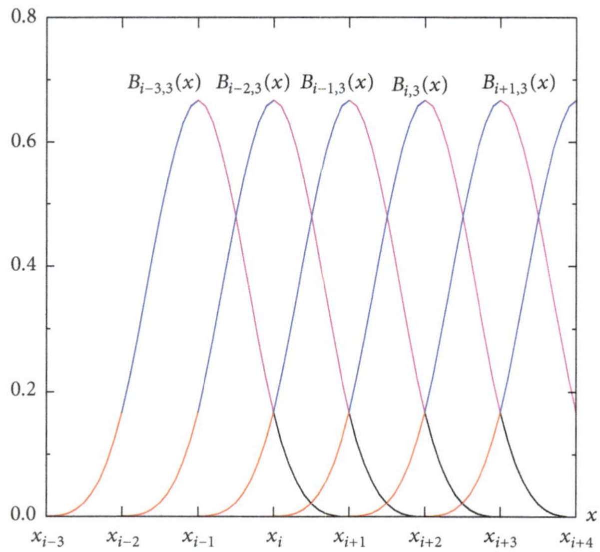

FIGURE 2: Cubic B-spline curve in piecewise form.

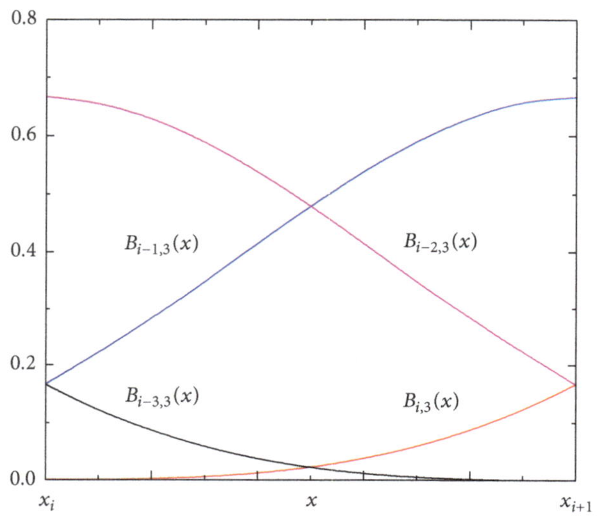

FIGURE 3: Cubic B-spline curve within the element i.

3.2. Two-Dimensional Manifold Element of the Cubic B-Spline. A method similar to bilinear interpolation is adopted to construct the B-spline overall interpolation function (as shown in Formula (8)) within the rectangular element domain of $x \in (x_i, x_{i+1})$ and $y \in (y_j, y_{j+1})$:

$$W_{i,j}(x,y) = \left[B_{i-3,3}(x) \quad B_{i-2,3}(x) \quad B_{i-1,3}(x) \quad B_{i,3}(x)\right] \cdot \begin{bmatrix} d_{i,j} & d_{i,j+1} & d_{i,j+2} & d_{i,j+3} \\ d_{i+1,j} & d_{i+1,j+1} & d_{i+1,j+2} & d_{i+1,j+3} \\ d_{i+2,j} & d_{i+2,j+1} & d_{i+2,j+2} & d_{i+2,j+3} \\ d_{i+3,j} & d_{i+3,j+1} & d_{i+3,j+2} & d_{i+3,j+3} \end{bmatrix} \times \begin{bmatrix} B_{j-3,3}(y) \\ B_{j-2,3}(y) \\ B_{j-1,3}(y) \\ B_{j,3}(y) \end{bmatrix} = \sum_i \sum_j B_{i-3,3}(x) B_{j-3,3}(y) d_{i,j} = \mathbf{T}_{i,j} \cdot \mathbf{D}^{i,j}. \quad (8)$$

In the formula, $\mathbf{D}^{i,j}$ is the generalized displacement vector of the manifold element (i, j), such that

$$\mathbf{D}^{i,j} = \left[d_{i,j} \;\; d_{i,j+1} \;\; \cdots \;\; d_{i+3,j+3}\right]^T,$$

$$\mathbf{T}_{i,j} = \left[B_{i-3,3}(x) B_{j-3,3}(y) \;\; B_{i-3,3}(x) B_{j-2,3}(y) \;\; \cdots \;\; B_{i,3}(x) B_{j,3}(y)\right]. \quad (9)$$

Each term of $\mathbf{T}_{i,j}$ is $(\mathbf{T}_{i,j})_k$, such that

$$\sum_{k=1}^{16} \left(\mathbf{T}_{i,j}\right)_k = \sum_{k_1=0}^{3} B_{i-k_1,3}(x) \sum_{k_2=0}^{3} B_{i-k_2,3}(y) = 1. \quad (10)$$

In the same way, cubic B-spline curves of $B_{j-3,3}(y)$, $B_{j-2,3}(y)$, $B_{j-1,3}(y)$, and $B_{j-2,3}(y)$ within $y \in (y_j, y_{j+1})$ can be obtained in piecewise form. The calculated B-spline functions are substituted into Formula (8) to obtain the final overall cover function of the manifold element.

We suppose that the numbers of manifold elements in the directions of x, y are, respectively, N_1, N_2, and the corresponding element nodes are $x_1, x_2, \ldots, x_{N_1+1}$. Through Formulas (7), a certain number of virtual nodes have to be added to construct the B-spline basis functions. We suppose that the added virtual nodes are x_{-1}, x_0, x_{N_1+2}, x_{N_1+3}; y_{-1}, y_0, y_{N_2+2}, y_{N_2+3}. The selected nodes only need to have the monotonically nondecreasing property.

Based on the continuity of the B-spline surface, the spline surface formed by k uniform B-spline basis functions has a $k-1$-order continuity [25] of x, y in the element boundary. The above manifold element adopts the model of the generalized node distribution, which is the same as that on the B-spline surface, such that second-order continuity exists in the element boundary. Figure 3 shows that the B-spline has a desirable property of local support, and its values are small within the interval to ensure the reasonability of the calculated condition number of the stiffness matrix. The number of generalized degrees of freedom required by the calculation is $(N_1+3)\times(N_2+3)$, which is significantly reduced compared with that of the conventional manifold method. Figure 2 shows that the B-spline basis functions have no δ property function in the nodes. Thus, these functions can reflect the dynamic property of the structure.

4. Manifold Scheme of the Bending Vibration of Simply Supported Thin Plates

4.1. Dynamic Equation of the Manifold Element Based on the Principle of Minimum Instantaneous Potential Energy.

Suppose that a thin plate has a thickness of h, a density of ρ, an elasticity modulus of E, a bending deflection (displacement) of $W(x,y,t)$, an actuating force in the unit area on the plate surface of $q(x,y,t)$, a normal direction for the simply supported boundary of n, and corresponding fixed deflection and bending moments of $\overline{W}_n$ and $\overline{M}_n$. According to the Kirchhoff plate theory [26] and based on the processing method of the boundary condition of the manifold element [1], the principle of the minimum instantaneous potential energy of the numerical manifold of the bending vibration of thin plates can be derived as

$$\begin{aligned}\prod\nolimits_G &= \iint_\Omega \frac{1}{2}\{\boldsymbol{\kappa}\}^T[\mathbf{D}_f]\{\boldsymbol{\kappa}\}\,d\Omega \\ &- \iint_\Omega \left(q - c\dot{W} - \rho h\ddot{W}\right) W d\Omega \\ &+ \frac{1}{2}\int_{S^1} \beta_2\left(M_n - \overline{M}_n\right)^T\left(M_n - \overline{M}_n\right) d\Gamma \\ &+ \frac{1}{2}\int_{S^1} \beta_1\left(W - \overline{W}\right)^T\left(W - \overline{W}\right) d\Gamma,\end{aligned} \tag{11}$$

where Ω is the area of the thin plate, S_1 is the simply supported boundary, c is the damping coefficient, and β_1 and β_2 are the stiffness coefficients of the penalty function. $[\mathbf{D}_f]$ is the bending stiffness matrix, and $\{\boldsymbol{\kappa}\}$ is the curvature and torsion of the elastic surface represented by vectors in the directions of x, y that meets the following equations:

$$\begin{aligned}\{\boldsymbol{\kappa}\} &= \begin{bmatrix}\kappa_x & \kappa_y & \kappa_{xy}\end{bmatrix}^T = \begin{bmatrix}-\frac{\partial^2}{\partial x^2} & -\frac{\partial^2}{\partial y^2} & -2\frac{\partial^2}{\partial x \partial y}\end{bmatrix}^T W \\ &= \{\mathbf{G}_2\} W,\end{aligned} \tag{12}$$

$$[\mathbf{D}_f] = \frac{Eh^3}{12(1-\mu^2)}\begin{bmatrix}1 & \mu & 0 \\ \mu & 1 & 0 \\ 0 & 0 & \frac{(1-\mu)}{2}\end{bmatrix}. \tag{13}$$

M_n in Formula (11) can be expressed as the following equation by the displacement function, where θ is the angle from the x direction to the n direction:

$$\begin{aligned}M_n &= \left[(\cos\theta)^2 \quad (\sin\theta)^2 \quad 2\cos\theta\sin\theta\right] \\ &\cdot [\mathbf{D}_f]\{\mathbf{G}_2\} W = \{\mathbf{G}_1\} W.\end{aligned} \tag{14}$$

The manifold element of Formula (8) is substituted into Formula (11) along with Formulas (12) and (14) to obtain the discretization form of the minimum instantaneous potential energy:

$$\begin{aligned}\prod\nolimits_G^* = \sum_{i=1}^{N_1}\sum_{j=1}^{N_2}\Bigg\{&\iint_{\Omega_{i,j}} \frac{1}{2}\left(\{\mathbf{G}_2\}\mathbf{T}_{i,j}\mathbf{D}^{i,j}\right)^T[\mathbf{D}_f]\{\mathbf{G}_2\}\mathbf{T}_{i,j}\mathbf{D}^{i,j} d\Omega \\ &+ \iint_{\Omega_{i,j}} \left(\mathbf{T}_{i,j}\mathbf{D}^{i,j}\right)^T \\ &\times \left(q - c\mathbf{T}_{i,j}\dot{\mathbf{D}}^{i,j} - \rho h\mathbf{T}_{i,j}\ddot{\mathbf{D}}^{i,j}\right) d\Omega \\ &+ \frac{1}{2}\int_{S^1_{i,j}} \beta_2\left(\{\mathbf{G}_1\}\mathbf{T}_{i,j}\mathbf{D}^{i,j} - \overline{M}_n\right)^T \\ &\times \left(\{\mathbf{G}_1\}\mathbf{T}_{i,j}\mathbf{D}^{i,j} - \overline{M}_n\right) d\Gamma \\ &+ \frac{1}{2}\int_{S^1_{i,j}} \beta_1\left(\mathbf{T}_{i,j}\mathbf{D}^{i,j} - \overline{W}\right)^T\left(\mathbf{T}_{i,j}\mathbf{D}^{i,j} - \overline{W}\right) d\Gamma\Bigg\}.\end{aligned} \tag{15}$$

The first variation is taken from Formula (15) that is the instantaneous variation that satisfies $\delta\dot{\mathbf{D}}^{i,j} = 0$ and $\delta\ddot{\mathbf{D}}^{i,j} = 0$. Therefore, the dynamic equation for any manifold element (i, j) is given by

$$\mathbf{M}_{i,j}\ddot{\mathbf{D}}^{i,j} + \mathbf{C}_{i,j}\dot{\mathbf{D}}^{i,j} + \mathbf{K}_{i,j}\mathbf{D}^{i,j} = \mathbf{R}_{i,j}, \tag{16}$$

where

$$\begin{aligned}\mathbf{M}_{i,j} &= \iint_{\Omega_{i,j}} \rho h\left(\mathbf{T}_{i,j}\right)^T\mathbf{T}_{i,j} d\Omega, \\ \mathbf{C}_{i,j} &= \iint_{\Omega_{i,j}} c\left(\mathbf{T}_{i,j}\right)^T\mathbf{T}_{i,j} d\Omega, \\ \mathbf{K}_{i,j} &= \iint_{\Omega_{i,j}} \left(\{\mathbf{G}_2\}\mathbf{T}_{i,j}\right)^T[\mathbf{D}_f]\{\mathbf{G}_2\}\mathbf{T}_{i,j} d\Omega \\ &+ \int_{S^1_{i,j}} \left(\beta_2\left(\{\mathbf{G}_1\}\mathbf{T}_{i,j}\right)^T\{\mathbf{G}_2\}\mathbf{T}_{i,j} + \beta_1\left(\mathbf{T}_{i,j}\right)^T\mathbf{T}_{i,j}\right) d\Gamma, \\ \mathbf{R}_{i,j} &= \iint_{\Omega_{i,j}} \left(\mathbf{T}_{i,j}\right)^T q d\Omega \\ &+ \int_{S^1_{i,j}} \left(\beta_1\overline{W}\left(\mathbf{T}_{i,j}\right)^T + \beta_2\overline{M}_n\left(\{\mathbf{G}_1\}\mathbf{T}_{i,j}\right)^T\right) d\Gamma.\end{aligned} \tag{17}$$

To facilitate the integration of the stiffness matrix, the transformation matrix $\mathbf{G}_{i,j}$ is introduced and can be defined by the following formula, where $\mathbf{D}$ is the total freedom of degree vector:

$$\mathbf{G}_{i,j}\mathbf{D}^{i,j} = \mathbf{D}. \tag{18}$$

The transformation matrix $\mathbf{G}_{i,j}$ corresponding to Formula (8) is

$$\mathbf{G}_{i,j} = \begin{bmatrix}\mathbf{0}_{4\times((i-1)\cdot(N_1+3)+j-1)} & \mathbf{I}_{4\times4} & \mathbf{0} \\ \mathbf{0}_{4\times(i\cdot(N_1+3)+j-1)} & \mathbf{I}_{4\times4} & \mathbf{0} \\ \mathbf{0}_{4\times((i+1)\cdot(N_1+3)+j-1)} & \mathbf{I}_{4\times4} & \mathbf{0} \\ \mathbf{0}_{4\times((i+2)\cdot(N_1+3)+j-1)} & \mathbf{I}_{4\times4} & \mathbf{0}\end{bmatrix}_{16\times((N_1+3)\cdot(N_2+3))}, \tag{19}$$

where $\mathbf{I}_{4\times4}$ is the identity matrix.

Formula (18) is used to integrate Formula (16) into the overall dynamic equation

$$\mathbf{M}\ddot{\mathbf{D}} + \mathbf{C}\dot{\mathbf{D}} + \mathbf{K}\mathbf{D} = \mathbf{R}, \tag{20}$$

where

$$\mathbf{M} = \sum_{i=1}^{N_1}\sum_{j=1}^{N_2} \mathbf{G}_{i,j}^T \mathbf{M}_{i,j} \mathbf{G}_{i,j}, \qquad \mathbf{C} = \sum_{i=1}^{N_1}\sum_{j=1}^{N_2} \mathbf{G}_{i,j}^T \mathbf{C}_{i,j} \mathbf{G}_{i,j}, \tag{21}$$

$$\mathbf{K} = \sum_{i=1}^{N_1}\sum_{j=1}^{N_2} \mathbf{G}_{i,j}^T \mathbf{K}_{i,j} \mathbf{G}_{i,j}, \qquad \mathbf{R} = \sum_{i=1}^{N_1}\sum_{j=1}^{N_2} \mathbf{G}_{i,j}^T \mathbf{R}_{i,j}. \tag{22}$$

It can be easily observed that (20) has an analogous form as the overall dynamic equation generated by FEM. Meanwhile, considering that **M** and **K** are symmetric positive matrices, we can solve the natural frequencies of the system by use of the techniques employed in FEM.

4.2. Initialization of the Dynamic Equation. Unlike the finite element method, $\mathbf{D}$, $\dot{\mathbf{D}}$, and $\ddot{\mathbf{D}}$ in the manifold method are no longer relevant to the nodes and thus need to be initialized. If the manifold element to be calculated is $N_1 \times N_2$ and the number of degrees of freedom is $(N_1 + 3) \times (N_2 + 3)$, the noncoincident points need to be regenerated in the directions of x, y to obtain the final nodes (x_i, y_j), $i = 1, 2, \ldots, N_1 + 3$, and $j = 1, 2, \ldots, N_2 + 3$. If the manifold element corresponding to node (x_i, y_j) is (I_i, J_j) and the real initial displacement is initialized with the velocity of $W(x_i, y_j, 0)$ and $\dot{W}(x_i, y_j, 0)$, we obtain

$$\begin{aligned} \mathbf{T}_0 \mathbf{D}_0 &= \mathbf{W}_0, \\ \mathbf{T}_0 \dot{\mathbf{D}}_0 &= \dot{\mathbf{W}}_0, \end{aligned} \tag{23}$$

where

$$\mathbf{T}_0 = \begin{bmatrix} \left(\mathbf{T}_{I_1,J_1}\right)\big|_{(x,y)=(x_1,y_2)} \mathbf{G}_{I_1,J_1} \\ \vdots \\ \left(\mathbf{T}_{I_{N_1+2},J_{N_2+2}}\right)\big|_{(x,y)=(x_{N_1+2},y_{N_2+2})} \mathbf{G}_{I_{N_1+2},J_{N_2+2}} \\ \left(\mathbf{T}_{I_{N_1+3},J_{N_2+3}}\right)\big|_{(x,y)=(x_{N_1+3},y_{N_2+3})} \mathbf{G}_{I_{N_1+3},J_{N_2+3}} \end{bmatrix}_{(N_1+3)\times(N_2+3)},$$

$$\mathbf{W}_0 = \left[W(x_1, y_1, 0) \ \cdots \ W\left(x_{N_1+2}, y_{N_2+2}, 0\right) \ W\left(x_{N_1+3}, y_{N_2+3}, 0\right)\right]^T,$$

$$\dot{\mathbf{W}}_0 = \left[\dot{W}(x_1, y_1, 0) \ \cdots \ \dot{W}\left(x_{N_1+2}, y_{N_2+2}, 0\right) \ \dot{W}\left(x_{N_1+3}, y_{N_2+3}, 0\right)\right]^T. \tag{24}$$

Formulas (23) can be used to obtain the generalized displacement $\mathbf{D}_0$ and the generalized velocity $\dot{\mathbf{D}}_0$, which are substituted into Formula (20) to obtain the generalized initial acceleration:

$$\ddot{\mathbf{D}}_0 = \mathbf{M}^{-1}\left(\mathbf{R} - \mathbf{K}\mathbf{D}_0 - \mathbf{C}\dot{\mathbf{D}}_0\right). \tag{25}$$

4.3. Integration Scheme of the Dynamic Equation. When the domain of area integration is a regular rectangular element, Gaussian integration can be adopted. When the domain is an irregular geometric configuration, it can be subdivided into several triangular elements. Then, the simplex integration or Hammer integration is used to obtain the calculation results with sufficient accuracy [27]. When the boundary integration is a regular curve, Gaussian integration can be directly performed. If the boundary integration is a complex curve, it should be subdivided into several, straight-line segments before Gaussian integration is performed.

For time integration, the Wilson-θ method can be selected in this paper. Formula (20) is then converted into the linear equation T at time $t + \theta\Delta t$. Consider

$$\widehat{\mathbf{K}}\mathbf{D}_{t+\theta\Delta t} = \widehat{\mathbf{R}}_{t+\theta\Delta t}, \tag{26}$$

where

$$\begin{aligned} \widehat{\mathbf{R}}_{t+\theta\Delta t} &= \mathbf{R}_{t+\theta\Delta t} + \mathbf{M}\left(\frac{6}{(\theta\Delta t)^2}\mathbf{D}_t + \frac{6}{\theta\Delta t}\dot{\mathbf{D}}_t + 2\ddot{\mathbf{D}}_t\right) \\ &\quad + \mathbf{C}\left(\frac{3}{\theta\Delta t}\mathbf{D}_t + 2\dot{\mathbf{D}}_t + \frac{\theta\Delta t}{2}\ddot{\mathbf{D}}_t\right), \\ \widehat{\mathbf{K}} &= \frac{6}{(\theta\Delta t)^2}\mathbf{M} + \frac{3}{\theta\Delta t}\mathbf{C} + \mathbf{K}. \end{aligned} \tag{27}$$

After solving $\mathbf{D}_{t+\theta\Delta t}$, the generalized total displacement, velocity, and acceleration are solved:

$$\begin{aligned} \ddot{\mathbf{D}}_{t+\Delta t} &= \frac{6}{\theta^3\Delta t^2}\left(\mathbf{D}_{t+\theta\Delta t} - \mathbf{D}_t\right) - \frac{6}{\theta^2\Delta t}\dot{\mathbf{D}}_t + \left(1 - \frac{3}{\theta}\right)\ddot{\mathbf{D}}_t, \\ \dot{\mathbf{D}}_{t+\Delta t} &= \dot{\mathbf{D}}_t + \frac{\Delta t}{2}\left(\ddot{\mathbf{D}}_{t+\Delta t} + \ddot{\mathbf{D}}_t\right), \\ \mathbf{D}_{t+\Delta t} &= \mathbf{D}_t + \Delta t\dot{\mathbf{D}}_t + \frac{\Delta t^2}{6}\left(\ddot{\mathbf{D}}_{t+\Delta t} + 2\ddot{\mathbf{D}}_t\right). \end{aligned} \tag{28}$$

Formulas (18) and (8) are used to obtain the function of the displacement, velocity, and acceleration of any manifold element (i, j) at time $t + \Delta t$:

$$\begin{aligned} W_{i,j} &= \mathbf{T}_{i,j}\mathbf{G}_{i,j}\mathbf{D}_{t+\Delta t}^{i,j}, \\ \dot{W}_{i,j} &= \mathbf{T}_{i,j}\mathbf{G}_{i,j}\dot{\mathbf{D}}_{t+\Delta t}^{i,j}, \\ \ddot{W}_{i,j} &= \mathbf{T}_{i,j}\mathbf{G}_{i,j}\ddot{\mathbf{D}}_{t+\Delta t}^{i,j}. \end{aligned} \tag{29}$$

5. Calculation Examples and Numerical Analysis

5.1. Calculation Example. As shown in Figure 4, the thin plate that is simply supported on the four edges has a length of $a = 6.3$ m, a width of $b = 2.1$ m, and a thickness of $h = 1 \times 10^{-2}$ m. The transverse dynamic load is $q(x, y, t) = Q\sin(\alpha t)$ Pa, $Q = 1 \times 10^4$ Pa, and $\alpha = 10$. In addition, the elasticity modulus is $E = 210$ GPa, the Poisson's ratio is $\mu = 0.3$, the stiffness

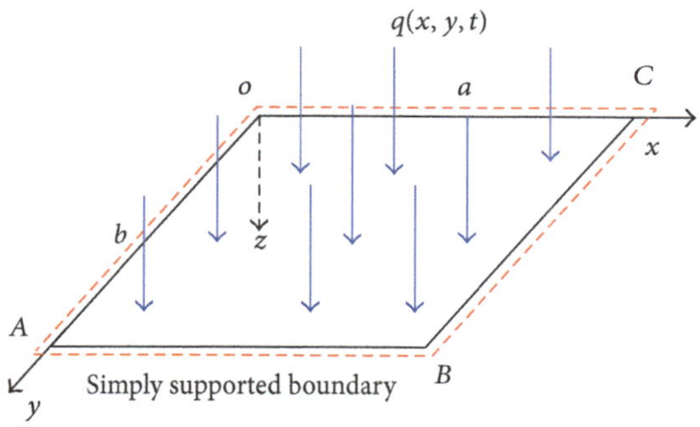

FIGURE 4: Scheme and configuration of the thin plate.

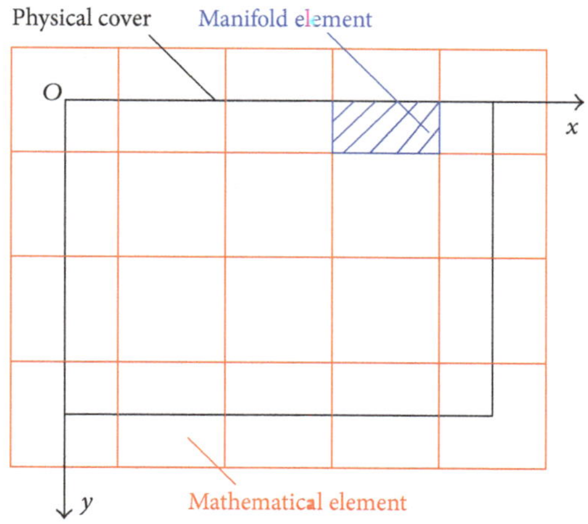

FIGURE 5: Cover system and manifold element of the thin plate.

coefficient of the penalty function is $\beta_i = 1.0 \times 10^9$, the material density is $\rho = 4\times10^4$ kg/m^3, the damping coefficient is $c = 0$, and the coefficient of the Wilson-θ method is $\theta = 1.4$. Different time steps Δt are selected for the calculation. As shown in Figure 5, the fixed rectangular mesh is used as the mathematical cover. The shaded area in the figure denotes any calculated manifold element.

With the theory illustrated in Section 4.1, the generalized instantaneous potential energy for this example can be expressed as

$$\begin{aligned}\prod_G &= \int_0^b\int_0^a \frac{1}{2}\{\kappa\}^T\left[\mathbf{D}_f\right]\{\kappa\}\,dx\,dy \\ &- \int_0^b\int_0^a \left(q(x,y,t) - c\dot{W} - \rho h\ddot{W}\right)W dx\,dy \\ &+ \frac{1}{2}\int_0^a \beta_1\left(W_{y=0}\right)^T\left(W_{y=0}\right)dx \\ &+ \frac{1}{2}\int_0^a \beta_2\left(W_{y=b}\right)^T\left(W_{y=b}\right)dx \\ &+ \frac{1}{2}\int_0^b \beta_3\left(W_{x=0}\right)^T\left(W_{x=0}\right)dy \\ &+ \frac{1}{2}\int_0^b \beta_4\left(W_{x=a}\right)^T\left(W_{x=a}\right)dy,\end{aligned} \tag{30}$$

where the latter four terms are the generalized energy form of boundary conditions.

After a large amount of discretization calculation, the matrices **M**, **C**, **K**, and **R** can be calculated by

$$\begin{aligned}\mathbf{M} &= \sum_{i=1}^{N_1}\sum_{j=1}^{N_2}\left(\mathbf{C}_{i,j}\right)^T\left\{\iint_{\Omega_{i,j}}\rho h\cdot\left(\mathbf{T}_{i,j}\right)^T\mathbf{T}_{i,j}dx\,dy\right\}\mathbf{C}_{i,j}, \\ \mathbf{C} &= \sum_{i=1}^{N_1}\sum_{j=1}^{N_2}\left(\mathbf{C}_{i,j}\right)^T\left\{\iint_{\Omega_{i,j}}c\cdot\left(\mathbf{T}_{i,j}\right)^T\mathbf{T}_{i,j}dx\,dy\right\}\mathbf{C}_{i,j}, \\ \mathbf{K} &= \sum_{i=1}^{N_1}\sum_{j=1}^{N_2}\left(\mathbf{C}_{i,j}\right)^T\left\{\iint_{\Omega_{i,j}}\left(\{\mathbf{G}_2\}\mathbf{T}_{i,j}\right)^T\cdot\left[\mathbf{D}_f\right]\right. \\ &\quad\left.\cdot\left(\{\mathbf{G}_2\}\mathbf{T}_{i,j}\right)dx\,dy\right\}\mathbf{C}_{i,j} \\ &+ \sum_{i=1}^{N_1}\left(\mathbf{C}_{i,0}\right)^T\beta_1 \\ &\quad\times\left\{\int_{S_{i,0}^1}\left(\mathbf{T}_{i,0}\big|_{y=0}\right)^T\left(\mathbf{T}_{i,0}\big|_{y=0}\right)dx\right. \\ &\quad\left.+\int_{S_{i,0}^1}\beta_2\left(\mathbf{T}_{i,0}\big|_{y=b}\right)^T\left(\mathbf{T}_{i,0}\big|_{y=b}\right)dx\right\}\mathbf{C}_{i,0}, \\ &+ \sum_{j=1}^{N_2}\left(\mathbf{C}_{0,j}\right)^T \\ &\quad\times\left\{\int_{S_{0,j}^1}\beta_3\left(\mathbf{T}_{0,j}\big|_{x=0}\right)^T\left(\mathbf{T}_{0,j}\big|_{x=0}\right)dy\right. \\ &\quad\left.+\int_{S_{0,j}^1}\beta_4\left(\mathbf{T}_{0,j}\big|_{x=a}\right)^T\left(\mathbf{T}_{0,j}\big|_{x=a}\right)dy\right\}\mathbf{C}_{0,j}, \\ \mathbf{R} &= \sum_{i=1}^{N_1}\sum_{j=1}^{N_2}\left(\mathbf{C}_{i,j}\right)^T\left\{\iint_{\Omega_{i,j}}q(x,y,t)\cdot\left(\mathbf{T}_{i,j}\right)^T dx\,dy\right\},\end{aligned} \tag{31}$$

where $\{\mathbf{G}_2\}$ and $[\mathbf{D}_f]$ are given in (12) and (13), respectively.

The frequency and the displacement for this case are as follows:

$$\begin{aligned}w_{ij} &= \pi^2\left(\frac{i^2}{a^2}+\frac{j^2}{b^2}\right)\sqrt{\frac{K_0}{\rho h}}, \\ W(x,y,t) &= \sum_{i=1,3,5,\ldots}\sum_{j=1,3,5,\ldots}\frac{16Q}{K_0\pi^2 ij\left(\pi^4\left((i^2/a^2)+(j^2/b^2)\right)^2-(\rho h/K_0)\,\alpha^2\right)} \\ &\quad\cdot\sin\frac{i\pi x}{a}\sin\frac{j\pi y}{b}\left(\sin(\alpha t)-\frac{\alpha}{w_{ij}}\sin\left(w_{ij}t\right)\right).\end{aligned} \tag{32}$$

TABLE 1: First five natural frequencies and their relative errors of manifold calculation.

Element	First frequency		Second frequency		Third frequency		Fourth frequency		Fifth frequency	
	w'_{11}/Hz	η	w'_{21}/Hz	η	w'_{31}/Hz	η	w'_{41}/Hz	η	w'_{51}/Hz	η
3 × 3	17.2564	0.084	22.5501	0.60	65.7306	111.79	70.9070	64.50		
5 × 5	17.2427	0.0046	22.4204	0.026	31.2165	0.58	44.9846	4.36	63.9337	9.06
7 × 7	17.2413	0.0035	22.4141	0.0022	31.0619	0.085	43.3865	0.65	60.3066	2.87
9 × 9	17.2411	0.0046	22.4130	0.0071	31.0406	0.016	43.1758	0.16	59.0368	0.71
11 × 11	17.2412	0.0041	22.4126	0.0089	31.0348	0.0026	43.1275	0.052	58.7638	0.24
Theoretical solution	17.2419	0.00	22.4146	0.00	31.0356	0.00	43.1049	0.00	58.6227	0.00

TABLE 2: Velocity and acceleration of the manifold calculation and their relative errors.

Elements	$x = a/3$, $y = b/3$, $\Delta t = 5 \times 10^{-3}$ s, $t = 50\Delta t$			
	W'_t ($\times 10^{-1}$ m/s)	η_t	W'_{tt} (m/s^2)	η_{tt}
3 × 3	6.2981	0.57	−33.705	1.81
5 × 5	6.2088	1.98	−33.693	1.85
7 × 7	6.2127	1.92	−34.199	0.38
9 × 9	6.2228	1.76	−34.437	0.32
Theoretical solution	6.3340	0.00	−34.328	0.00

5.2. Numerical Analysis. Table 1 shows the frequency of the manifold method, the accuracy of which will improve along with the increasing density of meshes. The relative error of frequency in the table is defined as $\eta = (|w - w'|/|w|) \times 100(\%)$. When the 11 × 11 mesh is adopted, the first-order frequency error is only 0.0041%, whereas the fifth-order frequency error is 0.24%.

W', W'_t, and W'_{tt} are defined as the numerical displacement, velocity, and acceleration, respectively. In the same way, W'_{xx}, W'_{xy}, and W'_{yy} are the partial derivatives of numerical displacement. The corresponding relative errors have the same definitions as in Table 1. Table 2 shows the velocities and acceleration of the given points at certain time point. The accuracy of the numerical solution is high. Considering that the given interpolation functions has the same continuity and coordination for x and y, Table 3 only shows part of the partial derivatives of displacement. The numerical solution shows that both stress and strain in the manifold method have good accuracy.

Figures 6, 7, and 8 are, respectively, the displacements, velocities, and acceleration at the points of $x = a/3$ and $y = b/3$ for each moment. With the reduction in time step Δt, the accuracy of the numerical solution improves. In particular, with the increase in time steps, smaller step lengths of Δt can significantly improve the accuracy of the numerical solution. Figure 9 shows some partial derivations at the points of $x = a/3$ and $y = b/3$. The figure presents the calculated values of the first several periods. As the meshes become denser, the calculation accuracy significantly improves. Therefore, the numerical solution is consistent with the theoretical solution.

Since the manifold method in this paper adopts an overall interpolation approximate, the integrated error norm has to be introduced to reflect the interpolation accuracy fully. $\varepsilon_0 = (\sqrt{\sum_{k=1}^{N_a} (W' - W)^2} / \sqrt{\sum_{k=1}^{N_a} (W)^2}) \times 100(\%)$ is defined as the integrated error of displacement, where N_a is the number of discrete points equally distributed in the x and y directions within the thin plate. During calculation, N_a is taken as 11 × 11 discrete points within the thin plate. The integrated errors of the other quantities in Tables 4 and 5 are defined by a similar method. Table 4 shows the bilinear Hermite finite element. The mathematical software MATLAB is strongly recommended for the calculation, similar to the manifold method. By comparing Tables 4 and 5, the overall degrees of freedom for the manifold method involving the same number of elements are significantly smaller than those for the finite element method, but its calculation accuracy is higher than that of the latter. Furthermore, Table 6 gives the global errors from the cubic B-spline finite element. By comparing Tables 6 and 5, we can safely conclude the cubic B-spline manifold element produces more accurate results than cubic B-spline finite element. All numerical results confirm the validity of the manifold method in solving structural dynamics problems.

6. Conclusions

In this paper, based on the nonuniform cubic B-spline basis function, a new numerical manifold method was derived. By using the same structural model as the B-spline surface, the second-order coordinate can be obtained at the boundary of the manifold element. The local support property of the B-spline basis function enables the good performance of the stiffness matrix, which facilitates the numerical calculation. The new method was used to solve the bending problems of thin plates. The linear elasticity dynamic equation was derived based on the principle of the minimum instantaneous potential energy. The method for the initialization of the dynamic equation and the time integration method were also derived. The fixed mathematical mesh was used to implement the manifold method. The numerical results agreed well with the theoretical solution. Compared with the finite element method, the overall degrees of freedom of the cover calculated with the new manifold method were significantly reduced. The calculation accuracy was obviously improved, which indicates the superiority of the manifold method in solving structural dynamics problems. The theoretical and application studies in this paper are preliminary, confined to linear elasticity problems, and limited to situations in which only the time integration method is adopted. Further research

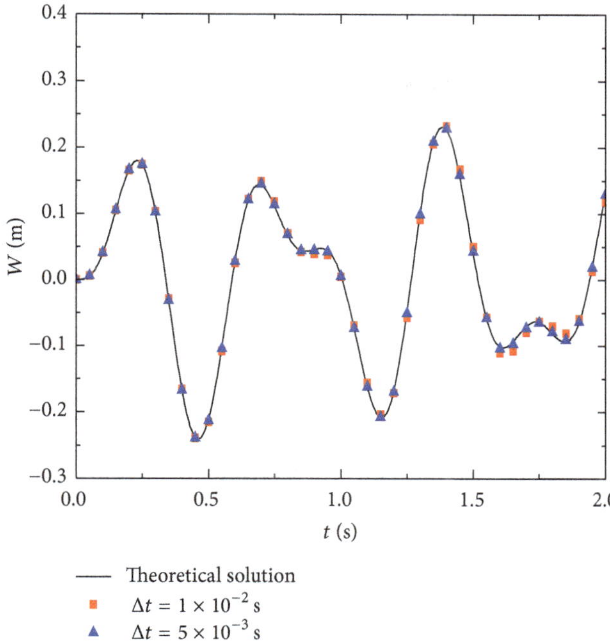

FIGURE 6: Numerical result of displacement W with 5×5 manifold elements.

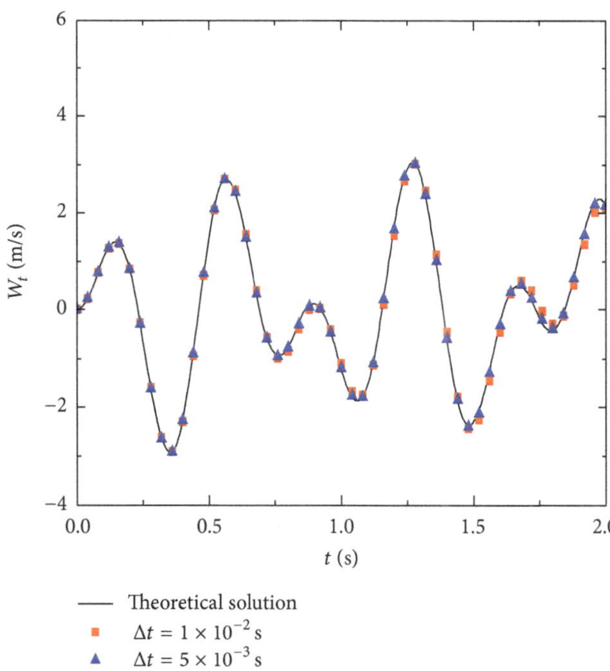

FIGURE 7: Numerical result of velocity W_t with 5×5 manifold elements.

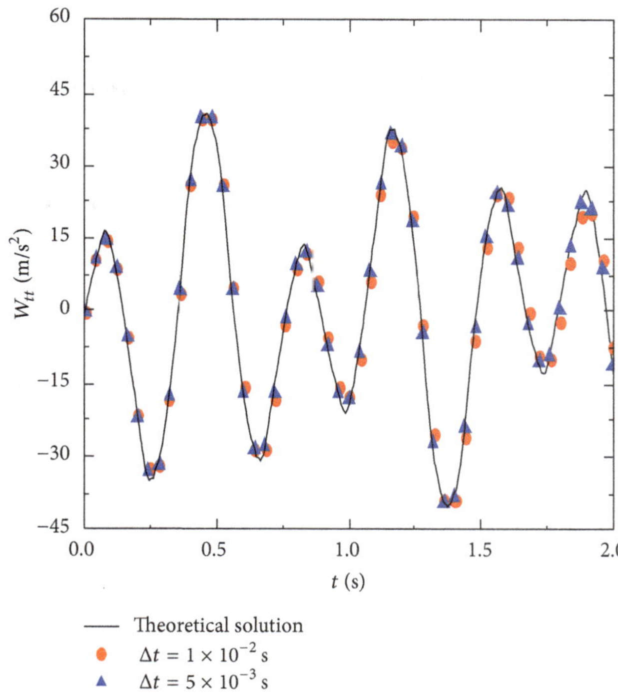

FIGURE 8: Numerical result of acceleration W_{tt} with 5×5 manifold elements.

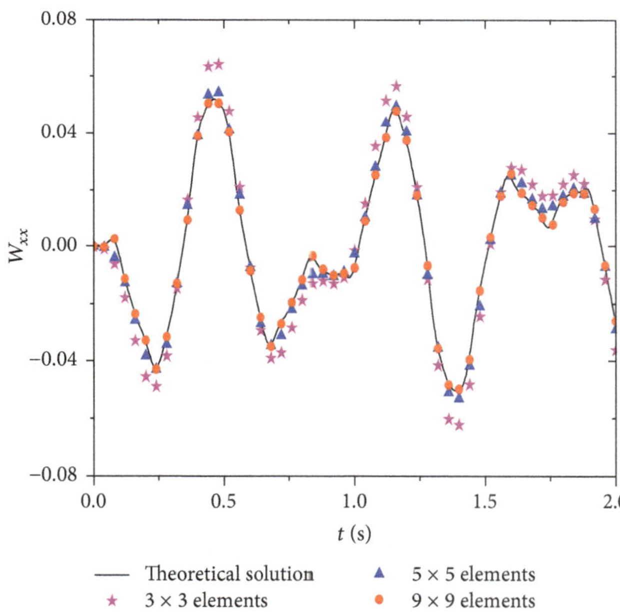

FIGURE 9: Numerical result of $\partial^2 W/\partial x^2$ with $\Delta t = 5 \times 10^{-3}$ s.

on nonlinear problems should be conducted in the future to perfect the theory.

Conflict of Interests

The authors declare that they have no conflict of interests.

Acknowledgments

This work is financially supported by the Natural Science Foundation of China (11302272), the Natural Science Foundation of China (11102169), and the Science and Technology Project Affiliated with the Education Department of Chongqing Municipality (KJ120801).

TABLE 3: Manifold results of the displacement partial derivatives and their relative errors.

Element	$x = a/3$, $y = b/3$, $\Delta t = 5 \times 10^{-3}$ s, $t = 50\Delta t$			
	W'_{xy} ($\times 10^{-2}$)	η_{xy}	W'_{yy} ($\times 10^{-1}$)	η_{yy}
3×3	4.2774	7.85	−4.2777	10.17
5×5	3.9085	1.45	−3.8569	0.66
7×7	3.9712	0.13	−3.8551	0.71
9×9	3.9721	0.15	−3.9208	0.98
Theoretical solution	3.9660	0.00	−3.8827	0.00

TABLE 4: Global errors of the finite method with Hermite element.

Time	$N_1 = 7$, $N_2 = 7$, DOFs = 256, $\Delta t = 5 \times 10^{-3}$ s					
	ε_0	ε_t	ε_{tt}	ε_{xx}	ε_{xy}	ε_{yy}
$20\Delta t$	1.22	0.66	3.21	23.26	9.59	1.52
$40\Delta t$	0.34	0.83	2.54	12.50	3.27	0.74
$60\Delta t$	0.47	0.47	4.11	6.67	1.41	0.88
$80\Delta t$	0.42	0.53	2.49	10.97	2.83	0.78
$100\Delta t$	0.50	1.02	2.26	10.25	2.50	0.77

TABLE 5: Global errors of the manifold method.

Time	$N_1 = 7$, $N_2 = 7$, DOFs = 100, $\Delta t = 5 \times 10^{-3}$ s					
	ε_0	ε_t	ε_{tt}	ε_{xx}	ε_{xy}	ε_{yy}
$20\Delta t$	0.55	0.51	3.18	7.51	7.02	0.99
$40\Delta t$	0.20	0.83	2.14	4.09	2.47	0.70
$60\Delta t$	0.46	0.41	4.12	4.23	1.14	0.84
$80\Delta t$	0.36	0.51	1.98	4.28	2.22	0.75
$100\Delta t$	0.45	0.94	2.23	2.66	1.87	0.74

TABLE 6: Global errors of the finite method with Cubic B-spline element.

Time	$N_1 = 7$, $N_2 = 7$, DOFs = 256, $\Delta t = 5 \times 10^{-3}$ s					
	ε_0	ε_t	ε_{tt}	ε_{xx}	ε_{xy}	ε_{yy}
$20\Delta t$	1.41	0.77	3.18	12.46	9.59	1.32
$40\Delta t$	0.37	0.88	2.14	8.37	3.27	0.72
$60\Delta t$	0.58	0.55	4.12	4.69	1.41	0.87
$80\Delta t$	0.49	0.56	1.98	6.85	2.83	0.75
$100\Delta t$	0.61	1.24	2.23	7.62	2.50	0.76

References

[1] J. M. Pei, "Numerical manifold method and discontinuous deformation analysis," *Chinese Journal of Rock Mechanics and Engineering*, vol. 16, no. 3, pp. 279–292, 1997.

[2] G. Ma, X. An, and L. He, "The numerical manifold method: a review," *International Journal of Computational Methods*, vol. 7, no. 1, pp. 1–32, 2010.

[3] G. Zhang, Y. Zhao, and X. Peng, "Simulation of toppling failure of rock slope by numerical manifold method," *International Journal of Computational Methods*, vol. 7, no. 1, pp. 167–189, 2010.

[4] Y. J. Ning, X. M. An, and G. W. Ma, "Footwall slope stability analysis with the numerical manifold method," *International Journal of Rock Mechanics and Mining Sciences*, vol. 48, no. 6, pp. 964–975, 2011.

[5] H. H. Zhang, L. X. Li, X. M. An, and G. W. Ma, "Numerical analysis of 2-D crack propagation problems using the numerical manifold method," *Engineering Analysis with Boundary Elements*, vol. 34, no. 1, pp. 41–50, 2010.

[6] S. C. Li, S. C. Li, and Y. M. Cheng, "Enriched meshless manifold method for two-dimensional crack modeling," *Theoretical and Applied Fracture Mechanics*, vol. 44, no. 3, pp. 234–248, 2005.

[7] S. C. Li, S. C. Li, J. W. Zhang, and Z. Q. Wang, "Numerical manifold method from mathematical theory and its application," *Engineering Mechanics*, vol. 24, no. 6, pp. 36–42, 2007.

[8] H. Gao and Y. Cheng, "A complex variable meshless manifold method for fracture problems," *International Journal of Computational Methods*, vol. 7, no. 1, pp. 55–81, 2010.

[9] G. Wei, K. Li, and H. Jiang, "Incompatible numerical manifold method for fracture problems," *Acta Mechanica Sinica*, vol. 26, no. 2, pp. 247–255, 2010.

[10] Z. Zhang, X. Zhang, and J. Yan, "Manifold method coupled velocity and pressure for Navier-Stokes equations and direct numerical solution of unsteady incompressible viscous flow," *Computers and Fluids*, vol. 39, no. 8, pp. 1353–1365, 2010.

[11] Q. H. Jiang, S. S. Deng, C. B. Zhou, and W. B. Lu, "Modeling unconfined seepage flow using three-dimensional numerical manifold method," *Journal of Hydrodynamics*, vol. 22, no. 4, pp. 554–561, 2010.

[12] J. Song and Y. Ohnishi, "High order rectangular element of manifold method," *Chinese Journal of Rock Mechanics and Engineering*, vol. 22, no. 6, pp. 932–936, 2003.

[13] S. Li, Y. Cheng, and Y.-F. Wu, "Numerical manifold method based on the method of weighted residuals," *Computational Mechanics*, vol. 35, no. 6, pp. 470–480, 2005.

[14] S. C. Li, Y. M. Cheng, and S. C. Li, "Meshless manifold method for dynamic fracture mechanics," *Acta Physica Sinica*, vol. 55, no. 9, pp. 4760–4766, 2006.

[15] S. Z. Lin, Y. F. Qi, and H. D. Su, "Improved local function of numerical manifold method and its application," *Journal of Yangtze River Scientific Research Institute*, vol. 23, no. 6, pp. 55–58, 2006.

[16] H. F. Gao and Y. M. Cheng, "Complex variable numerical manifold method for elasticity," *Acta Mechanica Sinica*, vol. 41, no. 4, pp. 480–488, 2009.

[17] S. C. Li and Y. M. Cheng, "Meshless numerical manifold method based on unity partition," *Acta Mechanica Sinica*, vol. 36, no. 4, pp. 496–500, 2004.

[18] H. Gao, Y. Cheng, and H. Jiang, "A meshless manifold method with complex variables for elasticity," *Chinese Journal of Applied Mechanics*, vol. 27, no. 1, pp. 15–19, 2010.

[19] W. B. Wen and S. M. Luo, "Polygonal manifold element for thin plate-bending analysis," *Engineering Mechanics*, vol. 29, no. 10, pp. 249–256, 2012.

[20] W. B. Wen, K. L. Jian, and S. M. Luo, "2D numerical manifold method based on quartic uniform B-spline interpolation and its application in thin plate bending," *Applied Mathematics and Mechanics*, vol. 34, no. 8, pp. 1017–1030, 2013.

[21] C. X. Guo and H. Zeng, "Study on linear dependence problem in high-order numerical manifold method," *Engineering Mechanics*, vol. 29, no. 12, pp. 228–232, 2012.

[22] H. D. Su, "Study on numerical manifold method with fixed meshes," *Chinese Journal of Theoretical and Applied Mechanics*, vol. 43, no. 1, pp. 169–178, 2011.

[23] S. C. Li and Y. M. Cheng, "Numerical manifold method and its applications in rock mechanics," *Advances in Mechanics*, vol. 34, no. 4, pp. 446–454, 2004.

[24] R. Sevilla, S. Fernández-Méndez, and A. Huerta, "NURBS-Enhanced Finite Element Method (NEFEM)," *Archives of Computational Methods in Engineering*, vol. 18, no. 4, pp. 441–484, 2011.

[25] L. Piegl and W. Tiller, *The NURBS Book*, Springer, London, UK, 1996.

[26] J. R. Barber, *Elasticity Book*, Kluwer Academic Publishers, New York, NY, USA, 2004.

[27] S. M. Luo, X. W. Zhang, W. G. Lu, and D. R. Jiang, "Theoretical study of three-dimensional numerical manifold method," *Applied Mathematics and Mechanics*, vol. 26, no. 9, pp. 1127–1132, 2005.

Permissions

The contributors of this book come from diverse backgrounds, making this book a truly international effort. This book will bring forth new frontiers with its revolutionizing research information and detailed analysis of the nascent developments around the world.

We would like to thank all the contributing authors for lending their expertise to make the book truly unique. They have played a crucial role in the development of this book. Without their invaluable contributions this book wouldn't have been possible. They have made vital efforts to compile up to date information on the varied aspects of this subject to make this book a valuable addition to the collection of many professionals and students.

This book was conceptualized with the vision of imparting up-to-date information and advanced data in this field. To ensure the same, a matchless editorial board was set up. Every individual on the board went through rigorous rounds of assessment to prove their worth. After which they invested a large part of their time researching and compiling the most relevant data for our readers. Conferences and sessions were held from time to time between the editorial board and the contributing authors to present the data in the most comprehensible form. The editorial team has worked tirelessly to provide valuable and valid information to help people across the globe.

Every chapter published in this book has been scrutinized by our experts. Their significance has been extensively debated. The topics covered herein carry significant findings which will fuel the growth of the discipline. They may even be implemented as practical applications or may be referred to as a beginning point for another development. Chapters in this book were first published by Hindawi Publishing Corporation; hereby published with permission under the Creative Commons Attribution License or equivalent.

The editorial board has been involved in producing this book since its inception. They have spent rigorous hours researching and exploring the diverse topics which have resulted in the successful publishing of this book. They have passed on their knowledge of decades through this book. To expedite this challenging task, the publisher supported the team at every step. A small team of assistant editors was also appointed to further simplify the editing procedure and attain best results for the readers.

Our editorial team has been hand-picked from every corner of the world. Their multi-ethnicity adds dynamic inputs to the discussions which result in innovative outcomes. These outcomes are then further discussed with the researchers and contributors who give their valuable feedback and opinion regarding the same. The feedback is then collaborated with the researches and they are edited in a comprehensive manner to aid the understanding of the subject.

Apart from the editorial board, the designing team has also invested a significant amount of their time in understanding the subject and creating the most relevant covers. They scrutinized every image to scout for the most suitable representation of the subject and create an appropriate cover for the book.

The publishing team has been involved in this book since its early stages. They were actively engaged in every process, be it collecting the data, connecting with the contributors or procuring relevant information. The team has been an ardent support to the editorial, designing and production team. Their endless efforts to recruit the best for this project, has resulted in the accomplishment of this book. They are a veteran in the field of academics and their pool of knowledge is as vast as their experience in printing. Their expertise and guidance has proved useful at every step. Their uncompromising quality standards have made this book an exceptional effort. Their encouragement from time to time has been an inspiration for everyone.

The publisher and the editorial board hope that this book will prove to be a valuable piece of knowledge for researchers, students, practitioners and scholars across the globe.

List of Contributors

T. Chinyoka
Center for Research in Computational and Applied Mechanics, University of Cape Town, Rondebosch 7701, South Africa

O. D. Makinde
Institute for Advanced Research in Mathematical Modeling and Computations, Cape Peninsula University of Technology, P.O. Box 1906, Bellville 7535, South Africa

Zhongwei Liang and Xiaochu Liu
School of Mechanical and Electrical Engineering, Guangzhou University, Guangzhou 510006, China

Bangyan Ye
School of Mechanical and Automotive Engineering, South China University of Technology, Guangzhou 510640, China

Richard Kars Brauwer
Department of Information Engineering, Indian Institute of Technology (IIT), Kanpur 208023, India

Ji Min Lee, Sung Hwan Park and Jong Shik Kim
School of Mechanical Engineering, Pusan National University, Jangjeon-dong, Geumjeong-gu, Busan 609-732, Republic of Korea

Hassan A. Wahab, M. Y. Noordin, S. Izman and Denni Kurniawan
Faculty of Mechanical Engineering, University Technology Malaysia, 81310 Skudai, Malaysia

R. Silva-Ortigoza, C. Márquez-Sánchez, M. Marciano-Melchor and D. Muñoz-Carrillo
Instituto Polit´ecnico Nacional, CIDETEC, ´ Area de Mecatr´onica, Unidad Profesional Adolfo L´opez Mateos, 07700 M´exico, DF, Mexico

M. Marcelino-Aranda
Instituto Polit´ecnico Nacional, UPIICSA, Secci´on de Estudios de Posgrado e Investigaci´on, 08400 M´exico, DF, Mexico

G. Silva-Ortigoza
Benem´erita Universidad Aut´onoma de Puebla, Facultad de Ciencias F´ısico Matem´aticas, 72001 Puebla, PUE, Mexico

R. Bautista-Quintero
Instituto Tecnol´ogico de Culiac´an, Departamento de Metal-Mec´anica, 80220 Culiac´an, SIN, Mexico

E. R. Ramos-Silvestre
Universidad Privada del Valle, Facultad de Inform´atica y Electr´onica, Tiquipaya, CBBA, Bolivia

J. C. Rivera-Díaz
Centro Nacional de Actualizaci´on Docente, ´ Area de M´aquinas, 13420 M´exico, DF, Mexico

Lizhong Xu and Fen Wang
Mechanical Engineering Institute, Yanshan University, Qinhuangdao 066004, China

Dong-yun Wang
School of Engineering, Zhejiang Normal University, Zhejiang 321005, China

Chen Guan
School of Mechanical Engineering, Zhejiang University, Zhejiang 321005, China

Jae-Hoon Lee, Changwan Han, Sang-Hu Park and Seonghun Park
School of Mechanical Engineering, Pusan National University, Jangjeon-dong, Geumjeong-gu, Busan 609-732, Republic of Korea

Dongsu Ahn and Jin Kyoo Lee
Suspension Systems Engineering Team, Doosan Corporation Mottrol, 456-3 Nae-dong, Seongsan-gu, Changwon, Gyeongnam 642-050, Republic of Korea

Qiang Gao, Jilin Chen, Li Wang and Yuanlong Hou
School of Mechanical Engineering, Nanjing University of Science and Technology, Nanjing 210014, China

Shiqing Xu
Research Institute, North Heavy Industries Group Corp. LTD, Inner Mongolia, Baotou 014033, China

Wenqing Zhang, Jie Li, Kun Zhang and Peng Cui
College of Mechatronics Engineering and Automation, National University of Defense Technology, Changsha, Hunan 410073, China

Jie Zhang, Zheng Liang and Chuanjun Han
School of Mechatronic Engineering, Southwest Petroleum University, Chengdu 610500, China

Waluyo Adi Siswanto
Department of Engineering Mechanics, Universiti Tun Hussein Onn Malaysia, Parit Raja, Batu Pahat, 86400 Johor, Malaysia

Agus Dwi Anggono
Department of Mechanical Engineering, Muhammadiyah University of Surakarta, Pabelan, Kartasura 57162, Indonesia

Badrul Omar
Department of Materials Engineering and Design, Universiti Tun Hussein Onn Malaysia, Parit Raja, 86400 Johor, Malaysia

Kamaruzaman Jusoff
Faculty of Forestry, Universiti Putra Malaysia, Serdang, 43400 Selangor, Malaysia

Jianping Wang
School of Mechanical Engineering, Henan Polytechnic University, Jiaozuo 454003, China

Kun Tao
Department of Orthopedics, Shanghai 10th People's Hospital, Tongji University School of Medicine, Shanghai 200072, China

Huanyi Li
Department of Orthopedics, Zhenjiang Jinshan Hospital, Zhenjiang 212001, China

Chengtao Wang
School of Mechanical Engineering, Shanghai Jiao Tong University, Shanghai 200240, China

Josip Sertić, Dražan Kozak and Ivan Samardžić
Mechanical Engineering Faculty in Slavonski Brod, Trg Ivane Brlic-Mazuranic 2, 35 000 Slavonski Brod, Croatia

Mohamed A. Omar
Mechanical Engineering Department, Taibah University, Almadinah Almonawwarah 42353, Saudi Arabia

Bin Huang and Heung Soo Kim
Department of Mechanical, Robotics and Energy Engineering, Dongguk University-Seoul, 30 Pildong-ro 1-gil, Jung-gu, Seoul 100-715, Republic of Korea

Ding Jun, Chen Song and Huang Xia
College of Mechanical Engineering, Chongqing University of Technology, Chongqing 400054, China

Wen Wei-Bin and Luo Shao-Ming
Zhongkai University of Agriculture and Engineering, Guangzhou 510225, China